Energy-Efficient Wireless Sensor Networks

Energy-Efficient Wireless Sensor Networks

Edited by
Vidushi Sharma and Anuradha Pughat

CRC Press
Taylor & Francis Group
Boca Raton London New York

CRC Press is an imprint of the
Taylor & Francis Group, an **informa** business

MATLAB® is a trademark of The MathWorks, Inc. and is used with permission. The MathWorks does not warrant the accuracy of the text or exercises in this book. This book's use or discussion of MATLAB software or related products does not constitute endorsement or sponsorship by The MathWorks of a particular pedagogical approach or particular use of the MATLAB software.

CRC Press
Taylor & Francis Group
6000 Broken Sound Parkway NW, Suite 300
Boca Raton, FL 33487-2742

First issued in paperback 2020

© 2018 by Taylor & Francis Group, LLC
CRC Press is an imprint of Taylor & Francis Group, an Informa business

No claim to original U.S. Government works

ISBN 13: 978-0-367-57328-7 (pbk)
ISBN 13: 978-1-4987-8334-7 (hbk)

Visit the Taylor & Francis Web site at
http://www.taylorandfrancis.com

and the CRC Press Web site at
http://www.crcpress.com

*I, Vidushi Sharma, **dedicate** this book to my lifeline and energy source—my husband, Rohit Sharma; my brother, Sadashiv Sharma; and my son, Shourya.*

*I **thank and acknowledge** my family especially my in-laws, Veena Sharma and Lt. Shri K.D. Sharma, for their blessings and motivation. I thank my parents, Shri K.N. Sharma and Saroj Sharma, for instilling the quest of knowledge and lovely people who brighten my life, Sharad Gaur, Shikha Gaur, and Shipra Sharma. My little angels, Shreya, Shourya, Vidip, Vidu, and Vasu, are my energy harvesters as they bring smile back in the moments of stress. Thank you all.*

*I, Anuradha Pughat, **dedicate** this book to my husband, Sumit Kumar, for standing beside me throughout my career and the writing of this book. He has been my rock and inspiration, motivation for continuing to improve my knowledge and move my career forward.*

I would like to thank my parents including my in-laws for allowing and supporting me to follow my ambitions throughout my career. I want to acknowledge my sisters (Manisha gg, Neelam, Reena, Moni, and Renu) who shared my happiness when starting this project and following with encouragement, when it seemed too difficult to be completed; my lovely children: HariOm Pughat, Prithveeraj S. Pughat, and new born Adhiraj S. Pughat, for always making me smile and for understanding on those mornings and nights when I was writing this book instead of playing with them. I hope that one day they can read this book and understand why I spent so much time in front of my laptop.

Above all, we thank GOD for being the guiding beacon and infusing us with aspirations, which he desires us to achieve.

Contents

Preface ix
Acknowledgments xv
Editors xvii
Contributors...xix

1   Introduction to Energy-Efficient Wireless Sensor Networks............................1
    VIDUSHI SHARMA AND ANURADHA PUGHAT

2   Medium Access Control in Wireless Sensor Networks..............................27
    VIDUSHI SHARMA, GAYATRI SAKYA, AND ANURADHA PUGHAT

3   Routing in Wireless Sensor Networks 43
    VINAY KUMAR SINGH

4   Clustering for Energy Efficiency in Wireless Sensor Networks....................69
    PUNEET AZAD

5   Power Management in Sensor Node...115
    ANURADHA PUGHAT AND VIDUSHI SHARMA

6   Energy Harvesting Issues in Wireless Sensor Networks 137
    GOURAV VERMA, VIDUSHI SHARMA, AND ANURADHA PUGHAT

7   Data Aggregation in Wireless Sensor Networks................................165
    PRASHANT SHUKLA, VIDUSHI SHARMA, AND ANURADHA PUGHAT

8   Sensor Network Security 179
    AARTI GAUTAM DINKER AND VIDUSHI SHARMA

9   Communication, Localization, Coverage, Error and Control, Time
    Synchronization, Naming and Addressing, and Cross-Layer Issues 213
    ANURADHA PUGHAT, PARUL TIWARI, VIDUSHI SHARMA, AND NEETA SINGH

10 Advanced Applications and Challenges 241
    VIDUSHI SHARMA

11 The Future for Sensor Networks—Cloud and IoT 249
    ARJUN K. SIROHI

Index 265

Preface

The world is changing as technology is taking a leap into the future. The traditional systems are gone and the plethora of applications has come up with the advent of sensor networks. These self-organizing miniature sensor nodes deployed in an area weave a magical network to sense, transmit, and communicate with the world through sink gateways and clouds. Wireless sensor networks (WSNs) comprised of a sensor node with sensing and communication capabilities have changed the business models of traditional enterprises such as healthcare industries, defense, supply chain management, and electronic commerce, to name a few, and are now permeating every sphere of our life. It is a paradigm shift where we can monitor and control remote locations through our computer, handheld devices, and Internet. It is a time when the Internet of Things (IoT) and sensor networks are closely integrated to create numerous other applications.

Energy is required by sensor networks for sensing, communication, synchronization, routing of information, and various other processes. Besides sensing and communication, energy consumption is there in hardware operations, software processes, and running various operative algorithms. Here one can say that the potential of WSNs is extensive, but one of the critical limiting factors out of main concern is the limited energy of the networks. Limited resource energy has led to research in the area of energy conservation: optimization, energy harvesting, and increasing the efficiency of sensor network will cease to exist if energy is depleted. Further, each component and process which consumes energy has to be critically analyzed and efficient mechanisms should be devised for optimizing resources.

This book aims to cover all the aspects of sensor networks with respect to energy conservation and optimization. There are several textbooks and reference books in the area of sensor networks. Researchers have also published their work in the form of articles and research papers in this domain. But one central idea to which we all arrive is to make WSNs more efficient. Efficiency metrics of WSNs include lifetime, cost, size, reliability, fault tolerance, etc., and most of them are culminating at the energy efficiency problem. Hence, the main objective of this book is to address energy-efficiency issues from diverse areas of low-level hardware design and communication to the high-level concept of data aggregation, routing, synchronization, localization, etc.

This book outlines the mechanisms, techniques, and algorithms of the physical layer, the media access control (MAC) layer, and the network layer in context with energy efficiency. It delves into energy-efficient security mechanisms and also presents open issues in the area of security. Special attention to MAC protocols is given and the state of the art has been presented. Advancements in energy-efficient algorithms using soft computing techniques and comparative analysis of these with traditional techniques have been brought up. The hierarchical network that improves lifetime has been discussed. Operational-level power management and energy harvesting have been explored. Other operational processes such as data aggregation, localization, time synchronization, and coverage have been presented along with open research issues. Finally, application and future trends in WSNs have been discussed at the end. This book is ideal to deal with the energy aspect of WSN. The information base presented in chapters is as follows.

Guided Tour

Chapter 1 gives an introduction of the sensor network starting with the basic architecture of WSNs and also fulfills the requirement of preliminary knowledge for the subsequent chapters. The protocol stacks structure for resource-constrained WSN is discussed along with the sensing, processing, and communication processes in the context of energy. An explanation on sensor node hardware and software use is presented along with the features of commercially available node components. Then, the types of networking environment and suitable operating systems for application-based WSNs are explained. Factors affecting the operation of WSNs such as hardware design, scalability, transmission medium, topology and control, power consumption in sensing, processing, and communication have been presented. Further, challenges for energy-constrained WSNs are explored.

Chapter 2 illustrates the medium access control protocols for WSNs and also how the existing mechanisms can be improved further for smart applications in WSNs. As WSNs are now used in many critical real-time applications for surveillance and monitoring, the main attributes of MAC protocols are discussed in this chapter, which are essential to design a good MAC protocol for WSNs. The existing MAC protocols are classified based on their access mechanism for acquiring the channel. This chapter explores the existing MAC models in WSN and also discusses them for their suitability in smart application scenarios. It further suggests the parameters for designing a protocol such as SMAC in NS-2.35 for its performance in mission critical scenarios. The case study on SMAC and ZigBee has been discussed. Finally, the open research issues for making WSN protocols suitable for real-time scenarios for smart functioning are also discussed.

Chapter 3 discusses the existing routing protocols used in WSNs and various challenges in the development of WSN routing protocols. Sensor nodes in a WSN are constrained in storage capacity, computation power, bandwidth, and power supply, so the routing schemes to be developed must take this fact into consideration while developing the routing protocols for WSNs. Several existing routing algorithms are presented. Soft computing algorithms are computationally heavy, but can be efficiently used for routing when they are deployed on the base station. GA-based algorithms and their hybridization with other existing algorithms have been discussed with simulation-based results to prove their efficiency.

Chapter 4 presents several clustering algorithms, which lead to the efficient use of energy, thereby increasing the overall network lifetime and improving other aspects such as average energy dissipation, deployment, data packets, and stability region. The techniques such as clustering and routing may be adopted for efficiently utilizing the energy of nodes using single or multipath communication to increase the lifetime of the network. In this chapter, several techniques for energy efficiency of WSNs have been presented and their performance has been compared with some existing protocols in terms of a number of metrics. Decision-making techniques such as Pareto optimal theory, technique for order of preference by similarity to ideal solution (TOPSIS), fuzzy TOPSIS, and the compromise ranking method VIsekriterijumskaoptimizacija I KOmpromisno Resenje (VIKOR) are utilized for selecting the cluster heads and making the clusters. The simulation results are given to prove their efficiency in decision-making. Improved cluster head selection protocols for increasing the lifetime of the network are also discussed at length with experimental results.

Chapter 5 illustrates the existing dynamic power management techniques, their implementation, and design issues with upcoming challenges in the wireless sensor node. Various techniques such as dynamic operation modes, dynamic voltage, and frequency scaling and scheduling are described at a more comprehensive level than before. Then, a stochastic approach for power management has also been depicted for workload modeling. The first approach is operational-level dynamic power management, which reduces the number of active states and their duration by operating the node components in low power modes or off. On the other hand, the second approach, i.e., dynamic voltage scaling (DVS), helps in adopting the desired voltage level based on the current workload/arrival rate and latency requirement during the active state of the processor. The low operating voltage and frequency of the microcontroller/processor also reduces the coupling noise of the chip. The third alternative is task scheduling, which enhances the energy efficiency of the node by ordering, organizing, and prioritizing the task before processing. This can be done at the operating system level. Finally, the cooperative power management technique gives the interplay between any two or more power management techniques on a single sensor node. The main advantage of the dynamic power management-based sensor node is its applicability with and without an energy scavenged node. However, these techniques require synchronization between node components and some guidance for desired application. Basic knowledge of sensor node architecture, application type, and task arrival rate at sensor input and/or preprocessor are the prerequisites for understanding the concepts of dynamic power management.

Chapter 6 explores the lifetime issues in WSNs and the concept of various energy harvesting techniques. The basic principle, architecture, presently available circuits, and storage technologies available for energy harvesting have been discussed. The current technology facilitated with energy harvesting and some applications have also been described. This chapter will help to find the efficient energy harvesting solutions for different kinds of environments such as habitat monitoring, agriculture monitoring, etc. The types of energy harvesting included in this chapter are photovoltaic energy harvesting, mechanical energy harvesting, thermoelectric generator, dynamic fluid, magnetic energy harvesting, solar energy harvesting, and radio frequency energy harvesting. The advantages and disadvantages of each are discussed with respect to the sensor network environment. Finally, the next section concludes this chapter with existing solutions and upcoming challenges in the energy harvesting arena.

Chapter 7 describes the concept of data aggregation in sensor networks, which has been viable in recent years because of the concept of energy optimization. As we all are aware of the fact that the energy harvesting in sensor network, data routing, and network lifetime are the most important aspects of sensor networks, a well-designed data aggregation algorithm covers the challenges of all these three areas. Based on transmission type, a complete review of the data aggregation algorithms and techniques has been discussed. The security aspects in data aggregation have been discussed. Finally, this chapter is concluded with possible future research possibilities in data aggregation.

Chapter 8 discusses different security risks and vulnerabilities to WSNs along with various security mechanisms to overcome the security issues. WSNs are generally deployed in an unattended and open environment which makes them vulnerable and attack prone especially in security-critical tasks. Security issues arise mainly due to network problems and various attacks on the nodes. First, the security solutions are classified in categories like attack detections and preventions, cryptography, key management, secure routing, secure location, secure data aggregation, etc. Second, the advantages and disadvantages of current security schemes are discussed with reference to the open research challenges in each area.

Chapter 9 describes the existing standards of the physical layer in WSNs. The energy efficiency in modulation and multiple access techniques have been compared and explained. Radio frequency and other technologies of physical layer are given. This chapter describes the miscellaneous terms associated with the effective operation of WSNs. Challenges in localization such as computational and energy constraints, range-based and range-free localization, etc., have been explained. The next section of this chapter explains the types of time synchronizations and their ground challenges associated with time sync protocols for sensor networks. Various coverage issues have been discussed. The classification and comparison of different error control schemes implemented and proposed for WSN have been illustrated. The naming and addressing issues have also been discussed in the next section of this chapter. The cross-layer interaction between MAC layer, physical layer, network layer, and application layer and their effect on performance and energy have been investigated.

Chapter 10 continues the range of applications in WSNs, which are based on the type of the sensing behavior and type of deployment environment. Periodic sensing applications such as monitoring, tracking and surveillance of borders, home automation, and environment to monitor the forest, military surveillance, healthcare monitoring, etc., have been discussed. The nonperiodic sensing applications such as event detection, query-based, and mobility-based systems have been discussed. To serve these different applications of WSNs, the protocol stack has not been standardized yet, and research is continuing with each layer to design energy efficient protocols suitable for specific applications. Terrestrial, underwater, and underground applications are also presented under the category of deployed environmental conditions. Then various other applications have been discussed to further broaden the scope of the WSN applications.

Chapter 11 introduces the advanced technologies such as big data, the IoT, Web 3.0, cloud computing and fog computing, etc. These technologies with WSNs produce solutions to next generation applications. The combination of WSNs and IoT has some infrastructure and data handling issues. The recent available data collection, integration, and analysis techniques have been discussed. Further, the security and privacy of the data collected from sensors and the distributed architecture to enable locally actionable intelligence have been elaborated. The ever-growing number of devices and sensors connected to any IoT system creates a big challenge to support heterogeneous devices and data formats while at the same time being scalable. Therefore, the performance and scalability challenges have been presented. After that, the intersection of artificial intelligence, data science, and machine learning with sensor network data in IoT is discussed. The factors which make the IoT successful have been categorized into two categories; first in technology and infrastructure-related factors and second business and consumer-related factors.

Target Audience

This book is mainly for students who are striving to understand the concepts and trying to discover new research areas in the field of WSNs. This book will also provide direction for budding researchers to explore a new area of research in WSNs. Finally, industry experts and technical managers will benefit from this book as it will help them to find new business ideas and models along with technological know-how.

A few words are necessary concerning chapter authors. The above topics in this book have been contributed by the researchers who have been working in this field for quite a long time. This is truly a team effort. In fact, this group of people has been working together for several years and they have covered almost all the domains of energy efficiency with respect to WSNs during their research. The chapters contributed by them are the outcome of their research in the field, hence the domain expertise is rich and varied. Authors have tried to cite the most relevant references to facilitate the beginners and researchers in their domains.

Far into the preface, we end by saying that we hope you will benefit from reading this book. If you have any comments or suggestions regarding this material, or any error detected, no matter how trivial, please write to us.

Vidushi Sharma
svidushee@gmail.com

Anuradha Pughat
anuradha.pughat@gmail.com

Acknowledgments

We extend our gratitude and thank our authors who arduously worked hard day and night to provide their domain expertise in the form of chapters. We are indebted to our reviewers, as their valuable suggestions have helped us improve the quality of the content. Heartfelt thanks to Honorable Vice Chancellor, Gautam Buddha University, our seniors Dean Academics, Prof. Arun Kansal; Dean Student Affairs, Dr. Anand Singh; Dean SoICT, Prof. A. K. Gautam; and our colleagues for their constant support. Compiling the book was a tedious task. Our sincere thanks to Rich O'Hanley, Publisher—ICT and Security, CRC Press/Taylor & Francis Group—who has been very patient and prompt at addressing the queries and giving guidelines for the completion of this book.

The editors would like to acknowledge the help of all the people who were involved in this project, especially Dr. Kriti Priya Gupta for her suggestions; Sonu Prasad, Vinay Yadav, and Abid Saifi for the artwork compilation.

The authors we will love to acknowledge who walked hand in hand in the journey of book compilation are Aarti Gautam Dinker, Arjun K. Sirohi, GouravVerma, Gayatri Shakya, Neeta Singh, Prashant Shukla, Puneet Azad, Parul Tiwari, and Vinay Kumar Singh.

Further, the editors will like to individually acknowledge the efforts of their respective supporters and motivators:

Editors

 **Vidushi Sharma** earned her MTech and PhD in computer science. She is presently working as a faculty at Gautam Buddha University, Greater Noida, India. She teaches network and IT-related subjects and has more than 50 international and national publications as research papers and articles. She has authored a book on information technology and her research interests include sensor networks, IT applications in management, and performance evaluation of information systems, which includes wireless systems, application software, E-commerce system, and scheduling in cloud computing.

 **Anuradha Pughat** is pursuing her PhD in wireless sensor networks in Gautam Buddha University, Greater Noida, India. She worked as a teacher while serving in engineering institutions from 2004 to 2013. She graduated in 2004 with a BTech degree in electronics and communication engineering. Along with working as a teacher and doing research in multidisciplinary areas, she earned her MTech degree in VLSI Design in 2010. She has published a number of research papers and books. Her research interests include wireless sensor networks, energy efficiency, system modeling, system on chip (SoC) design, and embedded system design.

Contributors

Puneet Azad
Maharaja Surajmal Institute of Technology
New Delhi, India

Aarti Gautam Dinker
SoICT
Gautam Buddha University
Greater Noida, India

Anuradha Pughat
SoICT
Gautam Buddha University
Greater Noida, India

Gayatri Sakya
JSS Academy of Technical Education
Noida, India

Vidushi Sharma
SoICT
Gautam Buddha University
Greater Noida, India

Prashant Shukla
eSecForte Technologies
Gurgaon, India

Neeta Singh
SoICT
Gautam Buddha University
Greater Noida, India

Vinay Kumar Singh
Anand Engineering College
Agra, India

Arjun K. Sirohi
Performance Engineering Kernel Cloud
Salesforce.com Inc
Bellevue, Washington

Parul Tiwari
Jaypee Institute of Information Technology
Noida, India

Gourav Verma
Northern India Engineering College
New Delhi, India

Chapter 1

Introduction to Energy-Efficient Wireless Sensor Networks

Vidushi Sharma and Anuradha Pughat

Contents

1.1 Introduction 2
1.2 Architecture of Wireless Sensor Nodes and Networks 4
1.3 Protocol Stack 4
    1.3.1 Physical Layer 5
    1.3.2 Data Link Layer 6
    1.3.3 Network Layer 6
    1.3.4 Transport Layer 6
    1.3.5 Application Layer 6
1.4 Features of Commercially Available Node Components 6
1.5 Available Operating Systems and Networking Environments for WSNs 9
    1.5.1 Operating Systems 10
        1.5.1.1 AmbientRT 10
        1.5.1.2 TinyOS 1.x 10
        1.5.1.3 TinyOS 2.0 10
        1.5.1.4 Contiki 11
        1.5.1.5 MANTIS 11
        1.5.1.6 LiteOS 11
        1.5.1.7 BTnut/NutOS 12
        1.5.1.8 MagnetOS 12
        1.5.1.9 SOS 12
        1.5.1.10 Nano-RK 12
        1.5.1.11 RIOT 13
        1.5.1.12 EYES OS 13

              1.5.1.13 NanoQplus...13
              1.5.1.14 EMERALDS..13
              1.5.1.15 SenOS 14
    1.5.2   Simulation/Emulation/Debugging Environments 14
              1.5.2.1 TOSSIM 14
              1.5.2.2 PowerTOSSIM 14
              1.5.2.3 Viptos 15
              1.5.2.4 GloMoSim 15
              1.5.2.5 ATEMU 15
              1.5.2.6 MATLAB/Simulink 15
               1.5.2.7 Avrora 16
               1.5.2.8 Network Simulator v2 (ns2).......................................16
              1.5.2.9 Network Simulator v3 (ns3).......................................16
              1.5.2.10 Cooja Simulator (Contiki OS) 17
              1.5.2.11 OPNET 17
              1.5.2.12 OMNeT++ 17
              1.5.2.13 QualNet 18
              1.5.2.14 J-Sim 18
              1.5.2.15 NetSim 18
              1.5.2.16 SensorSim 18
              1.5.2.17 SENSE..19
              1.5.2.18 EmStar 19
1.6   Factors Affecting Design of Energy-Efficient Sensor Nodes and Networks 19
    1.6.1   Hardware Constraints 19
    1.6.2   Network Traffic 20
    1.6.3   Quality of Service 20
    1.6.4   Fault Tolerance (Reliability).. 20
    1.6.5   Self-Management... 20
    1.6.6   Scalability and Network Density 20
    1.6.7   Production Costs 21
    1.6.8   Network Topology...21
    1.6.9   Operating Environment (Applications).................................21
    1.6.10 Security..21
    1.6.11 Transmission Media...21
    1.6.12 Power Consumption (Lifetime) 22
1.7   Energy Conservation Mechanisms in WSNs.................................... 22
1.8   Challenges for Energy-Constrained WSNs 23
References 24

1.1 Introduction

Wireless sensor networks (WSNs) have revolutionized the network technologies of today and tomorrow. What we will see in the future is the intricately woven infrastructure of feeder networks that will disseminate the information of their environment to a remote area where a receptor network will exist. The receptor network may be in the form of cloud or database server farms or monitoring and control centers.

The feeder networks of WSNs comprise of densely distributed sensor nodes distributed in a geographical area for carrying out the monitoring of physical activities such as temperature, pressure, humidity, vibration, seismic events, etc. Sensor nodes may be randomly deployed (Figure 1.1). A node can collect and transfer the data to the sink with the help of multihop communication. Data from the sink may be transferred to the gateway that is linked by the Internet or to a task manager where an appropriate decision can be taken by the Internet or satellite.

Sensor networks are different from the other wireless networks in several aspects:

- Sensor networks are densely deployed and their number can be in multiple orders of magnitudes higher than other wireless networks.
- WSNs are susceptible to failures.
- The node topology changes frequently and randomly. In mobile sensor nodes, the topology change is due to mobility, but for static nodes the topology frequently changes due to a node failure (physical damage or death of node) and different sleep–wake-up cycles of nodes.
- Broadcast communication is used in sensor nodes, whereas the other networks such as ad hoc networks use point-to-point communication paradigm.
- Sensor nodes are power constrained. They have limited processing capacities and limited memory.
- Unlike other networks, sensor nodes cannot have a global identification as the numbers of nodes are huge and global naming will lead to large overheads.
- Sensor nodes do not have a predefined topology, so the nodes should have self-organizing and autoconfiguring capabilities to enable communication.

These unique characteristics of WSNs have led to varied research avenues and challenges such as energy conservation, power management (PM), self-organized routing protocols, localization, development of unique naming and addressing schemes, an efficient sensor deployment mechanism to overcome the coverage problems, exploring schemes at various layers to reduce communication overheads, and above all developing mechanisms for lifetime enhancement. To understand and explore these aspects, one should have a clear understanding of the node architecture and the protocol stack of WSNs.

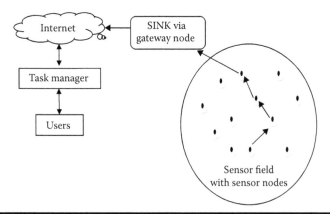

Figure 1.1   WSN network.

1.2 Architecture of Wireless Sensor Nodes and Networks

The sensor nodes are deployed in millions and billions to monitor a larger and more complex system. Hence, they should be small, self-configuring, low-cost devices (Kahn et al. 1999).

The basic function of these sensor nodes should be to measure the physical quantities such as light, temperature, and pressure and convert them into a form of signal that can be further interpreted by humans and communicating devices. To enable this functionality, a sensor node comprises of three basic components:

1. A sensing system for data acquisition
2. A processing system for data processing
3. A communication system for disseminating information

The power supply unit is present to provide energy to the sensor node. Normally this unit comprises of a battery, which is a limited source of energy, and hence the lifetime of the node is constrained as per the battery life. Figure 1.2 depicts a typical architecture in which the sensing unit consists of a sensor and analog-to-digital converters (ADCs). The function of ADCs is to convert the analog input received into a digital signal. The processing unit has a small storage unit and it not only processes the information but also coordinates with other nearby nodes. A communication unit comprises of the transceiver, which communicates with the network. In addition, this sensor node also has to carry out the task of location finding, mobility management (MM) through a mobilizer in case it is a dynamic node. There may be other units depending upon the application of the sensor node. Sometimes, the power unit also has an energy scavenging source such as solar cells.

1.3 Protocol Stack

The functioning and efficiency of WSNs depend on the protocol stack developed. In fact, the research challenges and the applications also can be outlined if one has a clear understanding of the protocol stack. The protocol stack for WSNs has five layers: physical layer, data link layer, network layer, transport layer, and application layer as depicted in Figure 1.3.

The physical layer addresses the need for signal modulation, transmission, and reception. The communication unit deals with the requirements of the physical layer. The data link layer, also

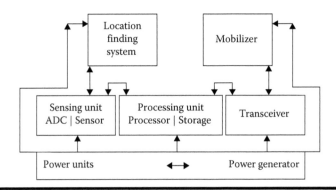

Figure 1.2   Architecture of wireless sensor network (WSN).

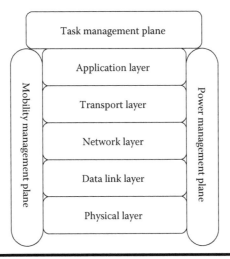

Figure 1.3   Protocol stack of WSN.

called the medium access control (MAC) layer, is primarily associated with the media access control. At various instances, the sensor network is mobile and its topology may also change; hence, the MAC protocols have to be power aware of minimizing the collision with a neighbor's broadcast. Routing of data from source to sink is carried out by the network layer. The flow of data is maintained by the transport layer. The sensor network is an application-specific network, so depending on the sensing, task-specific applications are built at the application layer. Besides, these five layers are management planes that help in controlling the sensor functions of the PM, MM, and task management (TM). The power levels of various functions of sensor nodes such as sensing, processing of information, transmission of data, and reception can be managed at the PM plane. The PM schemes are run at different layers as in the MAC layer the nonfunctional nodes can be turned off. At the network layer, the PM node makes an efficient selection of a neighbor node to transmit data so that the overall network lifetime increases. The MM plane tracks the neighboring nodes and changes in topology so that the route is maintained. The TM plane manages the sensing task in the network in such a way that not all the nodes perform the sensing at a time. This helps in conserving the energy of the system and improving the overall lifetime of the system. Various tasks are distributed among the multiple sensors to improve efficiency. All the management planes are required to ensure that the sensor nodes work in a coordinated way to bring about efficient PM, sharing of resources, and routing of data in WSNs.

1.3.1 *Physical Layer*

The physical layer converts a bitstream from the data link layer and makes it suitable for transmission. The transmission requires the communication channel. So far, the 915 MHz industrial, scientific, and medical bands have been used for the sensor network. Modulation, power efficiency, and efficient signal propagation are some areas that are being dealt with by researchers. Simple and low-power modulation schemes need to be developed. Further work is being done on communication strategies to overcome the signal propagation effects. At the physical layer, the major challenge is to develop hardware that is low power, small, and has low-cost transmitters, receivers, and sensing units.

1.3.2 Data Link Layer

The data link layer provides MAC, data stream multiplexing, error checking, and control. The packet frames are created at this layer and their transmission provides reliable point-to-point or multipoint communication. The MAC layer creates a network infrastructure by establishing a communication link between scattered nodes of the sensor network. This gives energy efficiency, high performance, high network throughput, and low latency in data delivery.

1.3.3 Network Layer

The network layer routes the data from distributed sensor nodes to their head or sink node. The deployed sensor nodes monitor and observe the phenomenon of their interest. Then the sensed data are transmitted to their neighbors for further processing. Basically, the data transmission can be either single-hop or multihop. The single-hop is used for long range while multihop for short-range communication. In single-hop communication, the nodes require higher power levels and complex implementation. On the other hand, the most effective multihop communication provides energy efficiency. This reduces the signal propagation and fading effects of channel. As far as the WSNs are concerned, the nodes are deployed in such a way that there is a short distance between them. Therefore, multihop communication is most suitable for these networks and the routing protocols provide them an optimum path for communication. The energy-efficient routing protocols for multihop communication are required for the WSNs to live long.

1.3.4 Transport Layer

The transport layer provides reliable and node-to-node data delivery. The traditional transport layer protocols for communication do not work well with WSNs. As WSNs are resource limited, high energy efficiency and performance are required for increasing the network lifetime. Therefore, the modified transport layer protocols that can work efficiently with WSNs need to be explored. They should provide modified error control, congestion control, and data transmission strategies for WSNs.

1.3.5 Application Layer

The application layer protocols are the sensor management protocols (SMPs) that are used for managing the sensor tasks, energy, etc. These protocols provide query dissemination, data security, localization and positioning, synchronization between nodes, etc. Operations such as transferring sensed data, synchronization between nodes, location-related information, node mobility, task scheduling, and node query are performed within the boundaries of these protocols, and they provide energy efficiency in the network too.

1.4 Features of Commercially Available Node Components

Some of the commercially available sensor nodes and their hardware platforms include onboard sensors; however, few of them require externally connected sensor modules for physical environment and microcontroller interaction. They can directly be used for any specific application or can be configured through programming. The day-to-day increase in the demand for WSN applications is motivating researchers, academicians, and industrialists toward this area. The criteria for selecting development, hardware, and debugging environment depend on the application requirement and energy goal. Minimum cost, complexity, time, space, memory, and energy are the basic and desired design parameters for resource-constrained WSNs. A sensor "node" is also termed as "mote" and it can differ in its size, cost, microcontroller, operating system, and battery type. The sensor nodes are customized for their hardware/software and prepared for the deployment of a WSN application. The commercially available sensor node constitutes hardware, operating system, the application code in a specified programming language, and a power supply to work unattended. The operating system provides support in terms of bug fixing, reconfiguration, program code update, and flexibility for energy and performance management. For example, Telosb motes work on "TinyOS" operating systems, applications programmed in a language nesC, and batteries for supplying power. This has ultra-low-power consumption, Universal Serial Bus (USB) programming, MSP430 microcontroller, and faster sleep and wake-up times. Table 1.1 depicts a list of sensor motes and their features. The microcontroller associated with the mote features such application requirements as operating voltage, clock frequency, real-time clock, memory, processing speed, internal/external interrupts, timers, and energy use. The MSP430 and Atmel ATmega128L are the most preferred microcontrollers for sensor networks because of their special features such as low-power consumption, varying operating voltage and frequency, high performance, and computational power. The operating system manages and schedules the available resources and data use of the services such as processing, transmission, and reception. It also provides hardware abstraction to application software.

Table 1.1 Comparative Analyses of Sensor Nodes

Sensor Mote	Microcontroller	Operating System	Clock Frequency	Battery Voltage
SunSpot	Atmel ARM920T	Java J2ME CLDC 1.1	180 MHz	3.7 V
SenseNode	TI MSP430F1611	GenetOS	8 MHz	1.8–3.6 V
S-Mote	TI/Chipcon CC2430 SoC	RETOS	32 MHz	2.0–3.6 V
Cricket	Atmel ATmega128L	TinyOS 1.1.6	8 MHz	2.1–3.6 V
Shimmer	TI MSP430F1611	TinyOS 2.x	8 MHz	1.8–3.6 V
EYES	TI MSP430F149	TinyOS, PEEROS	5 MHz	3.0 V
FireFly	Atmel ATmega1281	Nano-RK RTOS	7.3728 MHz	3.3–10.0 V
AVRaven	Atmel ATmega1284p	Atmel Studio	4–20 MHz	5–12 V or External power
BTnode	Atmel ATmega128L	BTnut, NutOS, and TinyOS	8 MHz	2.1–3.6 V

(Continued)

Table 1.1 (*Continued*)   Comparative Analyses of Sensor Nodes

Sensor Mote	Microcontroller	Operating System	Clock Frequency	Battery Voltage
iMote2	Intel PXA271 Xscale	SOS, Linux	13–416 MHz	3.2–4.5 V
TelosB/TmoteSky	TI MSP430F149, TI MSP430F1611	Contiki, TinyOS	4–8 MHz	2.1–3.6 V DC
Tmote Mini (Plus)	TI MSP430F1611	Moteiv's OS	8 MHz	2.1–3.6 V
Mica2/MicaZ	Atmel Atmega 128L	TinyOS	8 MHz	2.1–3.6 V
SHIMMER	TI MSP430F1611	TinyOS 2.x	4–8 MHz	1.8–3.6 V
IRIS	Atmel ATmega 1281	MoteWorks/ MoteRunner	8, 16 MHz	3.0 V or External power
Lotus	Cortex®M3	Free RTOS, MoteRunner, MEMSIC Kiel	10–100 MHz	3.0 V or External power
Sun SPOT	Atmel AT91RM9200	No	180 MHz	3.7 V rechargeable
EZ430-RF2480/2500	TI MSP430F2274	TinyOS	16 MHz	3.6 V
ANT	TI MSP430F1232	Ant	8 MHz	1.8–3.6 V
WiSense	TI MSP430	NA	8/16 MHz	1.8–3.6 V
Arduino BT	Atmel ATmega328	Arduino IDE/ Java	20 MHz	2.5–12 V
Neomote	Atmel ATmega128L	SOS, MantisOS, RTOS	8 MHz	3.0 V
µAMPS	Intel StrongARM SA-1100	µOS (adaptation of the eCOS microkernel)	59–206 MHz	NA
XYZ sensor node	OKI Semiconductor ML67Q5002	SOS	1.8–57.6 MHz	Ni-MH rechargeable batteries—3.6 V
PowWow—CAIRN	TI MSP430F1612	PowWow	8 MHz	2.7–3.6 V
Waspmote	Atmel ATmega1281	Libelium OTA	14.7456 MHz	3.3–4.2 V or solar (6–12 V)

1.5 Available Operating Systems and Networking Environments for WSNs

A sensor node typically has four basic software modules such as sensor/communication drivers, communication code processors, data processing applications, and an operating system (Pughat and Sharma 2015a). These modules together deal with the sensing, communicating transceivers, routing, encryption, topology control, encoding, synchronization, data processing, manipulation, reconfiguration, and architecture utilization for WSN application (Hill et al. 2000). This improves the performance and energy efficiency in event-driven, data-centric, and resource-constrained wireless sensor nodes.

There are many inexpensive, commercially available, and open source operating systems for the design of WSNs. Some of them are component based, event driven, power aware, design constrained, and work in distributed and multiple targeted environments. The salient features of a sensor network operating system are listed below:

■ It requires minimum memory space (in kilobytes) and less time complexity of scheduling.
■ It requires the memory-sharing feature that can reduce the memory space on the sensor node.
■ The sensor nodes should provide the finite state model for Kernel and application program interfaces for application to hardware access.
■ It should provide task processing within the deadline specified and meet the real-time application requirements (Pughat and Sharma 2015c).
■ The available hardware and software resources should be managed to reduce energy consumption, time, and memory use (Pughat and Sharma 2015b).
■ The sensor node should be power aware to improve the power and performance trade-offs (Pughat and Sharma 2016). It should be capable of implementing such PM policies as time-out, breakeven time, scheduling, dynamic PM, dynamic voltage scaling, etc.

Ideally, these operating systems are designed to aim for minimum space, small code size, faster implementation, and flexibility in modification. There are several issues related to the design and development of operating systems in WSNs.

One of the major issues is the required architecture of an operating system. It can be any of the monolithic types, microkernel type, layered type, and virtual machine type. It should be flexible, small size, and require less memory, but extendable as per the requirement of the application.

The second issue is the programming model, which should be lightweight and suitable for resource-constrained WSNs. There can be event-driven and multithreaded programming models. The multithreaded model is more flexible but resource intensive. Developers are searching for a way to event driven and multithreaded to make the light-weight multithreaded programming model for WSN.

The third important issue is resource allocation and sharing. It is provided by the operating system for efficient use of available resources to deal with the multiple programs for execution. The process requires scheduling, synchronization, and memory.

The fourth issue is the communication protocol support used by the operating system. It should provide the support for all protocol layers such as network layer, transport, data link, and MAC layers. It should be flexible irrespective of the homogeneous or heterogeneous environment.

The fifth issue is related to memory management in the operating system. Memory management can be a static type or a dynamic type. The later gives online memory management while the former does not support run-time memory allocation. The flexibility provided in dynamic memory allocation reduces the size of memory and ensures real-time execution. Apart from the above-mentioned points, the cross-layer design issues for WSN applications require the attention of both academicians and industrialists.

1.5.1 Operating Systems

In this section, the operating systems of WSNs that are suitable for the selection of any application are discussed.

1.5.1.1 AmbientRT

AmbientRT is a real-time operating system for the design and development of a mobile environment in WSNs and nodes such as µnode. The AmbientRT operating system has been proved energy efficient for real-time environments, even with limited available resources such as processing, memory, energy, etc. Hofmeijer et al. (2004) present the earliest deadline first (EDF) scheduling for AmbientRT and found it very efficient for resource-constrained WSNs. The dynamic features of this operating system also provide dynamic memory allocation and reconfiguration at runtime where other operating systems lack. The dynamically loadable module (DLM) written into the program memory supports the dynamically reconfigurable facility for sensor node. It also provides the solution for better processor utilization, saving energy, and central processing unit (CPU) cost.

1.5.1.2 TinyOS 1.x

TinyOS is an open source, static memory allocation-based, and popularly used operating system for resource-constrained WSNs (e.g., random access memory [RAM] of 512 bytes and program memory of 8 kB). It is written in nesC programming language and gives the component-level execution model and interfacing for event-driven applications. It has three layers of abstraction: hardware presentation, hardware abstraction, and hardware-independent layers. A first-in–first-out (FIFO) mechanism is used for queuing and scheduling the tasks for the events. It has a set of components and modules that are interfaced through wires. The currently executing task cannot be terminated on the occurrence of any other task (i.e., preemption of tasks). The TinyOS has small code size and provides less functionality in operation. Thus, the sensor node gives the faster execution of tasks and saves energy when it goes to sleep after completing the queued events.

1.5.1.3 TinyOS 2.0

This open source operating system was developed to overcome the limitations of TinyOS1.x and provides more flexible, efficient tasks and interfaces. The task scheduler is written in C language. It assigns task parameters by dividing the state into bytes for each task, thus reducing the number of POST failures. This way, different types of new tasks and their interfaces (such as PM interface for sleep scheduling) have been introduced. The different interface provides the concept of reuse and cooperative threading model for event-based sensor nodes (Chen 2010). The code for multiple tasks can run synchronously in TinyOS1.x and asynchronously in TinyOS2.x. The scheduler can differentiate between high-priority tasks and lower priority tasks and schedules the execution accordingly. The system file in TinyOS contains a basic scheduler for parameterizing tasks and a default scheduler that can wire the basic scheduler with the sleep-scheduling policies. Thus, it is more suitable for applications in which energy efficiency is the primary concern. It supports platforms such as MicaZ, TmoteSky, Telosb, iMote, etc.

1.5.1.4 Contiki

This is an open source, multitasking, portable, embedded operating system that is very useful for resource-limited (i.e., low power, processing, communicating, and less memory) devices such as sensor nodes and rules the Internet of Things (IoT). It is written in C language and provides a network simulation environment with the simulator "Cooja." The code size and event kernel require a space larger than that for TinyOS but smaller than that for the Mantis operating system. The developers like to work on the most promising, small, lightweight, and mature operating system like Contiki and give competition to the Windows operating systems for IoT. It has a hybrid execution model. The sensor nodes with TI MSP430 and Atmel AVR controllers can be emulated with Contiki 2.6 version. The operating system provides special features such as multitasking, multithreading, Internet protocol networking, Web browser, and graphical user interface (GUI). The new version, Contiki 3.0, was released in 2015 with more stability, flexibility, and low-power mesh networking protocol features (Contiki: The Open Source Operating System for the Internet of Things n.d.; Dunkels et al. 2004). It supports the Texas Instruments Sensortag and the Zolertia ReMote but not with the memory protection.

1.5.1.5 MANTIS

The classical, layered, and multithread architecture of the MANTIS operating system enhances the capability of event processing and priority scheduling. It is a developer-friendly operating system written in C-based application programming interface (API). It has a four-layer architecture with a hardware layer, a communication layer, a MANTIS API layer, and a network layer. The multiple numbers of threads can be handled simultaneously and coordinated for different types of tasks and their processing. This makes the task execution faster and improves the energy efficiency of the WSNs. It gives information such as task priority, pointer, stack, and state (Bhatti et al. 2005). To reduce energy consumption, the sensor node goes to idle or sleep when no task is processing and no thread is working. The faster execution time for the MANTIS operating system results in faster processing, while the average queue length decides the processing speed in TinyOS (Duffy et al. 2006). It has a more complex scheduling approach than TinyOS and its sleep duty cycle introduces more overheads. It supports Mica2, MicaZ, Telos, and Mantis nymph.

1.5.1.6 LiteOS

This Unix-like operating system provides ease to program developers and uses object-oriented programming languages, i.e., Lite C++. It can work with MicaZ and Atmel AVR motes as a modular architecture. It supports dynamic memory management, loading, file-based communication, synchronization, and process protection. It works well with the network simulator "Avrora." This adjusts the separation and size of dynamic memory for kernel and application requirement. It provides a nonpreemptive priority scheduling in which a task executes until completion and then the next task with higher priority is executed. It uses priority-based round robin scheduling and a file-based communication support system. This can miss the task deadline and does not guarantee real-time task execution (Cao et al. 2008). It also gives a provision to register call back function and avoid a race condition. The authors have presented the performance of LiteOS in 21 examples of applications (Cao et al. 2008) and described its implementation. The telecom company Huawei will launch an ultra-light, approximately 10 kB, LiteOS for smart sensors and IoT.

1.5.1.7 BTnut/NutOS

This real-time operating system is specially designed for BTnodes. This is a nonpreemptive, cooperative multithreaded operating system that is written in C language. In the cooperative multithreaded operating system, a program runs within limited resources and it can handle the external interrupt with the help of an event handler. This OS is designed for implementation in Atmel ATmega128 microcontrollers. It provides a reprogramming architecture environment and dynamic memory allocation in sensor nodes. This supports the multihop networking in WSNs. The BTnut operating system is open source under Berkeley Software Distribution (BSD) license and ready for modification in the source code, binary code, and kernel code (Embedded Ethernet n.d.).

1.5.1.8 MagnetOS

This distributed operating system provides power-aware support in WSN applications. It uses a Java language interface and virtual machine as a programming paradigm or execution model. The multihop routing protocol works on this adaptive operating system. It has the capability of modifying and application code statically and task creation/migration at the runtime. The power-aware support reduces the risk in manual programming and insufficient local optimization whenever required from the application development point of view (MagnetOS n.d.). This also makes the system stable and more energy efficient by managing the code and available resources. It does the partitioning in the application and then assigns the node components according to the immediate need (Barr et al. 2002). It gives easier development of WSN applications.

1.5.1.9 SOS

The dynamic operating system SOS has several advantages over TinyOS. It uses the application programming language C that is fairly well known to the users working in this area. The features such as dynamic memory allocation, loading of modules, message handling and ease in reprogramming, etc., enhance the flexibility of the operating system (Han et al. 2005). It has a module-based execution model. It is ported to the Imote2, Mica2, MicaZ, XYZ, emu, and Telos sensor nodes. It supports unattended monitoring and controlling, and updates regularly on the resource-limited sensor nodes. The operating system works in the cooperative scheduling mechanism and uses event-based programming without memory protection. The components of the sensor node can be modified after deployment of the entire network.

1.5.1.10 Nano-RK

The Nano-RK provides the monolithic architecture of kernel, lightweight network protocol, and real-time environments with multitasking. It has a task-based execution model. The user can change the task deadlines, CPU, and network bandwidth reservation for soft real-time applications. It is used for FireFly and MicaZ sensor motes. It can provide process-level priority scheduling as well as network-level priority scheduling. The preemptive multitasking scheduling in this operating system consumes energy and degrades the system performance. Thus, a task control block (TCB) provides the memory and synchronization for that (Farooq and Kunz 2011; Overview—Nano-RK n.d.). The research requires the new approach for making the dynamic memory management-based Nano-RK as it only supports static memory management. This operating system ensures the execution of the highest priority event first. It also uses rate monotonic (RM) and rate-harmonized scheduling, which is used to save energy during the idle period of the CPU. It is useful in multimedia sensor network applications.

1.5.1.11 RIOT

This is an open source, developer- and resources-friendly operating system for real-time applications. The application program uses C or C++ language flexibility. RIOT is very useful for applications where low power and less memory are required. It can run code on controllers such as MSP430 and Atmel ARM, etc. The microkernel architecture, multithreading with fewer overheads, memory allocation, and priority scheduling are the key features for the use of this operating system (Will et al. 2009). Apart from the salient features mentioned, the RIOT operating system lacks the energy efficiency if the sleep time duration is smaller comparatively and possesses low memory footprints. It provides fast task switching and a less complex scheduler for higher energy efficiency. This operating system can work in a heterogeneous environment such as overlapping of the IoT and WSNs. Thus, it bridges the operating system gap between WSNs and, for the next generation, Internet-based applications (Baccelli et al. 2013).

1.5.1.12 EYES OS

This is event-driven, distributed, and power-aware operating system for low-power wireless sensor nodes. It requires smaller code size and low energy for application support. The resource management and reconfigurability are the special features of EYES operating systems. The remote procedure call operation is useful when resource sharing from other nodes is required to perform any urgent computation. Although it does not guarantee real-time task execution and features such as resource and memory management, real-time scheduling has been incorporated in it to develop the PEEROS operating system. The PEEROS supports the priority event handling and multitasking capabilities and guarantees real-time application environment. The PEEROS operating system is more fault tolerant and less prone to errors due to node failures.

1.5.1.13 NanoQplus

The NanoQplus operating system has a real-time and thread-based execution model (Cheol Kim et al. 2008). The priority scheduling guarantees real-time application behavior and lower power consumption with a PM scheme. The limitations of this operating system are the dynamic reprogramming and memory management.

1.5.1.14 EMERALDS

It has a microkernel architecture and works on resource-constrained embedded systems like wireless sensor nodes. The distributed and object oriented language, i.e., C++, simplifies the distributed programming of the operating system. The OS performs a real-time operation with the EDF and RM scheduling techniques. The operating system provides features such as multithreading and memory protection. The kernel handles the interrupts and memory sharing (Zuberi and Shin 2001; Black et al. 2007).

1.5.1.15 SenOS

The SenOS contains elements such as the event queue for event management, state transition flow-table to represent actions between states and the callback library to access the hardware components of the node. The callback library is stored in ROM and the state table in RAM. Thus, it has a state-based execution model. It works on the concept of finite state machines in which the state table represents the behavior of any application used (Hong and Kim 2003). The change in the state and its transition can reconfigure any application on the SenOS. The monitor associated with the operating system helps in changing the transition table. It has lesser flexibility as the transition table can only be changed by changing an application. The finite state machine can represent dynamic PM on the SenOS and it helps in energy saving on the sensor node.

1.5.2 Simulation/Emulation/Debugging Environments

Several simulation/emulation/debugging environments are present, which can simulate different scenarios of WSN. These help in setting the functionality of WSNs.

1.5.2.1 TOSSIM

TOSSIM (TinyOS SIMulator) provides discrete event, high performance, and homogeneous simulation environment. It is used to configure, compile, and run discrete event simulations in high-scale sensor networks (up to 1000 nodes). This simulator supports the compilation environment of Mica2, MicaZ node, and radio CC2420. Python and C++ are the programming interfaces that TOSSIM uses for the application program. The generated XML files give information about component variables after compilation of TOSSIM (TOSSIM—TinyOS Wiki n.d.). TOSSIM runs the TinyOS codes and gives the simulation replacement to hardware requirement. The source code is modified and compiled in the TinyOS environment with less effort and more accuracy. It gives the flexibility of packet injection and scheduling of packets for the node. The limited scalability, no preemption in task handling, and invisible energy consumption on TinyOS environments restrict its use at a broad level. The data collection protocol under different network scenarios is analyzed and its performance is evaluated using a TOSSIM simulator (Nath 2013).

1.5.2.2 PowerTOSSIM

PowerTOSSIM is an effective tool for analyzing the power of sensor node components at the time of running application. A comparative power analysis of the four routing protocols (gossiping, flooding, gradient-based routing [GBR], and LEACH) using PowerTOSSIM shows the least power consumption in the LEACH protocol (Kumar et al. 2006). The LEACH protocol is referred to as a low-energy adaptive clustering hierarchy. It uses the component library present in TinyOS and TOSSIM for measuring the power state of components and a TinyViz tool to visualize the runtime date. The version PowerTOSSIM-2 gives power simulation for TinyOS/TOSSIM 1.x and MICA2 motes, while another version, i.e., PowerTOSSIM-Z, for TinyOS/TOSSIM 2.x and MicaZ motes (Perla et al. 2008). It has the ENERGY_HANDLER and several calls to evaluate the energy of the node components and debug the functions. For PM, TinyOS 2.x classifies node components in one class of microcontroller and another class of peripherals.

1.5.2.3 Viptos

This gives a heterogeneous, graphical, and simulation development tool for TinyOS-based or non-TinyOS WSNs. A diagram of blocks and arrows created in Viptos (Visual Ptolemy and TinyOS) can be converted to a TinyOS program. The tool can model network and communication channels and multinode routing, etc. This provides a visual interfacing between TinyOS and Ptolemy II. It can convert the diagram into a program in nesC. It gives the switching between high-level and low-level programs. The version of Viptos 1.0.2 can work with both Linux and Windows platforms. The network parameters such as delay, packet trace, physical quantities, corruption, etc., can be simulated using Viptos (Joint Modeling and Design of Wireless Networks and Sensor Node Software | EECS at UC Berkeley n.d.; Lee et al. 2006).

1.5.2.4 GloMoSim

GloMoSim (Global Mobile Information System Simulator) is an open source simulator (GLOMOSIM Simulator | Network Simulation Tools n.d.). It has a library developed in a C-based discrete event simulation language, i.e., Parsec (Parallel Simulation Environment for Complex systems). The GloMoSim and Parsec directories can be installed on a Windows environment using Visual Studio and Java SDK. The protocol and programs are compiled using a Parsec compiler. It is a scalable network analysis protocol that gives sequential and parallel simulation environments for wireless and wired networks. The visual configuration of the simulation model is done using PAVE (Parsec visual environment). It represents a node by its entity and simulates up to thousands of heterogeneous entities in one simulation scenario. The GloMoSim models can be found for any network layer such as physical (radio), data link (MAC), routing, transport, and application layers.

1.5.2.5 ATEMU

This is a heterogeneous network tool for emulation, simulation, and debugging purposes. This open source tool supports C-based AVR microcontrollers and TinyOS-based Mica2 nodes. Its visualization and graphical debugging can be done using XATDB. Authors have designed and implemented the ATEMU to rectify the gap between real simulation and deployment (Polley et al. 2004). It gives control of the operation of sensor nodes in real-time communication in a sensor network environment. The latest version is 0.4-mf4 (Atemu Sensor Node Simulator/Debugger 0.4—Free Download n.d.). It is compatible with operating systems such as Solaris and Linux. Thus, the accuracy and emulation in ATEMU ensure the deployment of new protocols on hardware. However, it requires research work on making faster simulation speed, more useful routing, and clustering functions.

1.5.2.6 MATLAB/Simulink

The MATLAB® (MATrix LABoratory), developed by MathWorks Inc., is used to design and simulate various research issues in WSNs (Wireless Sensor Network Simulation—Tutorial for Matlab | Our Work | Wireless System Laboratory of Brno n.d.). The built-in library in MATLAB/Simulink provides flexible and effective simulation for designing network layer architecture, communication channels, energy models, and node architecture for homogeneous and heterogeneous environments. It has the capability of programming, solving computational mathematics, and graphical visualization. Researchers have given important program codes for algorithms such as LEACH for routing (Low Energy Adaptive Clustering Hierarchy Protocol (LEACH)—File Exchange—MATLAB Central n.d.) and localization (Mobile User Localization in WSN—File Exchange—MATLAB Central n.d.).

A MATLAB-based PiccSIM (Platform for integrated communications and control design, simulation, implementation, and modeling) tool provides a network environment in NS-2 and control blocks in Simulink. Thus, it is a co-simulation tool for networked control systems, which gives the data exchange capability between Simulink and NS-2 (PiccSIM Simulation of Wireless Control Systems—Wireless Sensor Systems Group—Aalto University n.d.). A TrueTime–PiccSIM extension (TrueTime—PiccSIM—Wireless Sensor Systems Group—Aalto University n.d.) was released in 2013 for linking embedded kernels to the networked control systems. The kernel extends the real-time features such as task allocation, scheduling, task execution, etc.

1.5.2.7 Avrora

Avrora (AVR Simulation and Analysis Framework) is an open source simulator for AVR microcontroller platforms such as Mica2 and MicaZ. This can provide single-node emulation and many-node simulations. Its latest version is Avrora [Beta 1.7.105], which is much easier than TOSSIM. This is an instruction-level and cycle-based network simulator that has higher fidelity and scalability (up to 10,000 nodes). It provides almost one-half of the processing speed of TOSSIM and is 20 times faster than ATEMU. It works in the Linux or Windows environment. The programs are simulated and tested on Avrora before deploying on the hardware module (Titzer et al. 2005). Thus, the simulator can provide program monitoring, dynamic behavior, source code debugging, and stack and flow controlling. The battery life can be monitored using energy analysis tools. The drift in clock frequency for nodes and lacking authenticity in validation of results are the two main lacunas for this simulator.

1.5.2.8 Network Simulator v2 (ns2)

This is an open source, discrete event simulator written in C++ and provides Tcl scripting for networking applications. The simulator can work in the Linux and Windows environments. The installation of Cygwin is required for ns2 to work on Windows operating system. It helps in model development, node configuration, program execution, performance evaluation, and visualization (XGRAPH) for WSNs. There are various ns2 programs available online for working on the physical layer and MAC layer. A recent version is ns-2.35 (The Network Simulator Ns-2: Building Ns Version 2 n.d.). The all-in-one package of ns2 requires a large memory space of 320 MB. The modeling in ns2 is time consuming, and the scripting language is difficult to understand. This does not give precise timing simulation. The Mannasim (MannaSim Framework—Main Page n.d.), an extension to ns2, provides a WSN framework. It is used to design, develop, and analyze advanced modules in WSNs.

1.5.2.9 Network Simulator v3 (ns3)

The ns3 is an open source, discrete event, network simulator for Internet-based systems. This is suitable for working on the network and transport layers in wireless networks. It was first released in 2008, and 25 versions of ns3 were released through September 2014. The most recent version, ns-3.25, was released in March 2016 (Download « Ns-3 n.d.). The program codes are written in C++ and optional Python bindings. This features additionally the traffic control, congestion control, Wi-Fi modules, larger channel width, and routing protocols for Internet module. The WSN modules such as 802.15.4 are integrated into ns3. The researcher working with ns3 faces compatibility issues in the codes written in ns2. Thus, the WSN protocol support is less in ns3.

1.5.2.10 Cooja Simulator (Contiki OS)

The Contiki OS-based Cooja simulator is open source, Linux- or Windows-compatible software, which runs in the execution environment of *VMware Player* (virtual machine). It is used for compiling and testing of codes for the simulation of sensor motes (e.g., TMote Sky, Telosb) before running on real-time hardware. The package "Instant Contiki" provides a complete tool for development, compilation, and simulation (see for details: Get Started with Contiki, Instant Contiki and Cooja n.d.). This complete set-up including GUI occupies only 30 kB of RAM.

The Cooja simulator folder can be found in the folder ("tool") of the Contiki OS. Its front-end interface is provided by Java language. The tool has five different windows: *network window* to visualize the location and status of sensor nodes; *simulation control window* for monitoring and controlling execution and simulation time; *notes window* for writing notes on simulation; *motes window* to trace motes output data; *timeline window* for showing messages and motes timeline. The developed code gives the precise but slower analysis of simulation. The latest version is Contiki 3.0, developed and released in August 2015 (Contiki: The Open Source Operating System for the Internet of Things n.d.).

1.5.2.11 OPNET

OPNET is a C, C++ interface-based network simulation tool that provides performance management at the application and network layers. It gives an analysis for event-based and packet-level simulation. Since October 2012, the OPNET has been a part of the company Riverbed SteelCentral. That facilitates to teachers and academic researchers a free version of OPNET Network Simulator. Compared to ns2, OPNET provides more reliable and accurate results to IEEE 802.15.4/ZigBee protocol of WSNs. The OPNET simulation tool has been developed for slotted CSMA/CA mechanisms, guaranteed time slot (GTS) mechanisms, cluster-tree topologies, hierarchical tree routings, etc. The battery module gives the remaining energy levels for Micaz and Telosb sensor motes. The tool can analyze performance parameters such as propagation delay, collided frames, radio interference, noise, received power, bit error threshold, bit error rate (BER), etc. (Li et al. 2016).

1.5.2.12 OMNeT++

OMNeT++ (Objective Module Network Testbed in C++) has a component-based library, which provides a framework for discrete event simulation (Varga and Hornig 2008). It is used to model and evaluate the performance of different scenarios on wireless networks. The tool can support MAC protocol, power control, and localization protocols in sensor networks. Its latest version, OMNeT++ 5, was released in 2016 (OMNeT++ Discrete Event Simulator—Download n.d.). In addition to the previous version of OMNeT++4.x, it has some improved features such as Canvas API for 2D graphics, OpenSceneGraph for 3D graphics, Qt-based runtime environment, and logging. The open source OMNeT++ tool and its codes are available for academic purposes but not for commercial use. The graphical environment can be visualized using Eclipse (IDE). This tool is compatible with Linux, Windows, and Mac OS X operating systems.

1.5.2.13 QualNet

QualNet (Quality Networking) is a commercially available simulation tool for performance evaluation of communication networks. It supports the Linux- and Windows-based operating systems. The source codes for WSN component library are available in the C/C++ language. It is used to design, develop, manage, optimize, test, and create animation on network-centric systems. The tool consists of components such as QualNet Architect for scenario design and visualization, QualNet Analyzer for statistical graphics and post analysis, QualNet Packet Tracer, QualNet File Editor, and QualNet Command Line Interface (Qualnet—Packet Trace | SCALABLE Networks n.d.). It features real-time speed, scalability up to thousands of nodes, high fidelity, portability, and extensibility to connect to other network models.

1.5.2.14 J-Sim

J-Sim (Java-Simulator) is an open source, component-based, process-driven, real-time simulator. The tool can compile mathematical as well as text-based languages for its flexible models. It provides the interface for other languages such as Tcl, Python, and Perl. It has a scalability of around 500 nodes. It also has the capability of autoconfiguration and online monitoring. A version of J-Sim is found as J-SIMv1.3 + patch4, which can simulate communication channels and power consumption (Download J-Sim—J-Sim Official n.d.). A minor change with simplified installation was released in 2013. Then, a recent open source JSimVersion 2.16—Beta was released in 2015. In addition to previous version features, this version provides optimal control parameters, "Advanced" menus, and bug fixed reports (JSim Download Page n.d.). Researchers in the field of WSN can work on the localization, routing, and data diffusion protocol issues using J-Sim. Compared with NS-2, J-Sim, gives more scalability and requires less memory at the cost of execution time.

1.5.2.15 NetSim

The NetSim (Network Simulator) is a discrete event simulator that uses C++/Java scripting language. It is used for protocol modeling and performance evaluation through simulation. It provides support for protocol design, agent modeling, energy efficiency at the physical and MAC layers, PM, routing, self-configuration, localization, and clustering in WSNs. The user easily works on the C-coded libraries available in the NetSim package. Its academic version provides technology coverage, performance reporting, and packet animator. However, it does not support packet/event tracing, source codes for protocol library, development environment, and dynamic metrics (Common Modules).

1.5.2.16 SensorSim

SensorSim provides a ns2-based simulation framework for WSNs. Apart from the features available in ns2, the SensorSim provides additionally the sensor and sensing channel models, protocol stack for microsensors, battery models, and hybrid simulation models. It has the capability of real and simulated node communication, new communication protocol development, and real-time user-to-GUI interaction. SensorSim can give in-depth design and analysis of sensor nodes, but the scalability is the major problem (Park et al. 2000). On the other hand, NRL SensorSim is a WSN simulator developed by the Naval Research Laboratory (NRL) group. It is used for the energy consumption model and analysis in sensor networks. SensorSim and NRL SensorSim are not currently under development and are no longer available.

1.5.2.17 SENSE

SENSE (Sensor Network Simulator and Emulator) is a component-based, discrete event simulator written in C++ language. It uses the iNSpect tool for result visualization and animation purposes. It provides components such as physical layer protocol, MAC layer protocol, application layer protocols, network layer protocols, wireless channels, simulation engine, battery models, and graphical interface (G-Sense) (Perez et al. 2010). It has some special protocols such as self-selective-routing (SSR) protocols, self-selective reliable path (SRP) protocols, and self-healing routing (SHR) protocols. The tool supports component reusability, parallelization for scalability, and simulation model interdependency. The intercomponent communication and simulation speed need to be improved.

1.5.2.18 EmStar

EmStar is a real-time, trace-driven emulator/simulator for iPAQ-class and Mica2 sensor nodes (Girod et al. 2007). It gives development, deployment, emulation, simulation, and graphical environment for heterogeneous networks. More often, it is used for the development and deployment of complex and high-capability nodes such as microservers. These nodes consume more power but operate for higher performance and complex sensor network applications. The tool is built in C language and supports module reusability. It is compatible with the Linux operating system. The limited scalability reduces the use of this tool for WSNs.

1.6 Factors Affecting Design of Energy-Efficient Sensor Nodes and Networks

The design of a limited resource WSN for a particular application requires the knowledge of multidimensional aspects, which affect its energy efficiency. Consequently, the existing hardware, protocols, architecture, and operating environmental conditions affect the lifetime and overall performance of the network. There are ample protocols/techniques available in the state of the art, which enhance the energy of a sensor node or the entire network. The cluster head mechanism is one example, which increases the energy efficiency of WSNs. Apart from this, the improper implementation of these low-power design techniques can increase the cost and energy overhead of the network. Therefore, a designer should understand commercially available hardware/software and the multidimensional aspects (e.g., network traffic and topology, fault tolerance, quality of service, self-management, power consumption, production cost, etc.) before designing energy-efficient WSNs. A few of them are given below.

1.6.1 Hardware Constraints

The hardware components of sensor node such as sensors, processor, memory, ADC, relays, mobilizer, and radio are the main sources of energy consumption and necessitate their choice for low-power consumption at design time. The hardware operating conditions should not change with time and location. Sometimes, a lab-tested sensor operates with different characteristics in the deployed region. The implementation of low-power design techniques on sensor nodes may require additional memory space and computational cost. The technology advancement can also create a barrier during the implementation of existing techniques. Therefore, one should look for a feasible solution to low power design on the sensor node.

1.6.2 Network Traffic

The network traffic can be a periodic or continuous monitoring type. The nature of the traffic and its distribution depends on the type of WSN application. The energy efficiency depends on the way traffic flows from source nodes to sink nodes and the choice of hopping between nodes. The adaptive duty cycle (adaptive on/off) is one effective way of improving energy efficiency in WSNs. However, it is not applicable to continuous monitoring applications as these applications do not require the node to be off. On the other hand, the uniformity in load distribution throughout the network is another solution for traffic distribution. This requires known future traffic and continuous updates in the routing table to reduce the chances of network failure. Thus, we can say that network density, type of network traffic, and the traffic flow should be considered during WSN design.

1.6.3 Quality of Service

Task execution or data delivery within deadline defines the quality of service in WSNs. Latency in data delivery is acceptable if the quality of service is maintained. Otherwise, sensing becomes useless and increases energy overheads. The introduced intelligence capability in the sensor node increases the quality of service at the cost of on-board processing or computation. Further, this increases the power consumption of the node. Quality of service and power consumption affect each other, and choice of their trade-off limit depends on the application. If the quality of service is more important, then we compromise with the short lifetime of the node; otherwise, lower power consumption at the cost of quality of service is considered for designing the network.

1.6.4 Fault Tolerance (Reliability)

How long a sensor network is able to maintain its functionalities depends on the reliability of that network. The node failure in a network is tolerable until it affects the task and life of the node and network. The event reliability, packet reliability, and source and destination node reliability help in making the energy-efficient WSNs.

1.6.5 Self-Management

Self-management of the node is required when the backup node replaces a faulty/dead node or the link of any node is broken from the coverage area of the network. The node should be capable of self-management and the network needs to be self-configurable to adjust to a new member in the family of its actual nodes. Thus, proper routing and topology of the network are the main design concerns. The dynamic routing techniques are more power consuming.

1.6.6 Scalability and Network Density

Scalability or network density is defined by the number of nodes per square meter within the network range. It depends on the transmission range of the node, network, and coverage area. Higher network density (number of nodes in a network) affects overall fault tolerance, data processing and accuracy, communication, congestion, routing, and PM issues. The increased number of nodes in a network should not affect the functionality of the entire network.

1.6.7 Production Costs

The production cost of a node affects the cost of the entire network. Thus, the individual node should be less expensive and less power consuming to sustain for a long time in deployed networks. This makes a dynamic system and considers the effect of factors that influence the sensor network design. These factors help in comparing several schemes, protocols, and algorithms in WSNs.

1.6.8 Network Topology

Topology management is required at predeployment, deployment, postdeployment, and redeployment phases. The densely deployed network improves the fault tolerance, sensing coverage, and multihop routing coverage. However, sampling data from fewer nodes improves the energy efficiency of the network. It depends on the type of application for which the network is deployed and the minimum required for a lifetime of the network. Therefore, critical topology management is to either keep in reserve a few of the sensor nodes from the dense network as backup nodes or elect a coordinator for a group of nodes to forward data collectively. The sleep schemes also contribute to the lifetime improvement, but increase the multipath latency of the network.

1.6.9 Operating Environment (Applications)

The design and deployment of a WSN depend on what environmental conditions it can support for any application. A critical understanding of the type of physical quantity (e.g., temperature, humidity, pressure, proximity, etc.) to be measured is required. The types of transducers for measurement and the knowledge of interfacing unit between sensing and communication help in network design. The operating conditions of sensor node components such as temperature tolerance, humidity effects, data precision, and calibration change with time and the surrounding environment.

1.6.10 Security

Network security before deployment and after network formation should be considered to protect it from threats such as denial of service, false node, message corruption, node malfunction, passive information gathering, etc. The secure data has the features of data confidentiality, data authentication, data integrity, data freshness, etc. The major attacks on sensor network can be hijacking, eavesdropping, rushing, disruption, etc. Therefore, a suitable security mechanism needs to be designed that can consider such factors as processor performance, energy, memory capacity, environment, and application use.

1.6.11 Transmission Media

Transmission media for many of the applications in WSN is radio, i.e., radio frequency (RF) communication. It can be through wire, infrared, optical, acoustic, Bluetooth, etc. A low-power global positioning system (GPS) is required in some location-finding applications. Sometimes, external energy harvesting through RF energy, photovoltaic, piezoelectric, and thermos-electric energy is used for sampling, charging, and utilizing the battery resource.

1.6.12 Power Consumption (Lifetime)

The lifetime of a sensor node and network depends on the battery resource only. The nodes either continuously or with sampled intervals consume energy in sensing, processing, and communication from that source. Communication and network management protocols consume most of the energy. Although energy harvesting has been proved as an important mechanism for charging the node battery, it has various limitations so far. Thus, there is a definite need for efficient energy harvesting techniques.

1.7 Energy Conservation Mechanisms in WSNs

Researchers are making incessant efforts to create energy-efficient WSNs. Looking at the various factors affecting energy efficiency (refer to Section 1.6), several dimensions of energy conservation have evolved that are being explored by researchers. Figure 1.4 depicts energy conservation mechanisms being researched in present times to provide energy-efficient WSNs.

One of the key areas of research is PM. Different components of a sensor node have their own power requirements, and if they are addressed properly, a huge amount of energy can be saved. The operating system needs to efficiently schedule the tasks and apply techniques of dynamic PM, dynamic voltage scaling, and coordinated PM. The details are given in Chapter 5. It is seen that the computation system requires lesser energy than the communication system. Energy consumed in transmitting one bit may be used to compute a few thousands of instructions (Pottie and Kaiser 2000). Hence, computation is preferred over unwanted communication. Data reduction techniques like data aggregation and coded data transmission are being preferred to reduce the communication overheads. Further energy-efficient routing protocols are being explored by researchers to enhance the efficiency of the system. Controlling transmission power during communication is another area of extensive research to minimize the overall energy consumption. Chapter 7 deals with data aggregation techniques, whereas Chapters 3 and 4 are related to efficient routing based on soft computing techniques and cluster-based approaches used for enhancing the lifetime of the network, respectively. At the MAC layer, duty cycling is the most important mechanism to control the energy consumption. Duty cycling can be achieved in two ways. The first way is to control the topology whereby a minimum subset of nodes is selected for maintaining the connectivity. Alternatively, another subset of nodes is in sleep mode. An optimal subset of nodes is

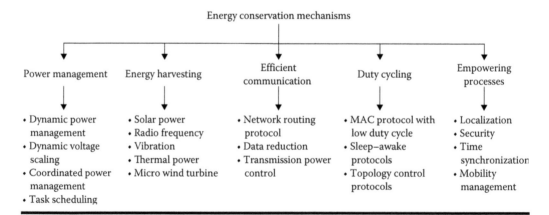

Figure 1.4   Classification of energy conservation mechanisms.

found to guarantee the connectivity. These subsets work alternatively and enhance the lifetime of the network. Second, duty cycling can be achieved by PM of nodes, i.e., the active nodes may not keep the radio in "on" state all the time. The protocols are designed to switch off the radio when the network is not active. There are several sleep and wake protocols and low-duty cycle MAC protocols to achieve low-duty cycle. Chapter 2 discusses MAC layer protocols and duty cycling concepts. Further, various processes need to be empowered and enhanced to address the energy consumption and provide solutions for energy conservation. The sensor network is unique due to its specific requirements such as self-organizing capabilities, low processing, small size, and memory space constraint. Several processes such as localization, time synchronization, MM, and security are concerns arising due to specific constraints of sensor networks; hence, mechanisms for efficiently managing them are different. Localization in a WSN is learning the location of the nodes autonomously or by assistance from any external source. The time synchronization in wake-up and sleep cycles of various nodes is required due to a duty cycling process at the MAC layer. Similarly, mobility can be either of a mobile sink or mobile node, which again is a unique require-ment. On top of all these processes, the basic requirement is energy efficiency, so these processes have to be empowered to make them energy efficient. Some of these techniques are discussed in Chapter 9. All these energy conservation schemes enhance lifetime but do not guarantee a per-petual lifetime to sensor node due to battery constraints. Hence, energy harvesting has become a key area of research in a WSN to actually capitalize its advantages by overcoming the constraints. Several energy-harvesting techniques are researched such as solar, wind, vibration, RF energy, thermal, etc. Chapter 6 gives an insight into these technologies that can be used for WSN. Besides these energy conservation mechanisms, there are still some challenges that need to be overcome to develop energy-efficient WSNs.

1.8 Challenges for Energy-Constrained WSNs

Although resource-limited sensor nodes have several excellent features such as low cost, small size, precise sensing, and accuracy, they require day-to-day research in terms of processor performance, radio range or coverage, memory capacity, and battery life. More often, the unattended nodes are destroyed by animals or tampered in natural disasters. Sometimes, they can harm the privacy of an individual and create undesired/unwanted effects too. Moreover, the highest energy consump-tion sources in a sensor node are sense, data processing, and communication. Network protocols such as the physical layer, data link layer, routing, clustering, security, packet forwarding, network communication, modulation schemes, traffic distribution, etc., are the main challenges which need to be immediately addressed. An energy-efficient MAC protocol for a mobile WSN is still a research problem. The proper routing algorithm for resource-limited (memory, energy) sensor networks is still finding a solution to make the network energy efficient. The security features and major attacks need to be considered for energy-efficient communication. Other issues such as user perceptions of reliability, competition from other protocols, the interconnection of WSNs and the Internet, networking issues, data aggregation, battery issues, DC–DC converter, overhead reduction, traffic distribution, and topology management should be taken care of. Energy-aware packet forwarding, energy-aware wireless communication, modulation schemes, PM of radios, energy-efficient routing, energy-efficient clustering, energy-aware software, power-aware comput-ing, coordinated PM, link layer optimizations, computation communication trade-offs, etc., are the immediate research areas in WSNs. A few of these issues are discussed in depth in the subse-quent chapters of this book.

References

Atemu Sensor Node Simulator/Debugger 0.4—Free Download. n.d. Available at: http://atemu-sensor-node-simulator-or-debugger.soft112.com/.

Baccelli, E., Hahm, O., and Günes, M. 2013. RIOT OS: Towards an OS for the internet of things. In *Proceeding of the 32nd IEEE International Conference on Computer Communications (INFOCOM 2013)*. Turin, Italy, 2453–2454, April 14–19, 2013, pp. 79–80. doi:10.1109/INFCOMW.2013.6970748.

Barr, R., Bicket, J.C., Dantas, D.S. et al. 2002. On the need for system-level support for ad hoc and sensor networks. *ACM SIGOPS Oper Syst Rev.* 36(2):1–5. doi:10.1145/509526.509528.

Bhatti, S., Carlson, J., Dai, H. et al. 2005. MANTIS OS: An embedded multithreaded operating system for wireless micro sensor platforms. *Mobile Netw Appl.* 10:563–579. doi:10.1007/s11036-005-1567-8.

Black, A.P., Hutchinson, N.C., Jul, E., and Levy, H.M. 2007. The development of the emerald programming language. In *Proceedings of the Third ACM SIGPLAN Conference on History of Programming Languages—HOPL III*. New York, NY: ACM Press, 11-1-11-51. doi:10.1145/1238844.1238855.

Cao, Q., Abdelzaher, T., Stankovic, J., and He, T. 2008. The LiteOS operating system: Towards Unix-like abstractions for wireless sensor networks. In *Proceedings of the 7th International Conference on Information Processing in Sensor Networks, IPSN 2008*. St. Louis, MO, 233–244. doi:10.1109/IPSN.2008.54.

Chen, Y.T. 2010. Enix: A lightweight dynamic operating system for tightly constrained wireless sensor platforms. In *Proceedings of the 8th ACM Conference on Embedded Networked Sensor Systems*. Zurich, Switzerland, 183–196. doi:10.1145/1869983.1870002.

Cheol Kim, S., Kim, H., Song, J., Yu, M., and Mah, P. 2008. NanoQplus: A multi-threaded operating system with memory protection mechanism for WSNs. 1st China-Korea WSN Workshop (CKWSN), 12–15 October, 2008 in Chongqing, China.

Contiki: The Open Source Operating System for the Internet of Things. n.d. Available at: http://www.contiki-os.org/.

Download «Ns-3. n.d. Available at: https://www.nsnam.org/ns-3-25/download/.

Download J-Sim—J-Sim Official. n.d. Available at: http://www.physiome.org/jsim/download/.

Duffy, C., Roedig, U., Herbert, J., and Sreenan, C.J. n.d. A performance analysis of MANTIS and TinyOS. Technical Report CS-2006-27-11, University College Cork, Ireland, November 2006.

Dunkels, A., Grönvall, B., and Voigt, T. 2004. Contiki—A lightweight and flexible operating system for tiny networked sensors. In *Proceedings—Conference on Local Computer Networks, LCN*. Washington, DC, 455–462. doi:10.1109/LCN.2004.38.

Embedded Ethernet. n.d. Available at: http://www.ethernut.de/.

Farooq, M.O. and Kunz, T. 2011. Operating systems for wireless sensor networks: A survey. *Sensors.* 11(6):5900–5930. doi:10.3390/s110605900.

Get Started with Contiki, Instant Contiki and Cooja. n.d. Available at: http://www.contiki-os.org/start.html.

Girod, L., Ramanathan, N., Elson, J., Stathopoulos, T., Lukac, M., and Estrin, D. 2007. Emstar: A software environment for developing and deploying heterogeneous sensor-actuator networks. *ACM Trans Sens Netw.* 3(3), August 2007, Article No. 13, 1–34. doi:10.1145/1267060.1267061.

GLOMOSIM Simulator | Network Simulation Tools. n.d. Available at: https://networksimulationtools.com/glomosim-simulator-projects/.

Han, C.-C., Kumar, R., Shea, R., Kohler, E., and Srivastava, M. 2005. A dynamic operating system for sensor nodes. In *Proceedings of the 3rd International Conference on Mobile Systems, Applications, and Services—MobiSys '05, 163*. New York, NY: ACM Press. doi:10.1145/1067170.1067188.

Hill, J., Szewczyk, R., Woo, A., Hollar, S., Culler, D., and Pister, K. 2000. System architecture directions for networked sensors. *ACM SIGOPS Oper Syst Rev.* 34(5):93–104. doi:10.1145/384264.379006.

Hofmeijer, T.J., Dulman, S.O., Jansen, P.G., and Havinga, P.J.M. 2004. AmbientRT: Real time system software support for data centric sensor networks. In *Proceedings of the 2004 Intelligent Sensors, Sensor Networks and Information Processing Conference*, Melbourne, Vic., Australia, 14–17 Dec. 2004, 61–66. doi:10.1109/ISSNIP.2004.1417438.

Hong, S. and Kim, T.H. 2003. Senos: State-driven operating system architecture for dynamic sensor node reconfigurability. In *International Conference on Ubiquitous Computing*, October 12–15, 2003 in Seattle, Washington, 201–203. http://citeseerx.ist.psu.edu/viewdoc/summary?doi=10.1.1.332.533.

Joint Modeling and Design of Wireless Networks and Sensor Node Software | EECS at UC Berkeley. n.d. Available at: https://www2.eecs.berkeley.edu/Pubs/TechRpts/2006/EECS-2006-150.html.

JSim Download Page. n.d. Available at: http://www.physiome.org/jsim/download/.

Kahn, J.M., Katz, R.H., and Pister, K.S.J. 1999. Next century challenges: Mobile networking for 'Smart Dust'. In *Proceedings of the 5th Annual ACM/IEEE International Conference on Mobile Computing and Networking*, Seattle, WA, USA—August 15–19, 1999, 271–278. doi:10.1145/313451.313558.

Kumar, G.S., Varghese, L.A., Mathew J.K., and Jacob, K.P. 2006. Evaluation of the power consumption of routing protocols for wireless sensor networks. In 2006 *International Symposium on Ad Hoc and Ubiquitous Computing*. IEEE, 20–23 December 2006, Mangalore, India, 194–195. doi:10.1109/ISAHUC.2006.4290675.

Lee, E.A., Zhao, Y., and Cheong, E. 2006. Viptos: A graphical development and simulation environment for tinyOS-based wireless sensor networks. In *Sensys '05: Proceedings of the 3rd International Conference on Embedded Networked Sensor Systems*, San Diego, CA, USA — November 02—04, 2005. http://citeseerx.ist.psu.edu/viewdoc/summary?doi=10.1.1.93.7444.

Li, X., Peng., Cai, J., Yi, C., and Zhang, H. 2016. OPNET-based modeling and simulation of mobile Zigbee sensor networks. *Peer-to-Peer Netw Appl*. 9(2):414–423. doi:10.1007/s12083-015-0349-8.

Low Energy Adaptive Clustering Hierarchy Protocol (LEACH)—File Exchange—MATLAB Central. n.d. Available at: http://in.mathworks.com/matlabcentral/fileexchange/40115-low-energy-adaptive-clustering-hierarchy-protocol–leach-.

MagnetOS. n.d. Available at: https://www.cs.cornell.edu/people/egs/magnetos/.

MannaSim Framework—Main Page. n.d. Available at: http://www.mannasim.dcc.ufmg.br/.

Mobile User Localization in WSN—File Exchange—MATLAB Central. n.d. Available at: http://in.mathworks.com/matlabcentral/fileexchange/29922-mobile-user-localization-in-wsn.

Nath, R. 2013. A TOSSIM based implementation and analysis of collection tree protocol in wireless sensor networks. *2013 International Conference on Communication and Signal Processing*, 03–05 April 2013, Melmaruvathur, Tamil Nadu, India, 484–488. doi:10.1109/iccsp.2013.6577101.

OMNeT++ Discrete Event Simulator—Download. n.d. Available at: https://omnetpp.org/omnetpp.

Overview—Nano-RK. n.d. Available at: http://nano-rk.org/projects/nanork.

Park, S., Savvides, A., and Srivastava, M.B. 2000. SensorSim: A simulation framework for sensor networks. *Proceedings of the 3rd ACM International Workshop on Modeling, Analysis and Simulation of Wireless and Mobile Systems–MSWIM '00*. Boston, MA, 104–111. doi:10.1145/346855.346870.

Perez, A.J., Labrador, M.A., and Barbeau, S.J. 2010. G-Sense: A scalable architecture for global sensing and monitoring. *IEEE Netw*. 24(4):57–64. doi:10.1109/MNET.2010.5510920.

Perla, E., Ó Catháin, A., Carbajo, R.S., and Huggard, M. 2008. PowerTOSSIM Z: Realistic energy modelling for wireless sensor network environments. In *Proceedings of the 3nd ACM Workshop on Performance Monitoring and Measurement of Heterogeneous Wireless and Wired Networks*. Vancouver, British Columbia, Canada, 35–42. doi:10.1145/1454630.1454636.

PiccSIM Simulation of Wireless Control Systems–Wireless Sensor Systems Group–Aalto University. n.d. Available at: http://wsn.aalto.fi/en/tools/piccsim/.

Polley, J., Blazakis, D., McGee, J, Baras, J.S., and Karir, M. 2004. ATEMU: A fine-grained sensor network simulator. In *2004 First Annual IEEE Communications Society Conference on Sensor and Ad Hoc Communications and Networks, 2004. IEEE SECON 2004.*, 145–152. doi:10.1109/SAHCN.2004.1381912.

Pottie, G.J. and Kaiser, W.J. 2000. Wireless integrated network sensors. *Commun ACM*. 43(5):51–58. doi:10.1145/332833.332838.

Pughat, A. and Sharma, V. 2015a. A review on stochastic approach for dynamic power management in wireless sensor networks. *Hum Cent Comput Inf Sci*. 5(1). doi:10.1186/s13673-015-0021-6.

Pughat, A. and Sharma, V. 2015b. Stochastic model for lifetime improvement of wireless sensor node. In *2015 8th International Conference on Contemporary Computing, IC3 2015*. Noida, India, 422–427. doi:10.1109/IC3.2015.7346718.

Pughat, A., and Sharma, V. 2015c. Queue discipline analysis for dynamic power management in wireless sensor node. In *2015 Annual IEEE India Conference (INDICON)*. IEEE, New Delhi, India, 1–5. doi:10.1109/INDICON.2015.7443670.

Pughat, A., and Sharma, V. 2016. Optimal power and performance trade-offs for dynamic voltage scaling in power management based wireless sensor node. *Perspect Sci.* 8:536–539. doi:10.1016/j.pisc.2016.06.013.

Qualnet—Packet Trace | SCALABLE Networks. n.d. Available at: http://web.scalable-networks.com/qualnet-network-simulator.

The Network Simulator Ns-2: Building Ns Version 2. n.d. Available at: http://www.isi.edu/nsnam/ns/ns-build.html.

Titzer, B.L., Lee, D.K., and Palsberg, J. 2005. Avrora: Scalable sensor network simulation with precise timing. In *2005 4th International Symposium on Information Processing in Sensor Networks, IPSN 2005*, Los Angeles, California, April 24–27, 2005, 477–482. doi:10.1109/IPSN.2005.1440978.

TOSSIM—TinyOS Wiki. n.d. Available at: http://tinyos.stanford.edu/tinyos-wiki/index.php/TOSSIM.

TrueTime—PiccSIM—Wireless Sensor Systems Group—Aalto University. n.d. Available at: http://wsn.aalto.fi/en/tools/piccsim/truetime-piccsim/.

Varga, A. and Hornig, R. 2008. An overview of the OMNeT++ simulation environment. In *Proceedings of the 1st International Conference on Simulation Tools and Techniques for Communications, Networks and Systems and Workshops*. Marseille, France, 1–10. doi:10.4108/ICST.SIMUTOOLS2008.3027.

Will, H., Schleiser, K., and Schiller, J. 2009. A real-time kernel for wireless sensor networks employed in rescue scenarios. In *2009 IEEE 34th Conference on Local Computer Networks*. IEEE, 20–23 October 2009, Zurich, Switzerland, 834–841. doi:10.1109/LCN.2009.5355049.

Wireless Sensor Network Simulation—Tutorial for Matlab | Our Work | Wireless System Laboratory of Brno. n.d. http://wislab.cz/our-work/wireless-sensor-network-simulation-tutorial-for-matlab.

Zuberi, K.M. and Shin, K.G. 2001. EMERALDS: A small-memory real-time microkernel. *IEEE Trans Softw Eng.* 27(10):909–928. doi:10.1109/32.962561.

Chapter 2

Medium Access Control in Wireless Sensor Networks

Vidushi Sharma, Gayatri Sakya, and Anuradha Pughat

Contents

2.1   Introduction 27
2.2   MAC Protocol Features for WSNs 28
2.3   Performance Parameters of MAC Protocols 29
       2.3.1   Total Residual Energy 29
       2.3.2   Latency and Delay 30
       2.3.3   Throughput 30
       2.3.4   Packet Delivery Ratio 30
       2.3.5   Bandwidth Utilization 30
2.4   Classification of MAC Protocols 30
       2.4.1   Scheduled MAC Protocols 32
       2.4.2   Unscheduled MAC Protocols 33
       2.4.3   Cross-Layer–Based MAC Protocols 35
2.5   The Simulations of WSN Protocols 36
       2.5.1   Case Study 1 36
       2.5.2   Case Study 2 37
2.6   Directions for Future Research 39
References 39

2.1 Introduction

Wireless sensor networks (WSNs) consisting of a dense network with heavy communication have to handle traffic efficiently with low-power constraints. Medium access control (MAC) decisions are made by MAC protocols. Since the nodes are deployed generally in unattended areas, replacing them is another challenging task. Hence the nodes should be tuned for providing an efficient network lifetime, which is again the task of the MAC protocol. For that, the on–off time (sleep–wake cycle) of the node should be managed as per the requirement of the applications so that energy consumption is minimized (Pughat and Sharma 2015a). The MAC protocol should also be adaptable to the changes in the communication pattern and topology since these are changing dynamically due to the node's energy depletion. Designing a good MAC protocol for WSNs should consider the behavior of the sensor nodes. The first consideration is that the area in which sensor nodes are deployed is not accessible at times such as in an underground sea-level Tsunami alert system or earthquake monitoring systems on mountains, etc. Second, nodes are self-configuring and controlling and, hence, there are special requirements for designing localization, time synchronization, and routing algorithms. Here the base station may not control the nodes; further nodes need to transfer information instantaneously when decision parameters cross the threshold value. The architecture of a sensor node is an important consideration as the data receiving and transmitting components in sensor nodes need lots of energy as compared to the sensing and processing components of the sensor node. One solution to reduce energy usage in a sensor node is to develop the decision rules based on which the transceiver states can be managed (Pughat and Sharma 2015b). The states of the transceiver (transmit, receive, idle, and sleep) need to be managed in such a way that the energy consumption is least and the objective of placing the sensor network is fulfilled, i.e., monitoring, control, and communication of environmental parameters. Another challenge of WSNs is that the network is prone to communication delay, which becomes more critical in disaster management or event-based systems. These considerations lead to the development of WSN MAC protocols, which are energy efficient with increased throughput, minimized packet delay, and packet loss.

In the proposed protocol stack for WSNs, the MAC protocol is responsible for handling the access mechanism of the channel with efficient energy-saving mechanisms. MAC protocols are categorized according to channel-accessing approaches into contention-based, time division multiple access (TDMA)–based (contention free), hybrid, and cross-layer MAC protocols. In this chapter, performance metrics for the evaluation of MAC protocols are presented after discussing different MAC protocols' performances. Mission-critical MAC parameters are also presented with a case study on an S-MAC protocol. Light is thrown on the efficacy of the ZigBee MAC module to be used in WSNs. The objective of this chapter is to discuss the concepts of MAC layers and various MAC protocols along with the future research directions.

2.2 MAC Protocol Features for WSNs

In general, MAC protocols are responsible for framing, medium access, reliability, flow control, and error control in the network. In a WSN, the MAC protocol focuses on medium access as the MAC protocol is directly responsible for controlling the transceiver, which consumes significantly higher energy. For medium access control, protocols are responsible for controlling the following activities in WSNs:

- ◾ Reducing communication overhead: extra transmission of control packets for handshaking should be minimized to reduce energy consumption.
- ◾ Collision avoidance mechanism: collision of data packets received from various nodes should be minimized.
- ◾ Overhearing of nodes is a situation when the nodes hear the transmitted packet due to the broadcast nature of the WSNs. Avoiding overhearing by sensor nodes may reduce energy to a considerable extent, but in some cases, overhearing is required to understand the traffic patterns of collecting some information for localization, time synchronization, etc.

- Retransmission of the data packets in conditions of failure or drop may consume energy. A limit on the optimal number of retransmissions has to be decided to reduce energy consumption. Idle listening of nodes is a state where the node is ready to receive but is not receiving. This is done in anticipation that a packet may arrive at any instant. TDMA-based protocols reduce idle listening, but in the case of high traffic or mission-critical applications the cost of packet loss will be high.
- Since the network is dynamic in nature, the MAC protocol needs to be adaptive toward the topology and network changes (Akyildiz et al. 2002).

There are also other issues that need to be addressed by MAC protocols such as the time synchronization algorithms should be simple as to increase the accuracy. This requires frequent message exchange, resulting in energy consumption. Further, to reduce cost, low-capability resources are required like oscillators and clocks. The scheduling algorithms need to be simple.

To strike a balance between all these is definitely a tough task. Several researchers have discussed MAC design issues like latency and throughput (Yick et al. 2008). Different approaches were used for designing MAC protocols like understanding the cause of energy absorption and traffic pattern to propose a suitable protocol (Ye et al. 2002). Some researchers used the traditional MAC protocols and tried to access their feasibility for WSNs such as contention-based and contention-free MAC protocols (Demirkol et al. 2006). The main objective is to reduce the duty cycle without affecting the efficiency of the system in terms of packet drop, throughput, and latency. Duty cycle is the ratio of the listen period to the wake-up period. A periodic sleep and wake-up cycle is used in most of the MAC protocols for managing the energy levels of the sensor nodes.

In recent research, it is pointed out that WSNs, when used in industrial control or monitoring applications, their energy is not a major concern as compared to latency and throughput. In other mission-critical applications such as on battlefields or in the oceans, both energy efficiency and data communication efficiency are required as high data traffic occurs due to the critical event and the information should be transmitted faster without energy depletion. The performance of MAC protocols is evaluated based on performance metrics such as lifetime, total residual energy, etc.

2.3 Performance Parameters of MAC Protocols

There are several performance metrics for sensor networks, but since MAC has the main function of media access, it is desirable that a trade-off is sought between the efficiency of the network, which is measured in terms of throughput, latency, bandwidth utilization, packet delivery ratio, etc., and the energy efficiency in terms of network lifetime. MAC protocol manages the duty cycle of the node, resulting in energy efficiency and thus enhancement of the lifetime, but at the same time, throughput, packet delivery, latency, and other performance parameters are not compromised. Hence, the performance of MAC protocols can be judged by the following parameters.

2.3.1 Total Residual Energy

It is the sum of the total energy at each node. The average energy at any instant of the system can be computed by dividing the total residual energy of each node by the number of nodes. The total residual energy is dependent on the duty cycle, which is given in Equation 2.1. If the duty cycle is more, then the transceiver will be in the ON state for a longer time and more energy is consumed, thus leading to lesser residual energy.

$$T_{\text{duty}} = \Delta t \text{ on } / (\Delta t \text{ on} + \Delta t \text{ off}) \qquad (2.1)$$

So, by designing a protocol with a comparatively low duty cycle, we can improve the energy efficiency of a protocol (Sakya et al. 2013). Total residual energy is an important parameter of lifetime of the network. Lifetime of the network has been calculated on various instances by researchers like the time for depletion of first node or the time at which half of the nodes die or the time at which the network partitions or the communication to sink breaks.

2.3.2 Latency and Delay

The latency and delay is a measure of the time taken by the MAC layer to send a packet from the sender node to the reception of a packet at the MAC layer. Latency may be a concern for mission-critical applications like a health monitoring system, gas pipeline leakage system, etc. Certain monitoring applications can tolerate some latency; hence, the duty cycle needs to be adjusted in such a way that the application requirements are satisfied. The average packet delay can be usually calculated for the performance evaluation of MAC protocols in WSNs.

2.3.3 Throughput

It is defined as the number of bits successfully delivered at the sink node at a specified time duration. It is the measure of the rate at which the channel serves the packets.

2.3.4 Packet Delivery Ratio

The ratio of the total packets received at the sink node to the packets generated at the source node can be defined as the packet delivery ratio.

2.3.5 Bandwidth Utilization

This parameter suggests the channel should be capable of sending maximum information with minimum delay and minimum bandwidth utilization as the bandwidth is limited.

In the mission critical applications, reliable and timely data delivery is of utmost importance. So the performance metrics should consider a trade-off between packet delivery ratio and throughput while managing the performance levels of the network.

2.4 Classification of MAC Protocols

In WSNs, the main aim is to provide an access to the channel. Since the WSNs are application specific, the MAC protocols have to be designed to suit the application's requirement; for example, environment-monitoring applications can be scheduled based on where the active period is to be manipulated, so that the energy consumption is decreased and the complete information about the environment is communicated. The concept of the threshold is also proposed for such applications. If at any point of time the threshold is crossed, their information is passed to the sink; otherwise, the sink considers that the environmental parameters are in control. Some researchers have classified MAC protocols as done in traditional networks, i.e., random, contention-based protocols, scheduled protocols. This classification does not consider the application-centric view of sensor networks. Some researchers also classified sensor networks on the basis of traffic conditions and protocol behavior.

Finding the most appropriate protocol for a particular application may be a difficult task, and hence one needs to understand the requirements of the application along with the constraints and then map the features of existing MAC protocols to it. Here, the requirement means that the essential features of applications in terms of permissible delay, mode of operation (event based or periodic monitoring), traffic conditions, the optimum packet delivery ratio to be achieved or the permissible number of packet drops in the system. These features are then compared to the existing MAC protocols, and then the protocol that fulfills the requirements is selected. Sometimes, some modifications are made in the existing protocols to map them to the application's requirement.

A general model for MAC protocol selection was proposed in WSNs, whereby the thumb rule of protocol selection was mathematically and analytically proved (Asudeh et al. 2016). According to the thumb rule, in low traffic conditions, preamble sampling protocols are desirable, whereas in the medium traffic conditions scheduled protocols perform better.

Researchers have critically analyzed contention-free and contention-based protocols. With increasing application complexity, a class of hybrid protocols has emerged, which combines the desirable facets of one or two protocols and overcomes the limitations. Cross-layer approaches have also been presented. In cross-layer protocols, the MAC layer uses the information from the upper layer and down layers for acquiring the channel. A taxonomy is presented in Figure 2.1.

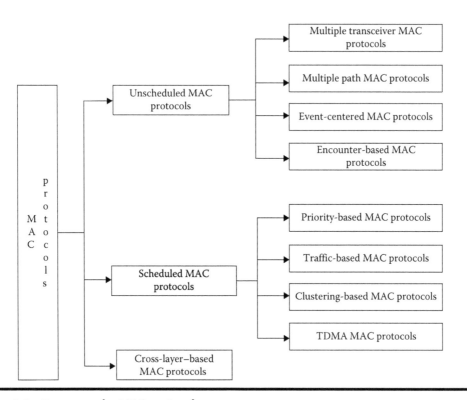

Figure 2.1　Taxonomy for MAC protocol.

2.4.1 Scheduled MAC Protocols

Contention-free protocols follow a schedule, which is common among the neighboring/all sensor nodes. Thus, they are also called scheduled-based MAC protocols. Most of the protocols use TDMA as the cost and energy requirements are considerably reduced. By following a schedule, problems such as idle listening, overhearing, and collisions are eliminated. Further, the nodes that are not communicating can go into sleep mode to save energy. In these protocols, the data-aggregating and data-prioritization schemes can be efficiently applied. But besides these advantages, there are some challenges such as to reduce the communication overhead. To enable synchronization control, messages are exchanged, which result in communication overhead. The dynamic nature of sensor networks (node redeployment, death, or node mobility) adds to the complexity and the schedule management becomes more complex. Synchronization is a major concern, as in the absence of proper synchronization latency increases and throughput decreases. The length of the time slot is another concern, as larger the time slot more is the wastage of sensor node's active state in low traffic conditions. During this period, if the transceiver is active, but not transmitting due to less traffic, the energy will be wasted. If the active time slot is decreased, then there are chances that the communication is missed, which increases the overheads in terms of latency and throughput. One option to avoid is to queue the messages or prioritize the retransmitted messages if they are dropped (Pughat and Sharma 2015c). The cost of latency and limited throughput can be overcome by increasing the duty cycle and by sharing additional information to properly synchronize between the nodes. Besides several TDMA-based protocols, they may be altered to suit the specific application requirements. Priority-based scheduling protocols are used to give priority to certain messages. Clustering-based protocols are used to meet the requirements of cluster-based networks and scheduled protocols, to meet the varying traffic demands.

Priority-based scheduling MAC protocols prioritize the links or the nodes, which are derived from a random function. The input to the random function is the sensor node ID and the time slot number, which after computation give access to the resource having the highest priority. Several protocols have been proposed such as node activation multiple access (NAMA) and link activation multiple access (LAMA). In NAMA prioritization of node occurs and in LAMA prioritization of link is done (Bao and Garcia-Luna-Aceves 2001). These protocols have to compute the priority of each neighboring resource for a time slot. These calculations add the energy requirements.

Traffic-based scheduling MAC protocols can operate on a wide range of conditions and are responsive to different traffic patterns such as an intermittent burst of high traffic. These protocols need to estimate the traffic pattern and share it with the neighboring nodes to work efficiently.

The traffic adaptive medium access (TRAMA) protocol (Rajendran et al. 2003) strikes a balance between the benefits of both the scheduled and the unscheduled protocols.

Here, the control packets are small and are sent on random access slots while the long data messages are provided with scheduled slots without any contention. A random access period followed by a schedule access period forms a cycle. TRAMA is supported by its three protocols. TRAMA learns the two-hop topology of the neighbor nodes using its neighbor protocols. The traffic queue information is exchanged by TRAMA's schedule exchange protocol and the adaptive election protocol of TRAMA helps in selecting the slots for data transfer based on traffic conditions and the topology. TRAMA is considered to be an efficient protocol for handling different traffic conditions, minimizing the transceiver energy consumption and limiting the message collisions. In case of inconsistent traffic conditions, TRAMA fails to optimize its performance and its resource utilization is high. Researchers have also proposed a pattern MAC protocol (P-MAC), which manages its duty cycle based on the traffic conditions. Sensor nodes having more data are given more slots than the other nodes. P-MAC also reserves some time slots for the exchange of pattern of node's sleep and awake time. P-MAC (Zheng et al. 2005) is a simple and adaptive protocol toward different traffic patterns. However, it faces limitations if the updated pattern schedules are not available due to channel errors.

2.4.2 Unscheduled MAC Protocols

Contention-based MAC protocols overcome the limitations of synchronization of nodes and overhead of control messages. They are simpler protocols and use less resources in terms of memory, processing, and message exchange for coordination. These protocols, however, encounter the problems of idle listening, overhearing, and collisions. To overcome these limitations, improvisations in contention-based protocols have been made. The concept of a duty cycle may also be incorporated, which can be made adaptive as per the traffic conditions. In low-traffic conditions, the transceiver must be put in a sleep mode for a longer time duration and thus the active period is reduced to save energy and other inherent problems of contention-based protocols (idle listening, overhearing, etc.). It may be controlled by proper adjustment of the duty cycle.

The concept of multiple transceivers was introduced with an objective that the transceiver may communicate simultaneously on different channels to increase the response time and throughput of the nodes. Mission-critical applications take advantage of these protocols to improve their response time. Additional hardware, i.e., a transceiver, adds to the cost of the network. Power-aware multiaccess with signaling (PAMAS) is an example of a multitransceiver protocol. PAMAS has two transceivers for handling data message and control message, which are sent by the respective transmitter on different channels. PAMAS has a mechanism for avoiding the problem of overhearing but presents no solution for idle listening. All messages of request-to-send (RTS), clear-to-send (CTS), and the busy tone are sent on the control channel and data packets are transmitted on data channels. Before sending data, the control channels of sender and receiver node exchange RTS and CTS packets. After a successful exchange, a busy tone propagates the control channel to indicate ongoing data transmission on the data channel, which is done after a successful RTS–CTS handshake. Power down of the transceiver occurs in the case when there is no data to transmit or in case the busy tone is received by the neighboring nodes. Though PAMAS is quite efficient, it adds to the complexity and cost due to two transceivers (Singh and Raghavendra 1998).

Other contention-based protocols were proposed, which used multiple paths to transmit the data packets. Using multiple paths for data transmission helps in reducing the collision as multiple copies of messages are propagated on different paths, thus increasing the probability of successful transmission (Chatzigiannakis et al. 2004, 2005). This probability of message delivery increases while eliminating the carrier sensing and control messages of synchronization. Simple random backoff protocol (SRBP), adaptive random backoff protocol (ARBP), and range adaptive random backoff protocol (RARBP) were proposed, each with an improvement in the earlier base protocol. These protocols used probabilistic forwarding protocol that considered the directional capability of transmission whereby the direction of the base station is known. The SRBP only considered the initial random backoff, while the ARBP took into account the traffic conditions and the sensor node density of a particular local region. The RARBP suggested improvement in the ARBP by reducing the number of hops and thus the latency. To handle the need of event-based applications, contention-based MAC protocols were modified. Event-based applications have very low or negligible traffic most of the time and whenever the event occurs the traffic intensity increases drastically. Hence, periodic-monitoring MAC protocols only add to the wastage of energy. MAC protocols that work on some threshold mechanism of triggering the media access serve the purpose of such applications. Energy consumption can be reduced in such applications by either putting the nodes in sleep in case of no event or by minimizing the message transfer by some aggregator mechanism. The correlation-based collaborative MAC (CC-MAC) protocol identifies the neighboring nodes and finds the correlate measurements. Filtering of nodes having high correlation measurement is done in such a way that the node data received by the highly correlated nodes is discarded as it will be duplicated or redundant, whereas the data originating from far-off node will have a low correlation coefficient and will be independent. There are two components of the CC-MAC: the event MAC and network MAC protocols. The event MAC protocol selects the data that come from least-correlated nodes, and highly correlated nodes sleep in that duration to save energy. The network MAC is responsible for forwarding the messages to the sink, which are considerably reduced by the event MAC protocol (Vuran and Akyildiz 2006).

Unscheduled protocols can be classified in another class, where the sensor nodes remain in sleep mode and whenever there is a requirement, they all synchronize and transmit the information. This class may be termed as encounter-based MAC protocols. Synchronization in event-based MAC protocols may be time consuming and futile if the event does not occur for quite some time and as the sensor network topology is inherently changing due to nodes dying or node failure. Then, synchronization will be a wasteful effort, so it is advised to use an encounter-based protocol. Traffic patterns that are random and occur rarely benefit from encounter-based protocols. A sparse topology and energy management (STEM) protocol can be considered the early unscheduled protocol in this category, as it attempts to wake up the neighboring nodes, which are in sleep mode by either a beacon message (STEM-B) or a wake-up tone, which is of considerable length, so that it can be heard by the neighboring node (STEM-T) (Schurgers et al. 2002).

Other protocols in this category are a transmitter-initiated cycled receiver (TICER) and a receiver-initiated cycled receiver (RICER). The functioning of a TICER is similar to STEM-B, whereby beacon messages are transmitted periodically by sensor nodes having data. They send the RTS packet, and then sense for CTS. The receiver also wakes up periodically to sense the data. The whole process is reversed in RICER, where the receiver periodically wakes up and sends the beacon. The sensor node wanting to send data listens to the beacon of the intended receiver (Lin et al. 2004a).

A B-MAC protocol introduces functions like the STEM-T protocol, as it uses a wake-up tone. In B-MAC, the target duty cycle of the network is considered and the nodes follow independent sleeping schedule. B-MAC uses long preambles for message transmission and senses the channel to avoid a collision before the transmission. However, long preambles add to the latency in B-MAC and are prone to the hidden terminal problem (Polastre et al. 2004).

WiseMAC (El-Hoiydi and Decotignie 2004) and CSMA-MPS (Mahlknecht and Bock 2004) were other encounter-based protocols which drew attentions of the researchers. WiseMAC further reduces energy consumption by remembering the sampling offset of the neighbors. By this extra field, the sensor node can know the sampling time of the neighboring nodes and attempt to transmit the preamble just before the receivers wake up. CSMA-MPS brought about the improvement by introducing minimal preamble sampling and thus brought about an improvement in energy saving and reduced latency over B-MAC.

Clustering-based MAC protocols were developed to take advantage of clustered networks, which makes the network scalable. With the help of clustering, the network clusters are made and hence the local control is administered instead of global. This arrangement is energy efficient and reduces the complexity in synchronization, localization and MAC protocols. Local and global traffic are also differentiated, thus enabling data aggregation at the local level, which is then transmitted to the next cluster head. The only drawback of clustering is that an added functionality of cluster head selection, reformation of the cluster, adds to the complexity; hence, for that one has to do a cost–benefit analysis. Normally, large and scalable sensor networks benefit from the clustering. Several protocols have been proposed in this. LEACH happens to be the most popular among them. The LEACH protocol rotates the role of the cluster head. For communication within the cluster nodes, use a direct-sequence spread spectrum (DSSS) to avoid interference from other clusters. Cluster heads use a reserved sequence to communicate with the base station. The LEACH protocol is discussed in Chapters 3 and 4.

The GANGS protocol is also a clustering protocol, but in contrast to LEACH it uses TDMA for transmission between cluster heads and an unspecified contention protocol for communication between nodes within a cluster. In both LEACH and GANGS (Ad et al. 2004), the cluster formation and reclustering consume energy. The basis on which clustering has to be done is again an important decision that will be explored in Chapter 4.

Another clustering MAC protocol is the Group TDMA, which divides the sensor nodes into various groups and assigns different TDMA slots to each group. Group TDMA reduces the collision and provides maximum channel utilization, but at the same time it is an energy consuming protocol with high latency due to its extensive setup phase.

S-MAC was proposed by Ye et al. (2002) and was further improved. Several variants of S-MAC have also been developed like DSMAC, T-MAC, etc. S-MAC clusters the sensor node by synchronizing the sleep schedules of the neighboring nodes. These clusters are virtual clusters. A sensor node usually adopts the schedule of the neighboring nodes. In case the nodes receive several schedules, it follows all the schedules to communicate between these virtual clusters. If the nodes do not receive any schedule, then it follows its own schedule. S-MAC uses RTS/CTS to transmit messages. S-MAC has advantages over other clustering protocols in terms of using fewer processing resources and better coordination and adequate synchronization. Scalability is also enhanced in S-MAC.

DSMAC (Lin et al. 2004b) is an improvement in S-MAC as it considers the dynamic duty cycle based on the energy and traffic conditions. T-MAC (van Dam and Langendoen 2003) extends the features of S-MAC by incorporating a timer to indicate the end of the active period and thus it overcomes the fixed schedule of the duty cycle. The main drawback of S-MAC is the latency introduced because of periodic sleep of nodes. Also the nodes at the border need to wake up more frequently due to following several schedules and so they die out early. Looking at the potential of the S-MAC research is still going on for its improvement.

Several other TDMA MAC protocols have been proposed like E-MAC, L-MAC, and AI-LMAC (Bhatia and Hansdah 2014; Van et al. 2004; Chatterjea et al. 2004). Slot assignment in all these protocols is same, i.e., random slot picking. These protocols differ on the basis of their node interaction.

2.4.3 Cross-Layer–Based MAC Protocols

The unscheduled protocols and the scheduled protocols mentioned above operate at a single layer. The WSN node has very few layers, which are defined fully as discussed in Chapter 1. The complex protocols and services are not used in these networks. At a single layer, efforts have been made to improve the performance parameters such as throughput, data transmission rate, delay, and energy efficiency. These performance features can be further enhanced if we apply cross-layer optimization as the performance of cooperative network improves substantially. Various protocols discussed like S-MAC and T-MAC use link-layer scheduling. B-MAC uses the physical layer for carrier sensing and thus all the protocols mentioned so far are restricted to a single layer. Researchers have proved the cross-layer protocol approach for designing the MAC protocol. It is more efficient as they can use the information of several layers to handle the situations. Some of the important cross-layer MAC protocols are BoX-MAC-1 (Moss and Levis 2008), BoX-MAC-2, P-MAC (Zheng et al. 2005), BulkMAC (Canli et al. 2010), CLMAC (Hefeida et al. 2013), XL-MAC (Hamid and Bashir 2013), XLP (Vuran and Akyildiz 2010), etc. BoXMAC-1 and BoX-MAC-2 use link layer and physical layer. At the physical layer, low-power mechanisms are combined with the information of the link layer. This considerably reduces the energy consumption. In BoX-MAC-1, which is based on B-MAC, data is transmitted continuously; however, no long preambles are required as in the case of B-MAC. BoX-MAC-2 is based on the X-MAC protocol where the receiver nodes remain awake to hear the incoming packet, but the channel sampling process is like a B-MAC. Both protocols have their own advantages: BoX-MAC-1 can work efficiently in low-traffic conditions and is very dynamic, while BoX-MAC-2 operates well in high-traffic conditions but is less dynamic.

2.5 The Simulations of WSN Protocols

2.5.1 Case Study 1

This is an NS-2–based case study for simulating the discrete event simulation scenarios in WSNs. The NS-2 is an open source tool for simulation of wired or wireless networks. It uses Tcl script at its front end for writing the network scenarios. The languages C++ and OTcl have been used for developing the objects and their handlers at its back end. The Tcl script file is used to interact with the C++ and OTcl functions. After creating the networked scenario and writing the scripts, two simulation files with extensions *.nam* and *.tr* are generated while executing the Tcl script in NS-2. The Nam tool provides visualization to the networked scenario, for example, packet traces, topology layout, node movement, etc. The popular Nam releases are Nam-1.11 and Nam 1.0a9 for Linux; Nam 1.0a11a and Nam 1.0a10 for Windows (32-bit) and Windows XP; Nam 1.0a8 for Solaris 2.5; Nam 1.0a6 for SunOS. The trace file (.tr file) carries information about all the events that occurred in network simulation. The following are the syntax to create a trace file in OTcl:

    set nf [open new.tr w]
    $ns trace-all $nf

This opens a trace file named "new" for writing and stores its data in the trace format. The file handler "nf" is used to handle the "new" file and "trace-all" traces all the packets. The syntax "$ns flush-trace" is used to flush the traced data and close the file. A trace string consists of the 12 fields, i.e., event type, packet tracing string time, from the node, to node, packet type, packet size in bytes, seven-digit flag, fid, source-id, destination-id, sequence number, and packet-id. The results are plotted on the XGRAPH or GNUPLOT. Refer to the Ns2 program tutorial for wireless topology 2016 for a list of wireless topology programs. We can implement a new S-MAC protocol using the NS-2 energy model given in the SCADDS project at USC/ISI (SCADDS: Scalable coordination architectures for deeply distributed systems 2016). An extended energy model for S-MAC provides the new parameters for power in a sleep state, transition time, and power from sleep state to active state (Energy model update in NS-2 2016). Let us discuss an S-MAC protocol for the multihop scenario, which is developed and simulated using NS-2.35. In this multihop scenario, 11 nodes have been taken, which are at least 250 m distant from each other (Figure 2.2).

Figure 2.2   Network topology for 11 nodes.

In this simulation setup, a fixed energy of 1000 J has been provided for each node. The source node sends packets of 80 bytes to the sink node at different time intervals or interarrival times. The interarrival times can vary from 0.1 to 50 seconds. The dynamic source routing (DSR) protocol has been used to provide an on-demand path from the source node to the sink node. The packets have been generated with a constant bit rate (CBR) and source node uses user datagram protocol (UDP). The varying duty cycle from 10% to 40% has been used for energy saving when a node is idle. Now, the trace file is created to record the residual energy of each node. The impact of varying the traffic load is analyzed by changing the packet transmission interarrival times at the source node for different duty cycles. The graph is plotted for each performance metric value against the traffic rate changing from 0.01 to 50 seconds. Figure 2.3 shows the variation of residual energy vs. CBR for various duty cycles. It has been observed that the residual energy at 40% duty cycle is minimized. It further decreases with the slower traffic rate. Other performance parameters such as system throughput and packet delivery ratio have also been analyzed (Sakya et al. 2013).

Similarly, one can analyze the performance of protocols in different scenarios and different topologies like grids, ring clusters, etc.

2.5.2 Case Study 2

This study shows how to configure a ZigBee module. The ZigBee module is very extensively used as a communication device or transceiver in several applications of WSNs. It is based on the 802.15.4 standard, which defines the characteristics of the physical and MAC layers for low-rate wireless personal area networks (LR-WPANs). The ZigBee nodes are low power consuming, communicate in the range of 100 m at the data rate of 250–500 kbps, use the bandwidth of 2.4 GHz, and connect in any of the star, mesh, or cluster tree network topologies. They provide the secure, reliable, convenient, and interoperable environment. These features make them suitable for resource-constrained WSNs. In WSNs, ZigBee can be used as a coordinator node, a router node, and a sink node. In a star topology, each node is directly connected to the coordinator node and a coordinator collects the information from surrounding nodes, then processes the information, and further routes it to the desired ZigBee

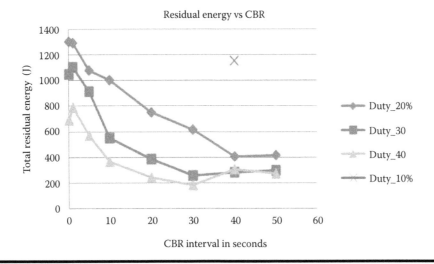

Figure 2.3   Residual energy vs. CBR.

device. This single coordinator-based topology increases the network reliability, but increases the chances of the network failure if the coordinator fails. Apart from the star topology, each node has been connected to the other nodes in a mesh topology. The mesh topology is more flexible and capable of handling high traffic, but has high costs for network setup and its maintenance. In the cluster tree topology, multiple parents in a network are used as clusters and form trees with their children. The cluster tree topology is mostly used in IEEE 802.15.4–based energy-constrained networks, but the ZigBee itself does not support a cluster tree topology. A ZigBee 3.0 is used for Internet of Things (IoT) applications such as home, building, office automation, smart homes, and health-care monitoring. The XBee is a device that uses the ZigBee protocol for its communication purpose (XBee: Connect devices to the cloud—Digi International 2016). The most common applications in which XBee is used are monitoring and control applications, but not limited to them. It provides high data throughput in the applications where the low duty cycle can be used for low power consumption.

The configuration of the ZigBee Modules on Firebird V BOTs is provided at http://www .e-yantra.org. The DigiMesh prepared the XBee® ZigBee for programmable application development, Python-based XBee® Gateway for scalable device management, connecting ports, XBee Smart Plug™ ZigBee energy manager, XBee-PRO® ZigBee Wall Router, XBee-Pro® ZigBee Adapters, USB to XBee Wireless Adapter, sensors with integrated ZigBee. These modules create the low-cost and low-power WSNs. The DigiMesh provides the simple architecture and mesh topology–based networks, which uses the sleep time synchronization to achieve low power consumption. It assists in the power optimization and network configuration for the network having more than a thousand nodes (https://www.digi.com/products/digimesh). These products have advanced security and networking features such as DSSS, self-routing, self-healing, retries, and acknowledgment.

The XBee DigiMesh (Sakya and Gautam 2016) modules are used for establishing a network in an ad-hoc topology. These modules are set up, configured, and managed using a software XCTU (see Figure 2.4). The XCTU is developed by the NEX Robotics (used Version 6.2.0.6). It provides simple GUI and interacts with firmware files for Digi's RF products.

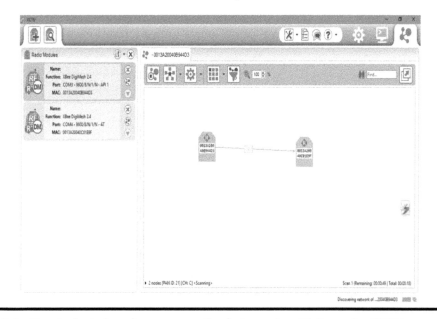

Figure 2.4   Terminal window shows connection between two ZigBee using XCTU.

On opening the XCTU software we can connect our XBee module to the PC via data cable. The first option from XCTU windows, ADD DEVICE OR DISCOVER DEVICES, is selected.

The software automatically detects every peripheral device connected to the PC. Then, select the XBee modules and click next. After the radio modules are added, select a radio module from the list to display its properties. Now, one can edit its properties according to use. For establishing an ad-hoc network, CH is the operating channel; ID is a network ID; MT, broadcast multitransmits; RR, unicast retries. Using the console log feature in the XCTU software, one can check whether the modules are communicating with each other or not.

The XCTU (6.2.0.6) also offers to observe the connection topology. This can be enabled by configuring modules on API mode operation and after that going to the network search tab in the XCTU software and start searching.

It has been found that in multihop and critical scenarios, the ZigBee protocol is not very efficient in WSNs. The S-MAC protocol can further be used in such scenarios, wherein delays should be less with better throughput. Therefore, a sensor node when designed should also take care of reliable data delivery and timely reporting to the base station. Research is also going on S-MAC to improve its performance further.

2.6 Directions for Future Research

- As the applications of WSNs are increasing day by day in implementing the projects of smart cities, building automation, industrial automation, and in alert systems, application-specific MAC protocols are needed to fulfill these needs. So existing protocols should be studied and according to the applications that should be modified.
- The optimization should be done among energy, latency, and throughput of the network as desired. The data delivery performance should be on time and reliable.
- Now we are in the era where each application demands intelligence and automation. For this purpose, we need to design smart protocols with intelligent algorithms so that they can dynamically change their behavior according to the scenarios. Certain approaches are suggested in this chapter.

References

Ad, S., Yawen, B., and Barowski, D. 2004. 'GANGS': An energy efficient MAC protocol for sensor networks. In *Proceedings of the 42nd Annual Southeast Regional Conference ACM-SE 42*. ACM New York, NY, 82–87. doi:10.1145/986537.986557.

Akyildiz, I.F., Su, W., Sankarasubramaniam, Y., and Cayirci, E. 2002. A survey on sensor networks. *IEEE Commun Mag.* 40(8):102–105. doi:10.1109/MCOM.2002.1024422.

Asudeh, A., Záruba, G.V., and Das, S.K. 2016. A general model for MAC protocol selection in wireless sensor networks. *Ad Hoc Netw.* 36:189–202. doi:10.1016/j.adhoc.2015.07.005.

Bao, L. and Garcia-Luna-Aceves, J J. 2001. A new approach to channel access scheduling for Ad Hoc networks. In *MobiCom '01 Proceedings of the 7th Annual International Conference on Mobile Computing and Networking*. Rome, Italy, 210–221. doi:10.1145/381677.381698.

Bhatia, A. and Hansdah, R.C. 2014. A TDMA-Based Energy Aware MAC (TEA-MAC) protocol for reliable multicast in WSNs. In *Proceedings—International Conference on Advanced Information Networking and Applications, AINA*, Victoria, Canada, May 13--16, 2014, 798–805. doi:10.1109/AINA.2014.97.

Canli, T., Hefeida, M., and Khokhar, A. 2010. BulkMAC: A cross-layer based MAC protocol for wireless sensor networks. In *Proceedings of the 6th International Wireless Communications and Mobile Computing Conference IWCMC '10*. New York, NY: ACM Press, 442–446. doi:10.1145/1815396.1815499.

Chatterjea, S., Van Hoesel, L.F.W., and Havinga, P.J.M. 2004. AI-LMAC: An adaptive, information-centric and lightweight MAC protocol for wireless sensor networks. *Proceedings of the 2004 Intelligent Sensors, Sensor Networks and Information Processing Conference, 2004*, 381–388. doi:10.1109/ISSNIP.2004.1417492.

Chatzigiannakis, I., Kinalis, A., and Nikoletseas, S. 2004. Wireless sensor networks protocols for efficient collision avoidance in multi-path data propagation. In *Proceedings of the 1st ACM International Workshop on Performance Evaluation of Wireless Ad Hoc, Sensor, and Ubiquitous Networks—PE-WASUN '04*. New York, NY: ACM Press, 8. doi:10.1145/1023756.1023759.

Chatzigiannakis, I., Nikoletseas, S., and Spirakis, P. 2005. Efficient and robust protocols for local detection and propagation in smart dust networks. *Mobile Netw Appl.* 10(1):133–149. doi:10.1023/B:MONE.0000048551.54039.f0.

Demirkol, I., Ersoy, C., and Alagoz, F. 2006. MAC protocols for wireless sensor networks: A survey. *IEEE Commun Mag.* 44(4):115–121. doi:10.1109/MCOM.2006.1632658.

El-Hoiydi, A. and Decotignie, J.D.(Eds.). 2004. WiseMAC: An ultra low power MAC protocol for multi-hop wireless sensor networks. In *Algorithmic Aspects of Wireless Sensor Networks*, 5005, ALGOSENSORS 2004, LNCS 3121, pp. 18–31, 2004. Springer-Verlag: Berlin, Heidelberg. doi:10.1007/978-3-540-27820-7_4.

Energy model update in NS-2. 2016. Accessed December 20. Available at: http://www.isi.edu/ilense/software/smac/ns2_energy.html.

Hamid, Z. and Bashir, F. 2013. XL-WMSN: Cross-layer quality of service protocol for wireless multimedia sensor networks. *EURASIP J Wireless Commun Netw.* 2013(1): 174. doi:10.1186/1687-1499-2013-174. Springer International Publishing.

Hefeida, M.S., Canli, T., and Khokhar, A. 2013. CL-MAC: A cross-layer MAC protocol for heterogeneous wireless sensor networks. *Ad Hoc Netw* 11(1):213–225. doi:10.1016/j.adhoc.2012.05.005.

Lin, E.-Y.A., Rabaey, J.M., and Wolisz, A. 2004a. Power-efficient rendez-vous schemes for dense wireless sensor networks. In *2004 IEEE International Conference on Communications (IEEE Cat. No.04CH37577)* 7(c), 20–24 June 2004, Paris, France, 3769–3776. doi:10.1109/ICC.2004.1313259.

Lin, P., Qiao, C., and Wang, X. 2004b. Medium access control with a dynamic duty cycle for sensor networks. In *2004 IEEE Wireless Communications and Networking Conference (IEEE Cat. No.04TH8733)* 3:1534–1539. doi:10.1109/WCNC.2004.1311671.

Mahlknecht, S. and Bock, M. 2004. CSMA-MPS: A minimum preamble sampling MAC protocol for low power wireless sensor networks. In *IEEE International Workshop on Factory Communication Systems 2004 Proceedings*, Vienna, 73–80. doi:10.1109/WFCS.2004.1377683.

Moss, D. and Levis, P. 2008. BoX-MACs: Exploiting physical and link layer boundaries in LowPower networking. Available at: http://citeseerx.ist.psu.edu/viewdoc/summary?doi=10.1.1.379.5537.

Ns2 program tutorial for wireless topology. 2016. Accessed December 20. Available at: http://www.ns2blogger.in/p/ns2-program-for-wireless-topology.html.

Polastre, J., Hill, J., and Culler, D. 2004. Versatile low power media access for wireless sensor networks. In *Proceedings of the 2nd International Conference on Embedded Networked Sensor Systems SenSys 04*. Baltimore, MD, 3(4): 95. doi:10.1145/1031495.1031508.

Pughat, A. and Sharma, V. 2015a. A review on stochastic approach for dynamic power management in wireless sensor networks. *Hum Cent Comput Inf Sci.* 5(1). doi:10.1186/s13673-015-0021-6.

Pughat, A. and Sharma, V. 2015b. Stochastic model for lifetime improvement of wireless sensor node. In *2015 8th International Conference on Contemporary Computing, IC3 2015*. Noida, India, 20–22 August. doi:10.1109/IC3.2015.7346718.

Pughat, A. and V. Sharma. 2015c. Queue discipline analysis for dynamic power management in wireless sensor node. In *2015 Annual IEEE India Conference (INDICON)*, 17–20 December, 2015 New Delhi, India. doi:10.1109/INDICON.2015.7443670.

Rajendran, V., Obraczka, K., and Garcia-Luna-Aceves, J J. 2003. Energy-efficient collision-free medium access control for wireless sensor networks. In *Proceedings of the First International Conference on Embedded Networked Sensor Systems SenSys 03* 12(1):181, Los Angeles, CA, USA. doi:10.1145/958511.958513.

Sakya, G. and Gautam, A. 2016. Smart agriculture system using adhoc networking among firebird V bots. *Int J Innov Adv Comput Sci* 5(10):2347–8616.

Sakya, G., Sharma, V., and Jain, P.C. 2013. Analysis of SMAC protocol for missioon critical applications in wireless sensor networks. In *Proceedings of the 2013 3rd IEEE International Advance Computing Conference, IACC 2013*, 488–492. doi:10.1109/IAdCC.2013.6514274.

SCADDS: Scalable coordination architectures for deeply distributed systems. 2016. Accessed December 20. Available at: http://www.isi.edu/scadds/.

Schurgers, C., Tsiatsis, V., Ganeriwal, S., and Srivastava, M. 2002. Optimizing sensor networks in the energy-latency-density design space. *IEEE Trans Mobile Comput.* 1(1):70–80. doi:10.1109/TMC.2002.1011060.

Singh, S. and Raghavendra, C.S. 1998. PAMAS—Power aware multi-access protocol with signalling for ad hoc networks. In *ACM SIGCOMM Computer Communication Review*. New York, NY, 28:5–26. doi:10.1145/293927.293928.

van Dam, T. and Langendoen, K. 2003. An adaptive energy-efficient MAC protocol for wireless sensor networks. In *Proc. 1st International Conference on Embedded Networked Sensor Systems (SenSys '03)*. Los Angeles, CA, 5–7 November, 2003, 171–180. doi:10.1145/958491.958512.

Van, L.F.W.H., Havinga, P.J.M., and Van Hoesel, L.F.W. 2004. A lightweight medium access protocol (LMAC) for wireless sensor networks: Reducing preamble transmissions and transceiver state switches. In *1st International Workshop on Networked Sensing Systems, INSS 2004*. University of Tokyo, Japan, 22–23 June, 2004, 205–208.

Vuran, M.C. and Akyildiz, I.F. 2006. Spatial correlation-based collaborative medium access control in wireless sensor networks. *IEEE/ACM Transac Netw.* 14 (2):316–329. doi:10.1109/TNET.2006.872544.

Vuran, M.C. and Akyildiz, I.F. 2010. XLP: A cross-layer protocol for efficient communication in wireless sensor networks. *IEEE Transac Mobile Comput.* 9(11):1578–1591. doi:10.1109/TMC.2010.125.

XBee: Connect devices to the cloud—Digi International. 2016. Accessed December 19. Available at: https://www.digi.com/lp/xbee.

Ye, W., Heidemann, J., and Estrin, D. 2002. An energy-efficient MAC protocol for wireless sensor networks. In *Proceedings of the Twenty-First Annual Joint Conference of the IEEE Computer and Communications Societies*, 23–27 June, 2002, New York,. 3:1567–1576. doi:10.1109/INFCOM.2002.1019408.

Yick, J., Mukherjee, B., and Ghosal, D. 2008. Wireless sensor network survey. *Comput Netw.* 52(12):2292–2330. doi:10.1016/j.comnet.2008.04.002.

Zheng, T., Radhakrishnan, S., and Sarangan, V. 2005. PMAC: An adaptive energy-efficient MAC protocol for wireless sensor networks. In *19th IEEE International Parallel and Distributed Processing Symposium*. Denver, CO, 3–8 April, 2005. doi:10.1109/IPDPS.2005.344.

Chapter 3

Routing in Wireless Sensor Networks

Vinay Kumar Singh

Contents

3.1   Introduction 44
3.2   Types of Routing Protocols for WSNs 44
       3.2.1   Data Centric Protocols (Flat-Based Routing) 44
       3.2.2   Hierarchical-Based Routing 45
       3.2.3   Location-Based Routing 46
       3.2.4   Mobility-Based Protocols 47
       3.2.5   Single Path and Multipath Protocols 47
       3.2.6   Heterogeneity-Based Protocols 47
       3.2.7   QoS-Based Protocols 48
3.3   Genetic Algorithms for Routing in WSNs 48
3.4   Energy-Efficient Routing Algorithm for WSNs Using Elitist GA 49
       3.4.1   System Model 50
       3.4.2   Elitist GA-Based Routing Algorithm 51
       3.4.3   Simulation and Analysis 53
3.5   Hybrid GA for Routing in WSNs 55
       3.5.1   The GASA Algorithm 56
       3.5.2   System Model 56
       3.5.3   Simulation and Analysis 59
3.6   Self-Organizing Migrating Algorithm for Routing in WSNs 60
       3.6.1   Self-Organizing Migrating Algorithm 62
                3.6.1.1   Perturbation 62
                3.6.1.2   Generating New Candidate Solutions 62
       3.6.2   System Model 63
       3.6.3   Simulation and Analysis 64
       3.6.4   Confidence Bounds 65
References 66

3.1 Introduction

Wireless sensor networks (WSNs) are nowadays being used extensively due to advancements made in the development of sensor node devices, wireless networks, and electronics engineering. Various communications such as text data, voice data, multimedia data, and others have been made ubiquitous by advancements in WSNs, thus affecting lifestyle in different ways. WSN applications with complex requirements face challenges like energy consumption by nodes in the network, bandwidth of the channel, acceptance by end user, and interoperability. Sensor network communication models are complex. The topology is dynamic as the node fails due to limited battery life, unpredictable link quality, etc. It is required that these challenges are overcome. The network administrator observes and reacts to events taking place in the network within the given environment. Sensor nodes provide collective information of events occurring in the network to the administrator in a reliable manner.

WSN (Akyildiz et al. 2002; Raghavendra et al. 2011; Heikalabad et al. 2011; Heinzelman et al. 2002) technology is growing at a very fast pace with new applications being developed day by day. It is widely used in the fields of military surveillance and monitoring, Internet, and various other scientific applications. Traffic from all directions is converged at the destination (sink), which leads to the exhaustion of the sensor nodes that are placed near the destination (sink). In order to make sure that these nodes do not deplete all their energy, the network traffic needs to be distributed evenly among nodes in the network. There have been a number of routing protocols proposed (Akkaya and Younis 2005; Al-Karaki and Kamal 2004; Shah and Rabaey 2002a) that are being used to enhance the performance of wireless networks.

Wireless networks are widely used because they are economically efficient, easy to deploy, and have relatively superior capabilities for monitoring the target area (Misra et al. 2008). However, there are many other problems associated with these networks such as congestion, loss of packet, improper bandwidth utilization, and unreliable data delivery to destination. The research work explained in this chapter focuses on reducing power consumption by sensor nodes and increasing network lifetime.

The research proposed various routing algorithms that address the issue of energy saving during data transmission in the network. Energy-efficient models were introduced that ensure data routing toward the sink utilizing a minimum number of active nodes by minimizing the distance from the source to the sink. Proposed protocols also ensure full coverage of the target area with a minimum number of active nodes. This increases the data transmission and the lifetime of the network with minimum energy consumption.

3.2 Types of Routing Protocols for WSNs

Classification of WSN routing protocols is shown in Table 3.1 (Akkaya and Younis 2005). Various routing protocols have been classified into different categories based on their characteristics such as data centricity, location, mobility, or quality of service, or having different structure like hierarchical or flat based.

3.2.1 Data Centric Protocols (Flat-Based Routing)

In this type of protocol, every node in the network is treated the same, and the sensor nodes work in collaboration to collect data and send it to the base station (BS). The largest number of

Table 3.1   Various Routing Protocols

Category	Representative Protocols
Data-centric protocols (flat-based routing)	SPIN, directed diffusion, rumor routing, COUGAR, ACQUIRE, EAD, information-directed routing, gradient-based routing, energy-aware routing, information-directed routing, quorum-based information dissemination, home agent-based information dissemination
Hierarchical protocols	LEACH, PEGASIS, HEED, TEEN, APTEEN
Location-based protocols	MECN, SMECN, GAF, GEAR, Span, TBF, BVGF, GeRaF
Mobility-based protocols	SEAD, TTDD, joint mobility and routing, data MULES, dynamic proxy tree-based data dissemination
Single and multipath-based protocols	sensor-disjoint multipath, braided multipath, N-to-1 multipath discovery
Heterogeneity-based protocols	IDSQ, CADR, CHR
QoS-based protocols	SAR, SPEED, energy-aware routing

sensors is deployed in the monitoring area, but is not assigned unique IDs. The BS sends queries regarding what kind of data it is interested in. The sensor nodes, after receiving the requested data, start sending data from one node to another node in the direction of the BS. Directed diffusion (DD) (Intanagonwiwat et al. 2000, 2003) is a protocol that reinforces the data transfer path, once the sensor knows the interest message. Sensor protocols for information via negotiation (SPIN) (Heinzelman et al. 1999) was proposed wherein the negotiation is done between the sensor nodes and then the transmission takes place. The GRAdient Broadcast (GRAB) (Ye et al. 2001) protocol helps in energy conservation by sending data after eliminating redundant data.

3.2.2 Hierarchical-Based Routing

These routing schemes are cluster based where high-energy nodes are selected as cluster heads; other nodes having data send it to the cluster while nodes without any data go into sleep mode to save energy. The scheme is scalable and is effective in saving network energy and increasing network lifetime. Some of the cluster-based protocols save data by data aggregation. These routing schemes work in two layers: the first layer is between the cluster head nodes and the other layer is between the cluster head and other nodes.

A low-energy adaptive clustering hierarchy (LEACH) (Haenggi 2004) is a cluster-based protocol that has one cluster head, and other nodes that have data send their data to the head, and then the head transmits the data to the BS as shown in Figure 3.1. Each and every node is given a chance to become the cluster head for balancing the power consumption. Each node selects a random number between 0 and 1; the node is made the cluster head if the random number is less than the following threshold:

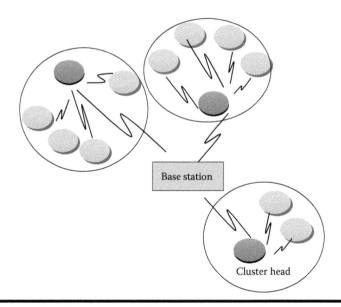

Figure 3.1 LEACH routing protocol.

$$T(n) = \begin{cases} \dfrac{p}{1 - p * (r \bmod 1/p)} & \text{if } n \in G \\ 0 & \text{otherwise} \end{cases} \tag{3.1}$$

where p = % of cluster heads (e.g., 5%), r = current round, and G = set of nodes that have not got a chance to become the heads till now.

Power-efficient gathering in sensor information systems (PEGASIS) (Younis and Fahmy 2004; Lindsey and Raghavendra 2002) is an enhanced version of LEACH. The data to be sent is aggregated in a chain and sent to the BS through direct transmission by any one node in the chain. The major drawback of PEGASIS is that it requires all the information about other nodes (Intanagonwiwat et al. 2000). Protocols like hybrid energy efficient distributed (HEED) can minimize the control overhead (Zou and Chakrabarty 2003). In this protocol, the cluster heads are well distributed over the network, and as a result more power is saved.

3.2.3 Location-Based Routing

This protocol works on the basis of the location of sensor nodes in the network. This location is used to calculate the distance between sensor nodes for sending the data and minimizing power consumption. Information about the location is obtained through a global positioning system (GPS) or some other positioning algorithms like the triangulation method (Chunlai 2009; Vural and Ekici 2007; Xu et al. 2000).

Geographic adaptive fidelity (GAF) (Xu et al. 2001) is a protocol that is energy efficient and is location based as shown in Figure 3.2. The network is divided into a number of zones and one sensor node is made to stay active for sending data to the BS after collecting it from other sensor nodes. Then, after some time, this node is turned off and some other nodes take its place. Through this process, the protocol saves power by turning off certain nodes that are not in use.

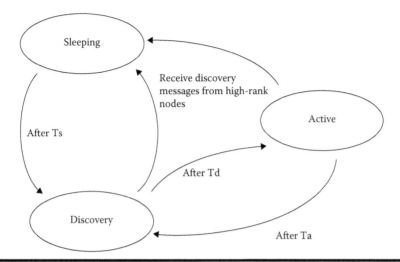

Figure 3.2   State transitions in GAF.

3.2.4 Mobility-Based Protocols

These protocols face new routing difficulties for routing protocols. When the sink is static, the nodes that are in the surrounding area of the sink get exhausted as they come in the path of data transmitting chains (Singh and Sharma 2011). Sink mobility requires energy-efficient protocols to guarantee data delivery from source sensors toward mobile sinks.

In a static sink, the sensors located near the sink transfer more data than those away from the sinks, and hence their energy depletion is more (Singh and Sharma 2011). This leads to loss in network connection to the link. To solve this problem we can have a mobile sink that may change its position from time to time and other nodes can come with a data chain to send the packets toward the sink.

3.2.5 Single Path and Multipath Protocols

A data chain through which the data is sent from a source to a destination is called a path. In single path routing, a single path is used for sending data between the source and destination. When a single explicit path is used, the nodes have information regarding the path delay and QoS of the data chain. Nodes may use this information to make local routing decisions. The shortest path routing is the best example of single path routing where the shortest path is used to send data from the source to the destination and it minimizes energy consumption. A multipath approach may be used, which provides multiple paths from the source to the destination for data transmission. This helps in even load distribution and effectively utilizes the network energy and minimizes queuing delay.

3.2.6 Heterogeneity-Based Protocols

In heterogeneous networks, there are two types of nodes, one that has more power line and energy supply as compared to the others having a battery power that has a limited power and lifetime.

The information-driven sensor query (IDSQ) protocol in applications of tracking and target localization reduces the latency and power consumption by dynamically querying the sensor and dynamically routing. The sensors sending the required information are kept active while other sensors are kept on sleeping mode to save energy. For improving tracking accuracy and detection, nodes should be kept alive in the vicinity of the object to be detected, but this increases the cost, and hence a subset of nodes are chosen to communicate while the others are put in sleep mode. These subsets alternatively go on active and sleep state for achieving the objective and for keeping the cost low. In this protocol, first, a leader is selected among the sensor nodes, which is responsible for selecting the optimal number of sensor nodes and classifying it into various subsets for alternative sleep and wake cycles.

Cluster-head relay routing (CHR): This routing scheme is used in heterogeneous networks, which use two types of sensors, L-sensors (low-end sensors) and H-sensors (high-end sensors). Both sensors contain information regarding their positions in the network. These sensors are randomly deployed in the network. L-sensors sense the occurrence of an event and transmit the data packets toward the cluster head through intermediate nodes. H-sensors perform data fusion inside their clusters. L-sensors use short-range data transmission, while H-sensors use long-range data transmission to the neighboring H-sensors in other clusters so as to transmit the message to sink sensors.

3.2.7 QoS-Based Protocols

Reduction of power consumption and providing quality of service (QoS) are two important features that must be kept in mind while designing a routing protocol. QoS ensures reliability, fault tolerance, and reduced delay.

The sequential assignment routing (SAR) protocol is a QoS-based protocol. It is table driven and uses multiple paths for data transmission. This protocol makes a tree of the neighboring nodes of the destination by a QoS metric. It takes into account the energy available in a particular path and the priority of the packets to be transmitted. The tree is created from sink to sensors having multiple paths, then one of the paths is selected for data transmission according to the available energy in the path and QoS.

There is another QoS protocol called SPEED. In this protocol, each and every sensor node has information about its neighbor and uses geographic forwarding to find the available paths. This protocol also tries to achieve a certain speed of packet delivery. The decision of admission of the packet to a particular path is made after dividing the distance to the destination (sink) by the speed of the packet. The path ensuring the highest speed is selected. This protocol also avoids congestion during data transmission.

3.3 Genetic Algorithms for Routing in WSN

Evolutionary algorithms are stochastic searching algorithms that use the process of natural selection for optimization. They are based on Darwin's theory of survival of the fittest. Various genetic operators such as selection, crossover, and mutation are used to achieve it. The algorithms work on a number of randomly generated solution strings called chromosomes/individuals. These chromosomes contain a number of genes that show various characteristics of the solution to the given problem. A fitness function calculates the fitness of these individuals, and those individuals that are fit compared to others are given more chances to reproduce. The selection operator ensures this by selecting the fittest parents for reproduction. Reproduction is the process of mixing the characteristics of two selected parents and producing new offsprings that have the characteristics of both the parents. Then, the mutation operator is applied to ensure that the diversity is maintained in the population. This process ensures the passing of characteristics of some fit chromosomes to the next generation and eliminating the rest. This process is repeated until a fit enough solution is obtained or a certain number of maximum iterations is reached.

As there are many sensor nodes deployed in the network area, nodes that take part in data transmission are kept active, while others are kept in inactive mode to save energy. The problem of selecting a set that is active and covers the monitoring area is an NP-complete problem.

In NP-complete programs, random deployment of nodes is done, which leads to the selection of an active node set of all the nodes. The conventional search algorithms cannot provide a solution to the problem in reasonable time, so evolutionary algorithms are used to find an effective solution (Singh and Sharma 2014). To minimize the power consumption in the network, a method is proposed to select a routing that has minimum distance from the source to the destination (sink). If the average path length of the data transmission is reduced, it will definitely reduce the power consumption (Singh and Sharma 2012b) and the solution is searched using a genetic algorithm (GA) approach. A simple GA can effectively find the solution to such problems, but the GA can be improved to converge faster. Such an improved GA using elitism was also proposed (Singh and Sharma 2012a). In elitism-based GAs, the best solution (elite) found so far is saved for the next generation during the search process; this ensures that the solution does not degrade in the next generation and always tries to get the better solution. This ensures faster convergence toward the global optima. Minimizing the distance from source to destination minimizes the power consumption, but this approach can be further improved by considering the residual energy of the selected path. Sending the data via the shortest path may result in a faster depletion of power in some of the nodes and this may break the network connectivity; therefore, another factor that considers the residual energy of the path further enhances the network lifetime (Singh and Sharma 2012c). When the energy of a certain node is depleting faster during data transmission, an alternate node can be used to replace that node in the data path as proposed in Wenliang et al. (2009). A GA is used to find the solution and the approach ensures that no node depletes its entire energy, thus increasing the network lifetime. An improved GA-based routing is discussed in the next section.

3.4 Energy-Efficient Routing Algorithm for WSNs Using Elitist GA

WSN routing is one of the main factors that should be considered for saving the power available in the sensor nodes. As the network operates for some time, the energy of some of the nodes is exhausted faster as compared to other nodes in the network. This is the main reason for unequal consumption of node energy. There are mainly two factors that cause unequal energy consumption:

1. The distance between the sensor nodes in the data transmission chain
2. The distance between the sensor nodes and the BS

The problem of finding the efficient routing is NP-complete and therefore GA is used to find the solution. As the WSN routing is a combinatorial optimization problem, GA can be effectively used to find the solution in reasonable time. This chapter uses three techniques for comparison with the proposed scheme: (1) Direct transmission between sensor nodes and the sink. (2) GA-based routing with least average energy consumption (ELGA) (Guo et al. 2008). A minimum spanning tree for shortest path routing is used for minimizing the average power consumption. (3) GA-based data routing with energy balance being taken into account (EBGA) (Guo et al. 2008). It is able to make those nodes having less residual energy not forward too much data from other nodes. If the energy of a sensor node is twice the average energy of the all the sensor nodes in the data path, then it is considered to be fit and the routing is selected. If the energy of a sensor node is less than half of the average energy of all the nodes in the data path, then it is considered less fit to be selected in the routing. It uses the mechanism of energy balance.

The proposed elitism-based GA routing considers both the cost and the distance between the nodes, i.e., the shortest path from the source node to the destination node as well as the energy balance of the sensor network that is given by the residual energy of the nodes divided by the total number of nodes in the network. This scheme uses the shortest path-based routing mechanism for data transmission in the network. Elitist GA is used to find energy-efficient routing by minimizing the path length by multi-hop routing. The proposed algorithm uses elitism to preserve the good solutions for the next generation to make the search algorithm converge faster. The routing scheme tries to minimize the path length; at the same time it tries to balance the energy consumption of the network to maximize the lifetime. The algorithm, when implemented, shows that GAs can be used effectively to get the routing solutions for WSNs faster than the conventional search techniques.

3.4.1 System Model

The model assumes that there is a BS that receives the data packets transmitted from the sensor nodes in the network. The BS has an adequate source of power supply. The nodes sense the events occurring in the network and send information toward the sink via selected energy-efficient routing. The network consists of homogeneous sensor nodes, i.e., nodes having the same structure. The first-order radio model is used to find the energy consumption during transmitting and receiving the data packets from the sensor nodes toward the sink. Power consumption by a sensor node to transmit a packet of size k bits from sender to receiver is situated at a distance of d meters is given in

$$ETX(k,d) = (\xi_{\text{elec}} + \xi_{\text{amp}} * d_i^n) * k \tag{3.2}$$

whereas power consumption during receiving of a packet of size k bit by a node is given in

$$ETX(k) = \xi_{\text{elec}} * k \tag{3.3}$$

where ξ_{elec} = energy utilized per bit to run the radio electronics and ξ_{amp} = energy utilized by the transmit amplifier.

The energy consumed by a sensor node for transmitting and receiving data packets to and from the source and destination is the same, i.e. power consumed from i to j is same as j to i. The computational energy is negligible in comparison to data transmission energy and is not taken into consideration. The path of data transmission is a chain of intermediate nodes from the source to the destination (sink) node. The time interval in which a node generates a packet and transmits it toward its neighbors is called the data collecting round. After receiving the packet from its previous node, every node sends the same packet to the node ahead in the path. Once the path is formed it should be ensured that the total energy consumed in the path for data transmission from the source to the sink is minimum. If there are N nodes, then the total power consumption is calculated by summation of the energy dissipated by all the nodes in the data path and if the packet is of size k, then the total energy is given in

$$E_{\text{TOTAL}} = \left\{ \sum_{i=1}^{N-1} (\xi_{\text{elec}} + \xi_{\text{amp}} * d_i^n) + \xi_{\text{elec}} \right\} * k \tag{3.4}$$

where d_i = distance between two adjacent nodes in the data path, n = path loss exponent (2.0 (free space) ≤ n ≤ 4.0 (multipath fading)), and dth = threshold value for communication radius.

3.4.2 Elitist GA-Based Routing Algorithm

GAs are stochastic search mechanisms. GAs are based on stochastic search and optimization methods and are based on the Darwinian theory of survival of the fittest (Guo et al. 2008). Dasgupta and McGregor (1992) proposed that GAs can easily find the optimum path for WSNs. The algorithm works on a number of alternate solution paths and gives the optimal result within a short period of time.

The algorithm works by storing useful information about the individuals from the current population implicitly through redundant representations.

Some important data about the population is stored by GA during the search process (Yang et al. 2010). The best solution found so far during the search is preserved because of elitism. The best solution replaces the worst solution during the iterations by crossover and mutation. As the data transmission goes on, during continuous data transfer, if the power of a particular node goes below a predefined level, then that node is replaced by some other neighboring node and this is based on some probability. This helps in the effective power utilization of the nodes and in increasing the lifetime of the network. A flowchart of the modified GA is shown in Figure 3.3.

The elitism-based GA proceeds via the following steps:

1. *Population initialization*: In the first step, N individuals (chromosomes) are produced randomly and the evolutionary generation starts with iteration 0. The threshold distance (communication radius) is initialized.
2. *Fitness calculation*: The data path with the shortest length is called the best path. The distance fitness of each routing is calculated as

$$C_f = \sum_{i=1}^{N-1} d_i^2 \tag{3.5}$$

The total energy consumed by the data path can be calculated in Equation 3.5. A data path is called chromosome C that contains N nodes (genes). The distance between adjacent nodes is denoted by d_i. If the length of the data path is less, it shows better solution as the total energy consumption will be less.

The energy balance for each individual is computed as given as

$$E_f = \frac{\sum_{i=1}^{N} P_i}{N} \tag{3.6}$$

where P_i is the residual power of a node and N is the total number of nodes. The fitness function that takes into account the cost (distance) between the nodes as well as the energy balance is

$$F(C,E) = C_f + E_f \tag{3.7}$$

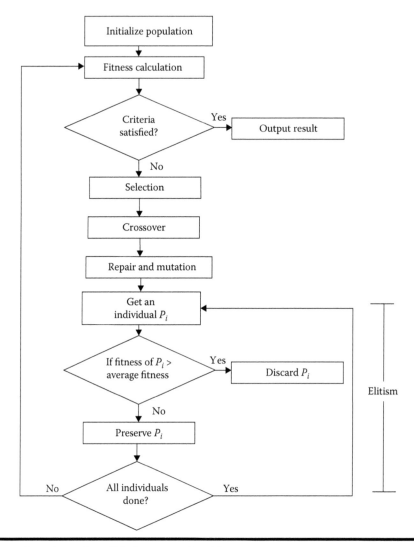

Figure 3.3 Flowchart of modified genetic algorithm.

3. *Selection*: This step selects the better individuals according to the corresponding selection operator. The selection operator used here is tournament selection whereby a mating pool of the individuals, the above average fitness value is maintained and two parents are randomly selected for crossover.

4. *Crossover*: This step produces new chromosomes by exchanging some part of one parent with other parents as per the corresponding crossover operator. The discussed method here uses two points for crossover.

5. *Repair*: This step is used to remove the cycle in the path. The gene values that are repeated in the chromosome are deleted and thus looping is removed. If a child violates the imposed constraints threshold distance, then it is rejected and crossover is performed again.

6. *Mutation*: This step produces new individuals and population diversity is maintained. It also helps to come out of local optima. A chromosome is selected randomly with some

probability and a certain gene number is changed to produce a new chromosome. In case the newly produced chromosome is not within the constraints, it is rejected and mutation is done again.

7. *Elitism*: It is used to find the best solution found so far and store it so that it is not lost in the next generation. This increases the convergence of the search.
8. *Checking the terminating criteria*: The algorithm is stopped if the maximum number of generations/iterations has been reached or if the power of a node is below a certain level.
9. *Result output*: The program outputs the best chromosome found so far as the result.

3.4.3 Simulation and Analysis

The elitism-based scheme shows much improvement over other existing schemes in terms of faster convergence. The sensor field in the experiment is considered to be 100 m × 100 m. The parameters for GA are

Population size = 100
Crossover (two-point crossover) probability $P_c = 0.6$
Mutation probability $P_m = 0.1$

In the simulation, various energy-saving methods are compared: (1) direct transmission, (2) ELGA, and (3) EBGA. Figure 3.4 shows a comparison of the four situations mentioned above.

Network lifetime has been improved using the proposed GA with elitism. There is 6% increase in lifetime with the proposed scheme for 50 nodes, as compared to EBGA, whereas the proposed scheme gives two times improvement over the direct method, which is more significant. The lifetimes of various networks before any single nodes expire are given in Table 3.2. The elitism-based GA scheme has shown considerable improvement as compared to other schemes for

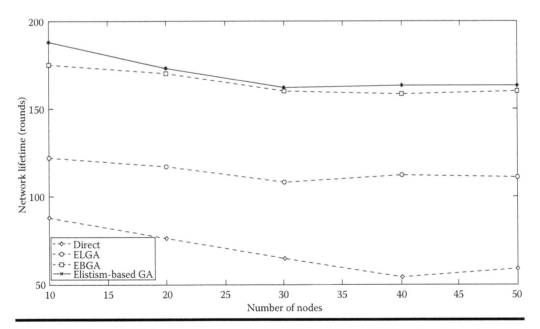

Figure 3.4   Network lifetime vs. number of nodes. Comparison of different routing schemes.

Table 3.2   Comparison of Network Lifetime for Various Schemes for the Number of Nodes During Data Communication

Schemes	Number of Nodes				
	10	20	30	40	50
Direct	88	76	65	54	59
ELGA	122	117	75	112	111
EBGA	175	170	160	158	160
Elitism-based GA	188	173	162	163	163

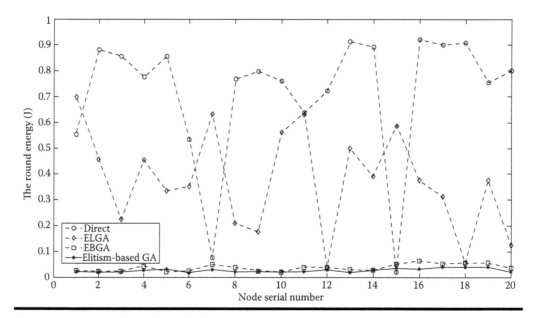

Figure 3.5   Comparison of residual energy of nodes under different routing schemes.

both 10 and 50 nodes. It is also evident from Figure 3.4 that in direct transmission, the network lifetime decreases with increase in the number of nodes, whereas in the proposed elitism-based scheme, the network lifetime is slightly more initially and gets stabilized as the number of nodes is increased.

Figure 3.5 shows the residual energy of 20 nodes after the termination of the algorithm for a 100 m × 100 m network area when the first dead node appears. The average residual energy of the network using the direct, ELGA, and EBGA methods is 0.6972, 0.3791, and 0.0184 J, whereas in the proposed algorithm, it is 0.0153 J, which is 16.8478% improvement over the EBGA as shown in Table 3.3.

It is inferred that the new algorithm exploits the remaining power of the nodes in a better way and increases the network lifetime significantly. When finding optimized routing, the elitism-based GA using the fitness criteria mentioned in Equation 3.6 converges much faster as compared to a simple GA for the same fitness criteria. This is because it preserves the best solution for the next generation, which improves the GA to a greater extent as shown in Figure 3.6. Simple GA

Table 3.3 Residual Energy of Various Schemes

Schemes	Average Residual Energy (J)
Direct	0.6972
ELGA	0.3791
EBGA	0.0184
Elitism-based GA	0.0153

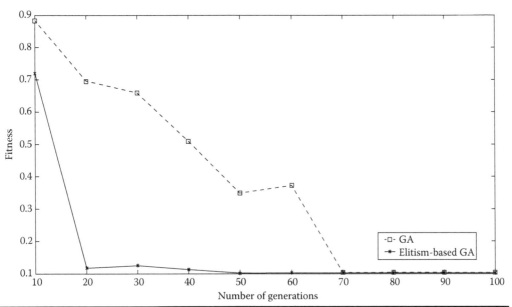

Figure 3.6 Comparison of convergence of simple GA and elitism-based GA.

converges in about 70 generations, whereas an elitism-based GA takes about 20 generations to converge, which is much faster.

The modified GA uses elitism because of which the best chromosome is preserved and transferred to the next generation. The data transfer takes place and as the power of any nodes goes below a certain predefined level, it is changed with some other neighboring node. Thus, the strategy minimizes power consumption during data transfer and also maintains energy balance in the network so that there is even energy distribution in the network. This increases the lifetime.

3.5 Hybrid GA for Routing in WSNs

The shortest path optimization is efficiently and effectively implemented using GAs in MANETs (Yang et al. 2010). They have shown that the algorithm can adapt the environmental changes quickly (i.e., the network topology change). It can output better results for dynamic shortest path problems. The next section presents the application of hybridization of GA with simulated annealing (SA).

3.5.1 The GASA Algorithm

GAs work well for NP-hard problems (Srinivas et al. 2000). They can easily find the global optimum solution. SA (Eglese 1990) is simulation of a physical annealing process with the help of a software implementation where a molten substance is gradually cooled down under a controlled environment. SA accepts better as well as inferior solutions which are determined by temperature parameter. During the initial stages of the algorithm, it is more likely to accept inferior solutions (high temperature) and during the final stages the probability of selecting an inferior solution is reduced with the reduction in temperature. *Genetic algorithm simulated annealing (GASA)* is a hybridization of the two techniques, GA and SA, for obtaining a better solution. It increases the convergence speed and makes the algorithm more consistent. Such hybridization (Bhagwan Das and Patvardhan 1999; Hui et al. 2010) has shown the use of SA for faster convergence and obtaining the global optimum because of GA. The GASA algorithm is shown in Figure 3.7.

3.5.2 System Model

The model considered in this section is the same as explained in Section 3.4.1.

Equations 3.8 and 3.9 (Bari et al. 2009) give the power consumption during transmitting and receiving, respectively. If the distance between the transmitting and receiving nodes is less than d_0 (threshold), the free space model ($d < d_0$) is used; otherwise, the multipath fading model ($d \geq d_0$) is used. The power required by the transmitter for sending a packet of size k bits to distance d is given in the following equation:

1. *Initialize final temperature T_f;*
2. *Initialize maximum number of iterations MaxIt;*
3. *Choose initial population;*
4. *Evaluate each individual's fitness;*
5. *Repeat {*
6. *Select individuals to reproduce;*
7. *Apply crossover operator to produce offsprings;*
8. *Repair the individual;*
9. *Select the best child as the parent for the next generation;*
10. *For each family, accept the best child as the parent for next generation if*
    $\Delta E < 0$ OR $\exp((\Delta E)/t) \geq \rho$
    t is the temperature coefficient.
    ρ is a random number uniformly distributed between 0 and 1.
    ΔE is as described in Equation 3.15
11. *Store the produced offspring for the next generation;*
12. *Mark the best offspring in the generation as Cfit*
13. *Increment generation number by $i = i + 1$*
14. *Decrement the temperature t according to cooling schedule*
15. *Calculate the individual fitness and store the best offspring in Cbest*
16. *} While $t \leq T_f$ or $i < $ MaxIt*
17. *Output the required route represented by Cbest*

Figure 3.7   GASA algorithm.

$$ETX_{k,d} = \begin{cases} \sum_{i=1}^{N-1}(kE_{\text{elec}} + k\xi_{\text{fs}} * d^2) & d < d_0 \\ \sum_{i=1}^{N-1}(kE_{\text{elec}} + k\xi_{\text{mp}} * d^4) & d \geq d_0 \end{cases} \tag{3.8}$$

Power consumption by the node during receiving k bit packet is given as

$$ETX_k = E_{\text{elec}} * k \tag{3.9}$$

The WSN is similar to a directed graph having a set of nodes N and a set of links A. The cost of each link is given by a cost matrix $C = [C_{ij}]$, where each entry in the cost matrix represents the cost of data transmission from the source S to the destination D. There is a link connection indicator that tells whether a link is present in the path or not, as shown in

$$I_{ij} = \begin{cases} 1, & \text{if the link } (i,\ j) \text{ is present in the route} \\ 0, & \text{otherwise} \end{cases} \tag{3.10}$$

It is clear from the above matrix that the diagonal elements are zero, i.e., $I_{ij} = 0$.

The problem of searching energy-efficient routing is given by a minimization objective function, as given in the following:

Minimize

$$f(C,I) = \sum_{\substack{i=S}}^{D} \sum_{\substack{j=S \\ j \neq i}}^{D} C_{ij}.I_{ij} \tag{3.11}$$

Subject to

$$\sum_{\substack{i=S \\ j \neq i}}^{D} I_{ij} - \sum_{\substack{j=S \\ j \neq i}}^{D} I_{ji} = \begin{cases} 1, & \text{if } i = S \\ -1, & \text{if } i = D \\ 0, & \text{otherwise} \end{cases}$$

and

$$\sum_{\substack{j=S \\ j \neq i}}^{D} I_{ij} \begin{cases} \leq 1, & \text{if } i \neq D \\ = 0, & \text{if } i = D \end{cases}$$

$$I_{ij} \in \{0,1\}, \quad \text{for all } i \tag{3.12}$$

The constraint (Equation 3.11) is used to make sure that the path searched by the algorithm does not have any loops or cycles. The system model given above is implemented using the GASA algorithm. The steps of the algorithm are as follows:

1. *Encoding*: An individual/chromosome contains a string of integer numbers, each integer representing a node in the network. The length of each such string is equal to the number of sensor nodes in the data routing path. In Figure 3.8a, a routing scheme having six nodes is shown as an example and the corresponding individual/chromosome is shown in Figure 3.8b. In this scheme, the node in position 1 is 2, which means node 1 is transmitting to node 2, and similarly the node at position 3 is 8, meaning that node 3 is transmitting data to node 8, and so on. The BS maintains a list of one hop neighbors of all the nodes. The greedy search is used to pick up a neighbor randomly from the available neighbors of a node. If there are d neighboring nodes for a node I and the total number of nodes is n, then the total number of paths that can be created is $O(d^n)$. This is very large when the n is large; therefore, to generate an optimal path for such a large network a nonconventional search and optimization technique such as GA is needed (Rana and Zaveri 2011). In the technique discussed here, we try to maximize the network lifetime. The network lifetime lasts till the first node is exhausted of all power. The function finds the lifetime (Shah and Rabaey 2002b) and is given in

$$L_{net} = \frac{E_{initial}}{E_{max}}$$ (3.13)

where L_{net} = lifetime (rounds)
$E_{initial}$ = initial energy of a node (same for all nodes at the time of initialization)
E_{max} = maximum energy dissipated by any node in the chromosome/routing in one round of data collection

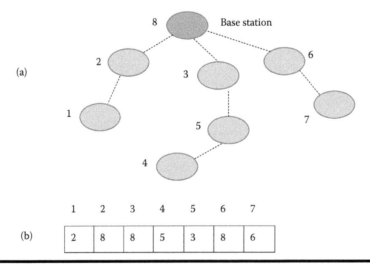

Figure 3.8 (a) and (b) Representation of network graph as chromosome.

2. *Fitness function*: The fitness function is critical in a GA search. The fitness function selected here tries to minimize the distance of the routing from the source to the sink and maximize the lifetime of the network as given in

$$\text{Fitness} = \min \; [\alpha. \, f(C, \, I) + \beta.(-L_{\text{net}})] \tag{3.14}$$

where $f(C,I)$ = tries to minimize the distance
$-L_{net}$ = tries to maximize the lifetime
α = weighting coefficient for distance
β = weighting coefficient for lifetime and
$\alpha + \beta = 1$

3. *Selection*: The elitist selection and tournament selection are used. Elitism maintains the good chromosomes in the population, and in tournament selection, one best chromosome out of two randomly selected parents is selected from the mating pool.

When Parent1 and Parent2 produce a new offspring as in

$$\Delta E = [Y_c - (Y_2 + Y_1)] * 0.5 \tag{3.15}$$

the above equation shows the difference in power of the two generations:

Y_1, Y_2 = parent's objective function value
Y_c= child's objective function value

If ΔE is negative, the child is better and it is not necessary that it will be selected always because of SA, which allows nonbetter solutions to be selected during initial search.

4. *Crossover*: A two-point crossover is used where two parents are selected for crossover. Some part of the parents between the crossover points are exchanged to get two new children. This is done with a crossover probability $P_c \leq 1$.

5. *Mutation*: It is used to diversify the search and to avoid the algorithm from being stuck in locala optimum. Mutation is done with mutation probability P_m.

3.5.3 Simulation and Analysis

The GA parameters are similar to those discussed in Section 3.4.3. The sensor fields of 200 m × 200 m are considered. This scheme is used to find the average power consumption, which is calculated as the summation difference between the initial and final energies of all the nodes divided by the total number of nodes as given in

$$E_a = \frac{\sum_{k=1}^{n} E_{ik} - E_{fk}}{n} \tag{3.16}$$

In Figure 3.9, it can be seen that the average energy of nodes is increasing as the number of nodes are increasing. In the GSA scheme, the overall network lifetime is increased as the fitness function chosen tries to increase the overall lifetime of the network.

It can be noted in Figure 3.9 that as the number of nodes keeps on increasing, the average energy consumption in broadcasting and clustering is increasing, whereas for direct diffusion and GASA there has not been a significant increase after 100 nodes.

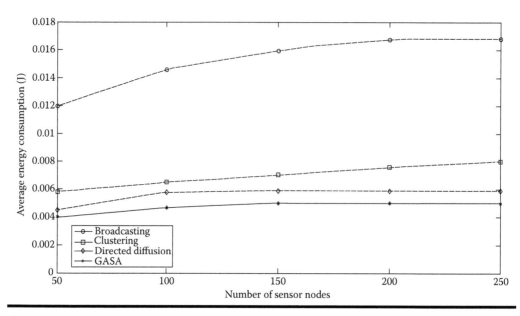

Figure 3.9 Average energy consumption per node.

Table 3.4 Average Energy Consumption (Joules) for Number of Nodes

	Number of Nodes				
Techniques	*50*	*100*	*150*	*200*	*250*
Broadcasting	0.0120	0.0146	0.0160	0.0168	0.0169
Clustering	0.0058	0.0065	0.0070	0.0076	0.0083
Directed diffusion	0.0045	0.0058	0.0059	0.0062	0.0067
GASA	0.0040	0.0047	0.0050	0.0057	0.0064

Table 3.4 shows that there are savings of 70%, 60%, and 53% in the average energy consumption of GASA as compared to broadcasting, clustering, and DD, respectively, for 200 nodes. Figure 3.10 shows the total energy consumption for the four schemes as a function of a number of sensor nodes. When the number of nodes is less, the clustering and DD have nearly the same power consumption as the GASA routing. There is not much difference in the total energy consumption of all the routing schemes up to 100 nodes. But after that the total energy consumption of the GASA is much less as compared to other routing schemes.

The total energy consumption of routing schemes is shown in Table 3.5. The GASA shows consistently better performance than other conventional routing schemes.

3.6 Self-Organizing Migrating Algorithm for Routing in WSNs

Engineering applications use conventional search and optimization techniques for local optimization. These techniques provide best-quality solutions for engineering applications. These local

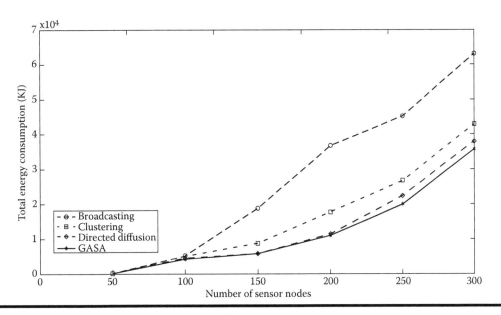

Figure 3.10 Total energy consumption of the network.

Table 3.5 Total Energy Consumption (Joules) for Different Numbers of Nodes

Techniques	Number of Nodes				
	50	100	150	200	250
Broadcasting	123	5102	18,782	36,754	63,214
Clustering	119	4973	8732	17,640	42,815
Directed diffusion	105	4519	5854	11,428	37,873
GASA	103	4232	5237	10,987	35,672

search techniques depend on the initial design and produce good-quality output if the initial design is good. If the starting point in finding the solution is out of the vicinity of the global optimum, then these local optimization algorithms get stuck and cannot find the global optimum. There are certain stochastic search and optimization algorithms that can be used to solve such problems like GAs, particle swarm optimization, ant colony optimization, and a self-organizing migrating algorithm (SOMA). The applications used nowadays require the use of nonconventional algorithms. These nonconventional algorithms are able to find the global optimum within a reasonable amount of time. These global search algorithms are allowed to go through the entire search space of the objective functions to obtain the global optimum. They work in parallel in various search regions of the solution space. Such algorithms are very good at solving NP problems. We have a large number of problems in science and technology that are NP-hard and they can be easily solved by nonconventional algorithms (Singh et al. 2014). Real-life problems are basically nonlinear or differential in nature. There are highly difficult combinatorial and permutative problems. The power-aware routing problem is one such real-life problem that has been addressed by most researchers recently (Singh et al. 2014).

Currently, a number of heuristics exist which deal with these problems; however, most arose as canonically real optimization tools, later modified for permutative problems. This chapter uses a next-generation optimization heuristic of SOMA, which has proven highly effective in solving real optimization problems.

3.6.1 Self-Organizing Migrating Algorithm

SOMA is a stochastic search algorithm that works on the principle of cooperation among the individuals in a social network. This algorithm has been extensively used in solving many such global optimization problems. They have proven to be fairly well at searching the solution for this type of problem. It works on a collection of individuals (random solutions) called the population and iterated for many generations (loops). The population is initialized in a random manner and distributed over the solution search space. The fitness of each individual in the population is evaluated and compared with the fitness of each other to get the solution with highest fitness.

The solution with the highest fitness is selected as the leader. In one generation loop, all the individuals in the population will traverse in the direction of the leader (Singh et al. 2014). A distance, called the path length, is traveled by an individual toward the leader in *n* steps, each step of size *step size*. The path length is chosen to be less than one, otherwise it will overshoot the leader.

3.6.1.1 Perturbation

Mutation is the process of making changes into an individual or string or chromosome in the search for the solution; the process of mutation randomly perturbs the individuals. This is an important feature of evolutionary search strategies. Mutation is done to maintain diversity among the individuals and it is also used to restore the information that is lost during the crossover over the generations. Mutation is quite differently applied in SOMA as compared to other search techniques. A PRT vector is used in SOMA for perturbation. The vector is defined in the range [0, 1] and is created according to

$$\text{if rand}_j < \text{PRT then PRT Vector}_j = 1 \qquad (3.17)$$

$$\text{else } 0, \qquad j = 1,....,n$$

The main advantage of this technique is the creation of a PRT vector that takes place before the search process starts and individuals forward in the direction of the leader.

The PRT vector generated is binary and is randomly obtained. The individual is allowed to change its position during the search process. The corresponding element in the PRT vector is one, and if it is set to zero the individual is not allowed to change its position.

3.6.1.2 Generating New Candidate Solutions

Once the population is generated randomly during initialization, the individuals are evaluated for fitness and the individual having the highest fitness value is selected as the leader, while the worst individual is called active. The active individuals travel in the direction of the leader in each migration loop. They travel a distance in n steps of a fixed length toward the leader. The travel path of an individual is perturbed by a PRT vector and is the same as a mutation operator. An individual moves in the direction of the leader as follows (Davendra and Zelinka 2009):

$$x_{i,j}^{\text{MLnew}} = x_{i,j,\text{Start}}^{\text{ML}} + \left(x_{L,j}^{\text{ML}} - x_{i,j,\text{Start}}^{\text{ML}}\right) t \text{PRTVector}_j \tag{3.18}$$

where $t \,\epsilon\, (0, \text{ by step to, Path})$, ML is the actual migration loop, $x_{i,j}^{\text{MLnew}}$ is the new position of an individual, $x_{i,j,\text{Start}}^{\text{ML}}$ is the position of active individual, and $x_{L,j}^{\text{ML}}$ is the position of leader

The computational steps of SOMA explained in [39], are as follows:

1. Perform population initialization.
2. Calculate the fitness of each individual.
3. For every individual, calculate the PRT using Equation 3.17.
4. Perform sorting of all of them.
5. The individual having the best fitness is the leader, and the worst is active.
6. The new position is calculated for an active individual according to Equation 3.18. The best position among them is found and this replaces the active one with the new.
7. Stop if the termination condition is met, otherwise go to step 2.
8. Output the individual, the best individual, as solution.

3.6.2 System Model

The system model used is the same as in Section 3.4.1. The first-order radio model is used for transmitting as in Equation 3.2 and receiving as in Equation 3.3. The fitness of each routing is calculated in Equation 3.5.

The energy load is balanced and a particular path is calculated by adding the residual energy of all the nodes in the path P_i divided by the total number of nodes as shown in Equation 3.6.

The fitness function, which takes into account the cost (distance) between the nodes as well as the energy balance, is thus given as

$$F(C,E) = \alpha C_f + \beta \frac{1}{E_f} \tag{3.19}$$

The distance is to be minimized and the residual energy of the selected route should be maximum, that is why the E_f is inverted, so that the function $F(C,E)$ becomes a minimization function. α and β are the weights for the cost (distance) from source to sink and energy balance of the network, respectively, and $\alpha + \beta = 1$.

3.6.3 Simulation and Analysis

The GA parameters are the same as discussed in Section 3.4.3. The sensor fields of 200 m × 200 m is considered. But the population size taken for this experiment is cut down to 30 for SAMA as it works very well on low population.

The value of the PRT is taken to be 0.1. The path length is set as 3, since the solution jumps two steps beyond the leader from its starting point. Path length longer than 3 is not deemed feasible due to the extended execution time required. The step size is the only variable that can generally be said to be problem dependent. A starting point of 0.1 and increments of 0.02 per execution would be sufficient to obtain the optimal operating value. The convergence of the elitism GA with immigrants and memory scheme is far better than the Simple GA elitist GA. The inherent advantage of elitism is that it keeps the elite solutions for the next generation of faster convergence.

The immigrants in each generation add diversity to the population so that the most optimal solution can be found in a smaller number of generations and memory is used to store some useful information to quickly converge in case of a topology change. As can be seen in Figure 3.11, the convergence of SOMA is much faster as compared to other algorithms. Table 3.6 shows that SOMA converges mainly 46 generations, whereas elitism-/immigrants-/memory-based GA converges in 83 generations, which is 45% faster.

If compared with a simple GA there is a 70% improvement. The network lifetime maximization with SOMA is slightly better as compared to the elitism GA with immigrants and memory scheme as shown in Figure 3.11.

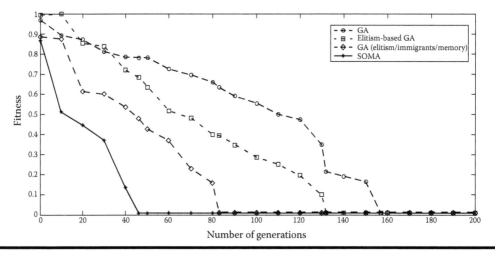

Figure 3.11 Comparison of convergence of various algorithms.

Table 3.6 Number of Generations Taken by Each Algorithm to Converge

Algorithm	Number of Generations	Population Size
GA	157	60
GA (Elitism)	132	60
GA (Elitism/immigrants/memory)	83	60
SOMA	46	30

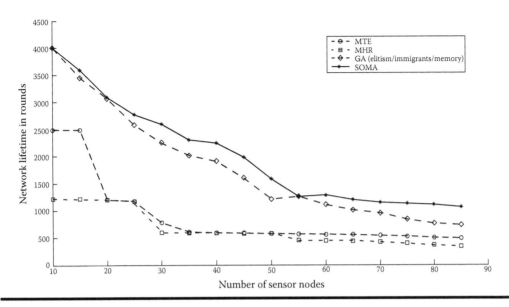

Figure 3.12   Comparison of network lifetime.

SOMA outperforms all the algorithms in terms of number of generations taken to converge the population size, as can be seen in Table 3.6. For nodes numbering up to 25, the network lifetime using SOMA is similar to elitism-/immigrants-/memory-based GAs. When the number of nodes is increased, SOMA performs better and nearly becomes stable as shown in Figure 3.12. The values of the weighted coefficients optimal route α and energy balance of the network β are taken to be 0.4 and 0.6, respectively, to obtain the best performance.

3.6.4 Confidence Bounds

As the nodes in the WSN are randomly deployed, for each simulation a different topology is obtained. The distance of the nodes from the BS and the distance between the nodes is different for each simulation. The GA is also a stochastic search and optimization technique, which may take different numbers of generations (iterations) to converge. Thus, it may result in different lifetime of the network for each run. The network lifetime for each run is noted that varies with the number of data collection rounds. The confidence bounds provide a measure of confidence that the network lifetime obtained lies within a definite range for different locations of nodes during each run. Confidence bounds give the range of values in the interval. The interval width shows how certain or uncertain the results are about the obtained network lifetime values. The bounds have a level of specified certainty. The certainty level is sometimes 95% but it can have any value. For example, a 95% confidence interval indicates that there is a 95% chance that the obtained network lifetime value shall lie within the specified bounds. The confidence bound is found by calculating the mean of the lifetime values obtained, which is given as

$$\bar{x} = \frac{\sum_{i=1}^{n} x_i}{n} \tag{3.20}$$

Table 3.7 Confidence Bounds 99% and 95% for Network Lifetime for SOMA

Confidence Bounds	99%	95%
LCL	1285	1288
UCL	1310	1307

The standard deviation of data values is represented

$$s = \sqrt{\frac{\sum_{i=1}^{n}(x-\bar{x})^2}{n-1}} \quad (3.21)$$

The estimated standard deviation is used to create a lower confidence limit (LCL) and an upper confidence limit (UCL) about the mean given as

$$\bar{x} \pm t\frac{s}{\sqrt{n}} \quad (3.22)$$

where t represents the Student's t cumulative distribution function and α is 0.05 for 95% confidence and 0.01 for 99% confidence. The Student's t distribution test is being used because the simulation is run for 25 samples at a time. For more samples, the z test can be performed.

In the experiments conducted, the nodes are randomly distributed and the standard deviation is not known; hence, the t test is used. The calculated values of the LCL and UCL for the overall network lifetime (in number of rounds of data communication) are shown in Table 3.7. The results obtained in the experiments follow the values presented in Table 3.7 and are within a specified range.

Network lifetime can be improved to a greater extent by the proposed technique. The energy of a node falling below a predefined level is restrained from the path based on probability, to increase the overall lifetime of the network. The energy load balancing strategy prevents uneven energy dissipation and thus the network lifetime is prolonged considerably. The proposed routing uses SOMA, which takes into account both the cost (distance) between the nodes, i.e., the shortest path from the source to the sink, and the energy balance of the network based on the residual energy of the nodes divided by the total number of nodes. Thus, the results show that the proposed algorithm improves the overall network lifetime by selecting the shortest path and the residual energy of the nodes is evenly utilized. The experimentations with WSN route optimization have validated the stated approach, with comparable results and optimal solutions. The results obtained especially for finding an energy-efficient route in a WSN highlighted the effectiveness of SOMA in terms of the small population size and faster convergence.

References

Akkaya, K. and Younis, M. 2005. A survey on routing protocols for wireless sensor networks. *Elsevier Ad Hoc Netw J.* 3:325–349.

Akyildiz, I.F., Su, W., Sankarasubramaniam, Y., and Cayirci, E. 2002. A survey on sensor networks. *IEEE Commun Mag.* 40(8):102–114.

Al-Karaki, J.N. and Kamal, A.E. 2004. Routing techniques in wireless sensor networks: A survey. *IEEE Wireless Commun.* 11:6.

Bari, A., Wazed, S., Jaekel, A., and Bandyopadhyay, S. 2009. A genetic algorithm based approach for energy efficient routing in two-tiered sensor networks. Ad Hoc Netw. 7(4):665–676.

Bhagwan Das, D. and Patvardhan, C. 1999. Solutions of economic load dispatch using real coded hybrid stochastic search. *Int J Elec Power.* 21:3165–3170.

Chunlai, C. 2009. Efficient intelligent localization scheme for distributed wireless sensor networks. *J Comput.* 4(11):1159–1166.

Dasgupta, D. and McGregor, D. 1992. Nonstationary function optimization using the structured genetic algorithm. In *Proceedings 2nd International Conference Parallel Problem Solving Nature.* Brussels, Belgium, 28–30 September 1992, 145–154.

Davendra, D. and Zelinka, I. 2009. Optimization of quadratic assignment problem using self organising migrating algorithm. *Comput Inform.* 28:1001–1012.

Eglese, R.W. 1990. Simulated annealing: A tool for operation research. *Eur J Oper Res.* 46(3):271–281.

Guo, L., Huichang, S., Jun, Y., and Yifei, Z. 2008. An application of genetic algorithm in energy efficient routing. In *Microwave Conference.* China-Japan, 10–12 September 2008.

Haenggi, M. 2004. Twelve reasons not to route over many short hops. In *the proceedings of the 60th IEEE Vehicular Technology Conference (VTC).* 26–29 September 2004, 3130–3134.

Heikalabad, S.R., Ghaffari, A., Hadian, M.A., and Rasouli, H. 2011. DPCC: Dynamic predictive congestion control in wireless sensor networks. *IJCSI Int J Comput Sci Issues.* 8(1):472–477.

Heinzelman, W.B., Chandrakasan, A.P., and Balakrishnan, H. 2002. An application-specific protocol architecture for wireless microsensor networks. *IEEE Trans Wireless Commun.* 1(4):660–670.

Heinzelman, W.B., Kulik, J., and Balakrishnan, H. 1999. Adaptive protocols for information dissemination in wireless sensor networks. In *Proceedings of IEEE/ACM International Conference on Mobile Computing and Networking.* Seattle, Washington, 15–19 August 1999, 174–185.

Hui, X., Zhigang, Z., and Xueguang, Z. 2010. A novel routing protocol in wireless sensor networks based on ant colony optimization. *Int J Intell Inf Technol Appl.* 3(1):1–5.

Intanagonwiwat, C., Govindan, R., and Estrin, D. 2000. Directed diffusion: A scalable and robust communication paradigm for sensor networks. In *Proceedings of the 6th Annual ACM/IEEE International Conference on Mobile Computing and Networking (MobiCom'00), August 06–11, 2000, Boston, MA, USA.*

Intanagonwiwat, C., Govindan, R., Estrin, D., Heidemann, J., and Silva, F. 2003. Directed diffusion for wireless sensor networking. *IEEE/ACM Trans Netw.* 11(1):2–16.

Lindsey, S. and Raghavendra, C.S. 2002. PEGASIS: Power-efficient gathering in sensor information systems. In *Proceeding IEEE Aerospace Conference*, 9–16 March 2002, 1125–1130.

Misra, S., Reisslein, M., and Xue, G. 2008. A survey of multimedia streaming in wireless sensor networks. *IEEE Commun Surveys Tuts.* 10(4):1553–1877.

Raghavendra, V.K., Forster, A., and Venayagamoorthy, G.N. 2011. Computational intelligence in wireless sensor networks: A survey. *IEEE Commun Surveys Tuts.* 13(1):68–96.

Rana, K.M. and Zaveri, M.A. 2011. Techniques for efficient routing in wireless sensor network. In *International Conference on Intelligent Systems and Data Processing (ICISD)-Special Issue published by International Journal of Computer Applications (IJCA) January 24–25, 2011, Vallabh Vidya Nagar, Anand, Gujarat, India.*

Shah, R.C. and Rabaey, J. 2002a. Energy aware routing for low energy ad hoc sensor networks. In *IEEE Wireless Communications and Networking Conference (WCNC),* 17–21 March 2002 , Orlando, Florida, USA.

Shah, R.C. and Rabaey, J. 2002b. Energy aware routing for low energy ad-hoc sensor networks. In *Proceedings of IEEE WirelessCommunications and Networking Conference (WCNC),* 17–21 March 2002, 1:350–355.

Singh, D., Agrawal, S., and Singh, N. 2014. A novel variant of self-organizing migrating algorithm for global optimization. In *Proceedings of the Third International Conference on Soft Computing for Problem Solving, Advances in Intelligent Systems and Computing,* Vol. 258, Springer, India.

Singh V. K. and Vidushi, S. 2011. On the hybridization of evolutionary algorithms and optimization techniques. In *Conference on Advancements in Communication, Computing & Signal Processing (COMMUNE CACCS 2011)*. Gautham Buddha Nagar, Uttar Pradesh, India, 16–17 April 2011.

Singh, V. K. and Sharma, V. 2012a. Elitist genetic algorithm based energy efficient routing scheme for wireless sensor networks. *Int J Adv Smart Sensor Netw Syst*. 2:2.

Singh, V. K. and Sharma, V. 2012b. Extending wireless sensor network lifetime through effective genetic algorithm based approach. In *COMMUNE Conference on Advancements in Communication, Computing & Signal Processing*.

Singh, V. K. and Sharma, V. 2012c. Lifetime maximization of wireless sensor networks using improved genetic algorithm based approach. *Int J Comput Appl*. 57(14):0975–8887.

Singh, V. K. and Sharma, V. 2014. Elitist genetic algorithm based energy balanced routing strategy to prolong lifetime of wireless sensor networks. *Chinese J Eng*. 1:6. Hindawi Publishing Corporation.

Srinivas, K., Patvardhan, C., and Bhagwan Das, D. 2000. A hybrid stochastic search technique for optimization of difficult functions. *J Inst Eng (India), Comput Eng Division*. 81:45–49.

Vural, S. and Ekici, E. 2007. Hop-distance based addressing and routing for dense sensor networks without location information. *Ad Hoc Netw*. 5(4):486–503.

Wenliang, G., Huichang, S., Jun, Y., and Yifei, Z. 2009. Application of genetic algorithm in energy-efficient routing. In *China-Japan Joint Microwave Conference, 10–12 September 2008*, 737–740.

Xu, Y., Heidemann, J., and Estrin, D. 2000. *Adaptive energy-conserving routing for multihop ad hoc networks*. Research Report. USC/Information Sciences Institute.

Xu, Y., Heidemann, J., and Estrin, D. 2001. Geography-informed energy conservation for ad hoc routing. In *Proceedings of the Seventh Annual ACM/IEEE International Conference on Mobile Computing and Networking (MobiCom 2001)*. Rome, Italy.

Yang, S., Cheng, H., and Wang, F. 2010. Genetic algorithms with immigrants and memory schemes for dynamic shortest path routing problems in mobile ad hoc networks. *IEEE Trans Syst Man Cybern C Appl Rev*. 40(1):52–63.

Ye, F., Lu, S., and Zang, L. 2001. GRAdient Broadcast: A Robust, Long-lived, Large Sensor Network. Available at: http://irl.cs.ucla.edu/papers/grab-tech-report.ps.

Younis, O. and Fahmy, S. 2004. HEED: A hybrid, energy-efficient, distributed clustering approach for Ad Hoc sensor networks. *IEEE Trans Mobile Comput*. 3(4):660–669.

Zou, Y. and Chakrabarty, K. 2003. Energy-aware target localization in wireless sensor networks. In *Proceedings of the IEEE International Conference on Pervasive Computing and Communications*. 26–26 March 2003, 60–67.

Chapter 4

Clustering for Energy Efficiency in Wireless Sensor Networks

Puneet Azad

Contents

4.1 Motivation and Basic Ideas 70
4.2 Use of Clustering for Energy Efficiency in WSNs 70
    4.2.1 Partitional Clustering 71
    4.2.2 Hierarchical Clustering.. 71
    4.2.3 Exclusive Clustering... 74
    4.2.4 Overlapping Clustering.. 74
    4.2.5 Fuzzy Clustering.. 74
    4.2.6 Complete and Partial Clustering... 75
4.3 Single Criteria-Based Algorithms 75
    4.3.1 Low-Energy Adaptive Clustering Hierarchy 75
    4.3.2 Energy-Efficient Clustering Hierarchy 78
    4.3.3 Distributed Energy-Efficient Clustering Algorithm 79
    4.3.4 Novel Clustering Approach for Extending the Lifetime 79
    4.3.5 Energy-Efficient Clustering Scheme 79
    4.3.6 Performance Evaluation of NCAEL and EECS 81
    4.3.7 Maximum Residual Energy-Based Clustering Scheme................................... 83
        4.3.7.1 Static Maximum Residual Energy-Based Clustering Algorithm 83
        4.3.7.2 Maximum Residual Energy-Based Clustering Algorithm 86
        4.3.7.3 Comparative Analysis 86
4.4 Multiple Criteria-Based Algorithms 90
    4.4.1 Pareto-Optimal Theory..91
    4.4.2 TOPSIS Approach 92
    4.4.3 Ranking Using Fuzzy TOPSIS 98

        4.4.3.1 Fuzzy TOPSIS Approach 98
        4.4.3.2 Performance Evaluation of Fuzzy TOPSIS 101
     4.4.4 Ranking Using VIKOR 102
        4.4.4.1 Performance Evaluation of VIKOR 105
4.5 Conclusion 112
References 112

4.1 Motivation and Basic Ideas

One of the recent challenges in designing a wireless systems is the efficient use of two key resources, i.e., energy and communication bandwidth. These restrictions require innovative techniques for efficiently utilizing the bandwidth. Communication protocols should be designed for suitability in current conditions than in worst-case conditions. One of the key challenges is the conservation of energy for the design and operation of wireless sensor networks (WSNs). All the sensor nodes should be designed so that they are small in size, low-cost, and have good communicating capabilities. There are many applications such as habitat monitoring, disaster management, etc., which demand battery replacement and remain unattended in many cases. For meeting the conflicting requirement of prolonged unattended operation, WSNs will have to depend on lightweight techniques of communication and protocols, which are capable of enhancing the lifetime of all the nodes.

Many algorithms have been designed for the best utilization of energy in the nodes present in the WSN. Thus, the main purpose is to develop an energy-efficient algorithm for increasing the lifetime of the network. The lifetime can be defined in terms of either the death of the first node or the death of all the nodes present in the network. In the present study, both criteria have been considered in different articles. Assuming all the nodes have similar data processing and routing capabilities, recent research has largely been concentrated on various routing and clustering techniques (Romesburg 1990; Yu and Chong 2005). In the last few years, several published research works on WSNs have dealt with the topic of "energy conservation." In these papers, the researchers presented minimization of energy usage of sensor nodes and prolonged the lifetime of WSNs by (1) reducing communication between various nodes or (2) increasing the sleep time of nodes. It is commonly agreed by the authors that WSNs of cluster-based organization have a good potential to combine both (1) and (2) effectively. Accordingly, clustered networks come in the category of most energy-efficient and long-lived sensor networks. If the nodes are not in range of each other, then a multihop communication network may be adopted. Hence, there is a need to develop clusters to enable communication to different remote locations in the sensor network. The strategy of clustering in the present work starts with the selection of cluster heads (CHs) and using various statistical methods. The methods include the best nodes to act as CHs based on a certain ranking technique. To achieve good performance in network lifetime, it is essential that the nodes be deployed either remotely or in a strategic manner. Manual deployments possess a difficulty in inaccessible terrains, where it is not possible to manually place the sensor nodes. Thus, placement of these nodes is another area that needs further attention and has been addressed and discussed in the literature for uniform and nonuniform placements. The above points of discussion have motivated the development of energy-efficient clustering schemes (EECSs) in the WSN.

4.2 Use of Clustering for Energy Efficiency in WSNs

Clustering is the classification (unsupervised) of patterns into certain groups, which are called clusters (Abbasi and Younis 2007). The analysis comprises of the organization of a pattern collection represented as a vector, or a point in a multidimensional space into clusters based on some similarity. In WSNs, sensor nodes are grouped into clusters. The similarity between these nodes can be in terms of a number of properties such as their location, residual energy, signal strength, connectivity, distance from the sink, etc. Clustering enables better resource allocation, saves energy, and reduces overhead communication. There are various types of clustering techniques, which are described as follows.

4.2.1 Partitional Clustering

A partitional clustering is a division of the set of data objects divided into nonoverlapping data sets (called clusters) such that all data objects are in exactly one subset. Figure 4.1 shows the collection of clusters for two different clusters. The first type of clustering is K-means, which is a partitional clustering technique in which the user specifies the number of clusters (K) with their centroids. It creates a one-level partitioning of the data objects. The partitioning of data into K clusters (mutually exclusive) returns the cluster's index. This clustering works on actual observations instead of sets of dissimilarity measures and creates clusters of a single level. In this method, cluster centroids are computed for each distance measure for the minimization of the sum with respect to the measure specified by the user. It uses an algorithm (iterative in nature) to minimize the sum of distances from all the objects to its cluster centroid. The resulting set of clusters is compact and well-separated as shown in Figure 4.2.

4.2.2 Hierarchical Clustering

If subclusters are permitted in the design, then hierarchical clustering is obtained. It is a set of nested clusters that are in the form of a tree. All the nodes in the tree are the union of their children and the root is the cluster, which contains all the objects. An important type of hierarchical clustering

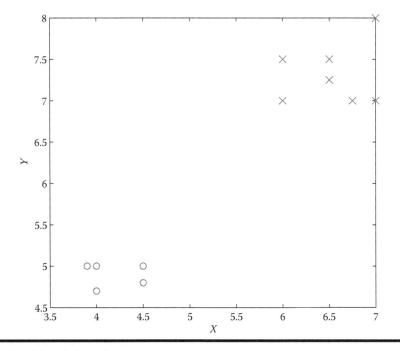

Figure 4.1   Partitional clustering.

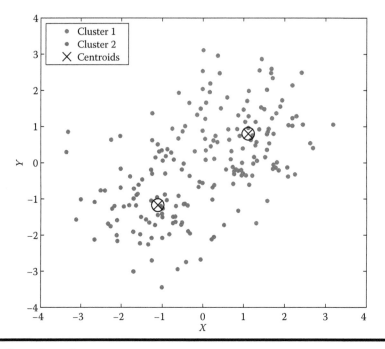

Figure 4.2 Clustering using *K*-means.

is hierarchical agglomerative clustering (HAC) (Akylidiz and Wang 2005), which is a mathematically and conceptually simple approach for data analysis. There are four types of HAC-algorithm methods:

1. Single LINKage (SLINK): It is also termed as the nearest neighbor method, which defines the measure of similarity between clusters as the lowest resemblance coefficient (minimum) among all the entities in the two clusters.

$$C_{\text{SLINK}} = \text{Min}(C_{(1,1)}, C_{(1,2)}, ..., C_{(i,j)}, ..., (m,n))$$ (4.1)

2. Complete LINKage (CLINK): It is also termed as the furthest neighbor method, which defines the similarity measure between clusters as the highest resemblance coefficient among all entities in the two clusters.

$$C_{\text{CLINK}} = \text{Max}(C_{(1,1)}, C_{(1,2)}, ..., C_{(i,j)}, ..., C_{(m,n)})$$ (4.2)

3. Unweighted pair-group method using arithmetic averages (UPGMA): It defines the similarity measure between clusters as the average (arithmetic) of resemblance coefficients among all entities in the two clusters. It is the most widely adopted clustering method.

$$C_{\text{UPGMA}} = \frac{1}{mn} \sum_{i=1, j=1}^{m,n} C_{(i,j)}$$ (4.3)

4. Weighted pair-group method using arithmetic averages (WPGMA). It is the arithmetic average of resemblance coefficients between clusters without taking into consideration the cluster size.

$$C_{\text{UPGMA}} = \frac{1}{mn} \sum_{i=1, j=1}^{m,n} C_{(i,j)}. \tag{4.4}$$

The results of this algorithm are usually presented by a binary tree or dendrogram. The root of the dendrogram presents the data set and each leaf is termed as a node.

The intermediate nodes describe the extent in which the nodes are close to each other and the dendrogram. An eight-link network as shown in Figure 4.3 is considered and the results of hierarchical clustering with the techniques SLINK, CLINK, and UPGMA are shown in Figure 4.4.

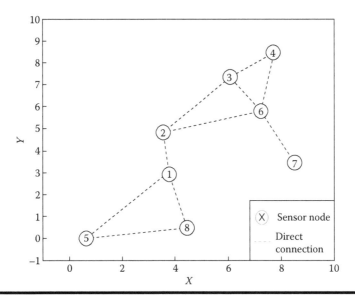

Figure 4.3 A simple eight-node network.

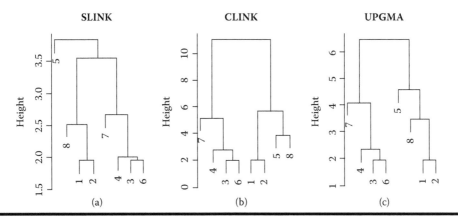

Figure 4.4 Dendrogram using different hierarchical agglomerative clustering algorithms with quantitative data.

4.2.3 Exclusive Clustering

The clusterings as shown in Figure 4.1 are all exclusive in which each object is assigned to a single cluster. Exclusive clustering suggests that each data object can only exist in one cluster. Thus, the given data are grouped exclusively, so that a datum belongs to only one fixed cluster. This separation is based on the characteristic that allows a data object to exist in one or more than one clusters. *K*-means clustering comes under the category of exclusive clustering algorithms and partitional clustering.

4.2.4 Overlapping Clustering

An overlapping clustering depicts that an object can belong to more than one group simultaneously. For example, a person can be simultaneously enrolled as a student of the university and an employee of the university. A nonexclusive clustering suggests that an object belongs to two or more clusters and can be reasonably assigned to any cluster.

4.2.5 Fuzzy Clustering

In fuzzy clustering, each object belongs to every cluster having a membership weight between 0 and 1. Thus, the clusters are identified as fuzzy sets, because the probabilities or membership sum up to 1. The most famous fuzzy clustering algorithms is the fuzzy c-means (FCM) algorithm developed by Dunn and Bezdek. This method obtains the best location of clusters in an optimum manner by minimizing an objective function as given by

$$J = \sum_{i=1}^{N} \sum_{j=1}^{C} u_{ij}^m \| x_i - c_j \|^2, \qquad (4.5)$$

where x_i is the set of data points, N is the number of data set, c_j is the center of the clusters, C is the number of clusters, u_{ij} is the membership degree of x_i in jth cluster, $\|*\|$ is any norm expressing the similarity between any measured data and the center (which is the distance here), and m is the weighing exponent on each fuzzy membership. Fuzzy partitioning is performed using an iterative optimization of the objective function as shown in Equation 4.5, with the membership u_{ij} and the cluster centers c_j given by

$$u_{ij} = \frac{1}{\sum_{k=1}^{C} \left(\frac{\| x_i - c_j \|}{\| x_i - c_k \|} \right)^{\frac{2}{m-1}}}, \text{ where } c_j = \frac{\sum_{i=1}^{N} u_{ij}^m . x_i}{\sum_{i=1}^{N} u_{ij}^m}. \qquad (4.6)$$

These iterations stop when $\max_{ij}\{| u_{ij}^{(k+1)} - u_{ij}^{(k)} |\} < \varepsilon$, where ε lies between 0 and 1, called the termination criterion, whereas k signifies the iteration steps.

Figure 4.5 shows two different clusters with their centers given by the FCM technique for a specific data set. The centers of the two clusters are approximately placed physically in the center only by this technique.

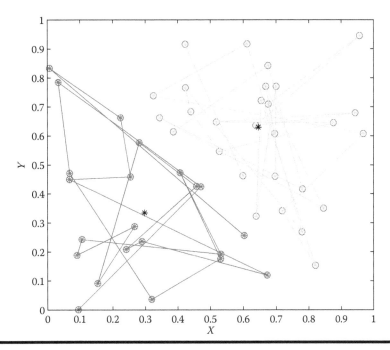

Figure 4.5   Clustering using fuzzy c-means.

4.2.6 Complete and Partial Clustering

In this clustering, every object is assigned to a cluster in which some objects in a data set may not belong to any defined group. In this clustering, many objects may represent outliers, noise, or uninteresting background in the data set. For example, some news channels may share a common theme, such as terrorism, while others can be more generic.

4.3 Single Criteria-Based Algorithms

To enhance the overall lifetime of the network, clustering algorithms may be designed by suitably selecting the CHs and forming the clusters. There are a number of methods, which are based on selecting the CHs using a single criterion for the formation of clusters. Many of them use a single parameter such as distance from the sink, residual energy, or the average energy of the node or received signal strength at any time to find out the CHs for a particular round. Some of the methods based on single criterion are as follows.

4.3.1 Low-Energy Adaptive Clustering Hierarchy

Low-energy adaptive clustering hierarchy (LEACH) (Heinzelman et al. 2002), a protocol architecture, presents the ideas of cluster-based energy-efficient routing with aggregation for achieving a satisfactory performance in terms of network lifetime. Several microsensor nodes are laid down for monitoring a remote environment by intelligently aggregating the data from all the nodes present in the network. The protocol adopts a method of formation of clusters enabling the self-organization of hundreds of nodes, which involves the rotation of position of CHs for uniform expenditure of energy of the nodes. In this method, CHs are selected randomly, whose positions are rotated to balance the energy dissipation of the sensor nodes present in the network. Each node becomes a CH on the basis of a predefined threshold given whereby

$$T(S) = \begin{cases} \dfrac{p_{opt}}{1 - p_{opt} * \left(c * \mathrm{mod} \dfrac{1}{p_{opt}} \right)} & \text{if } S \in G, \\ 0 \end{cases} \qquad (4.7)$$

where c is the present cycle, p_{opt} is the percentage of CHs, and G is the set of those nodes that have not become CHs in the last $1/p_{opt}$ cycles.

A radio energy model for transmitting an l bit message over a distance d is presented in this section using the simulation parameters of practical interest shown in Table 4.1. A free space model and a multipath fading model are used for the data transmission of packets from nodes to CH and further to sink. The threshold value of distance of transmission is given by

$$d_t = \sqrt{\frac{\varepsilon_{fs}}{\varepsilon_{mp}}}, \qquad (4.8)$$

where ε_{fs} and ε_{mp} are the radio energy parameters (Table 4.1) for the free space and multipath fading models. If the transmission distance d is less than the d_t threshold, the free space model is adopted, otherwise the multipath model is used. The energy used to send a l bit data at a distance d is given by

$$E_{TX} = \begin{cases} l * E_{elec} + l * \varepsilon_{fs} * d^2 & \text{if } d \le d_t \\ l * E_{elec} + l * \varepsilon_{mp} * d^4 & \text{if } d \le d_t. \end{cases} \qquad (4.9)$$

Table 4.1 Simulation Parameters

Description	Symbol	Value
Number of nodes in the system	N	100
Size of data packet	–	500 bytes
Initial energy	E_{Init}	2/1.5 J
Broadcast/Hello/CH_Join message	–	25 bytes
Energy consumption in the amplifier to transmit at a short distance	ε_{fs}	10 pJ/bit/m²
Energy consumption in the amplifier to transmit at a longer distance	ε_{mp}	0.0013 pJ/bit/m⁴
Energy consumption in the electronics circuit to transmit or receive the signal	E_{elec}	50 nJ/bit

The two expressions are equated at $d = d_o$, and we have

$$d_o = \sqrt{\frac{\varepsilon_{fs}}{\varepsilon_{mp.}}} \qquad (4.10)$$

For receiving an l bit message, the energy consumed is

$$E_{RX} = l * E_{elec}, \qquad (4.11)$$

where E_{elec} is the energy consumption in the electronic circuit for transmitting or receiving the signal, and ε_{fs} and ε_{mp} are the energy consumptions in the amplifier for transmission at shorter or longer distances.

The CH nodes aggregate the data arriving from all the nodes and send the aggregated data to the sink to reduce the total amount of data. This collection of data is performed periodically. When clusters are formed, each node decides to become the CH on the basis of a probability. LEACH-C, an improvement over LEACH, finds clusters using the simulated annealing algorithm and finds k-optimum clusters with minimum average energy dissipation per round. Since the distribution of CH is random, the algorithm does not guarantee a uniform distribution of CH nodes all over the field. This results in a low network lifetime and inappropriate cluster distribution. Figure 4.6 shows the total number of nodes that remain alive over the simulation time.

Results from this experiment show that LEACH provides the high performance needed under the tight constraints of the wireless channel. An assumption has been made that all the nodes are considered within the range of communication with other nodes and the base station (BS), which limits the scalability of the protocol. But this limitation can be relaxed by collision-avoidance methods.

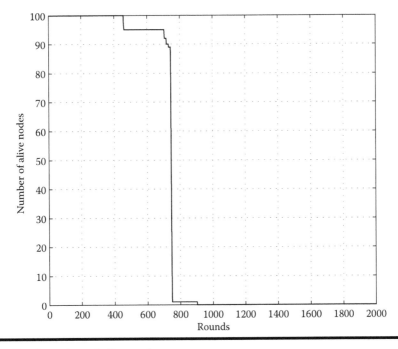

Figure 4.6   Total number of alive nodes.

4.3.2 Energy-Efficient Clustering Hierarchy

The energy-efficient heterogeneous clustered scheme (EEHC) (Kumar et al. 2009) adopts the heterogeneity of the nodes in terms of their initial energy, i.e., a percentage of nodes are equipped with more energy than others. This method adopts the heterogeneity in which few nodes have more energy than others present in the network. To improve the performance of the system, this work reports the weighted probability of the election of CHs given by

$$p_n = \frac{p_{opt}}{1 + m * (\alpha + m_o \beta)},$$ (4.12)

where p_{opt} is the predetermined percentage of the CH, m is the fraction of the total number of nodes, and m_o is the percentage of m having β times more energy than others. The probability of the selection of a node (as a CH) is more when the initial energy of the node is higher. It is reported that the lifetime of the network increased by 10% as compared with the results obtained by using the LEACH method. The performance of the EEHC protocol is evaluated by considering the first-order radio model as shown in LEACH and the simulation parameters for our model are mentioned in Table 4.1. In Figure 4.7, the network lifetime of EEHC is illustrated, which shows the number of alive nodes. For the same number of rounds, the number of nodes that die in LEACH is more than that in EEHC. The normal nodes die faster in comparison to the advanced and super nodes and as a result the sensing field becomes sparse very fast. This is because advanced and super nodes have higher energy than normal nodes, and hence live for a longer duration. It is observed that there are no CHs present in many rounds, which results in nonreporting of data by various nodes.

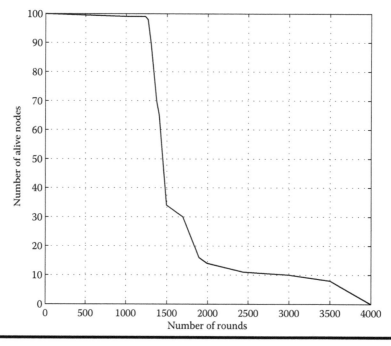

Figure 4.7 Network lifetime of energy-efficient heterogeneous clustered scheme (EEHC) under heterogeneity.

Also, the number of data packets received from CHs at the BS are much higher than LEACH during the lifetime of the network. One of the disadvantages is that the actual energy consumption is not available in this algorithm for a varying number of clusters and if the CHs are not chosen optimally.

4.3.3 Distributed Energy-Efficient Clustering Algorithm

Distributed energy-efficient clustering algorithm (DEEC) (Qing et al. 2006) is another energy-efficient clustering protocol that elects CHs on the basis of a probability, which depends on the ratio of residual energy of the node and the total average energy of the network. Thus, the nodes having high initial and residual energy will become CHs. The algorithm works in a two-level heterogeneous mode in which there are two types of sensor nodes, namely normal and advanced nodes. The probability threshold to become a CH in each round is

$$T(S) = \begin{cases} \dfrac{p_{opt}}{1 - p_{opt} * \left(c^* \mathrm{mod} \dfrac{1}{p_{opt}} \right)} & \text{if } S \in G, \\ 0 \end{cases} \qquad (4.13)$$

where G is the set of nodes that are eligible to be CHs at round r. The performance of DEEC protocol using MATLAB® is done by considering a 100 m × 100 m field and the results are compared with some well-known protocols such as LEACH and the stable election protocol (SEP) (Smaragdakis et al. 2004). SEP is based on weighted election probabilities in which each node becomes a CH on the basis of the residual energy of the node. It always prolongs the stability period compared to (and that the average throughput is greater than) the one obtained using current clustering protocols. The network lifetime of DEEC, SEP, and LEACH is shown in Figure 4.8.

The simulation results show that the lifetime in DEEC is significantly longer than other important clustering protocols in heterogeneous environments.

4.3.4 Novel Clustering Approach for Extending the Lifetime

Novel clustering approach for extending the lifetime (NCAEL) (Azad et al. 2011) works in a homogenous environment in which all the nodes present in the network have the same initial energy. Consider a system model in a 100 m × 100 m sensor field containing 100 nodes placed as shown in Figure 4.9. Each CH is responsible for collecting the data from the nodes present in the cluster and finally sending it to the BS. The criterion of selection of CHs is the highest residual energy. Clusters are formed on the basis of relative distances between the nodes.

The radio energy dissipation model (Heinzelman et al. 2002) (Figure 4.10) is same as given in Equations 4.9 and 4.10.

Finally, the lifetime of the network is plotted in terms of "number of alive nodes" per round.

4.3.5 Energy-Efficient Clustering Scheme

The EECS (Ye et al. 2006) works under a heterogeneous environment, where a fraction of all the nodes is equipped with more energy than other nodes in the network. These nodes which have

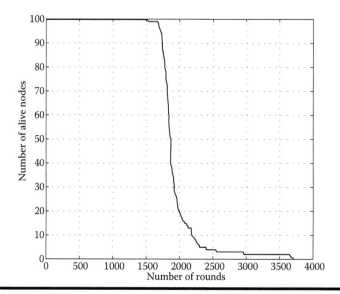

Figure 4.8 Network lifetime of distributed energy-efficient clustering algorithm under heterogeneity.

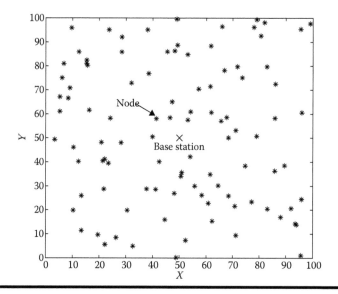

Figure 4.9 Node placement in the homogeneous model.

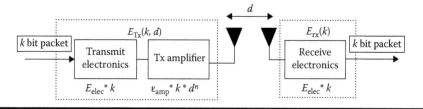

Figure 4.10 Radio energy model used in a wireless sensor network.

higher energy than others are called advanced nodes as shown. A fixed number of CHs are chosen from the set of advanced nodes in the entire area initially having maximum energy in each round. All the nodes will check their distance from all the CHs and send their data to the nearest CH. Thus, each CH will now be associated with a group of sensor nodes which are sending their data in their respective time division multiple access (TDMA) schedule.

4.3.6 Performance Evaluation of NCAEL and EECS

In this section, we evaluate the performance of algorithms *NCAEL* and *EECS* in a homogeneous and heterogeneous environment in MATLAB. We have considered a first-order radio model simulation similar to LEACH and the simulation parameters for our model are mentioned in Table 4.2. The performance of the proposed protocol in a homogeneous environment is calculated in terms of the network lifetime. The performance of NCAEL is compared with two existing protocols, LEACH and distributed hierarchical agglomerative clustering (DHAC) (Lung and Zhou 2009) as shown in Figure 4.11. It is clear that the present method of selecting CH works efficiently than the reported protocols (DHAC and LEACH) for similar input parameters in a similar environment. The performance of NCAEL is better in terms of lifetime and is 28% better than DHAC and 70% better than LEACH. Also, the stability region (when all the nodes are alive) is also the best in NCAEL. The extension in the stability region proves that there is a uniform consumption of energy till 800 cycles and nodes start dying only after that. In other algorithms, the stability region is around than 50% or less as compared to NCAEL.

In a heterogeneous environment, we evaluate the performance of EECS in MATLAB as shown in Figure 4.12. To check the performance, a heterogeneous clustered WSN with 100 sensor nodes is chosen. The normal and advanced nodes are distributed over the field in a random fashion. The energy of the advanced is higher than the normal node keeping the overall energy constant in the setup. The size of the message is 2000 bits.

Interestingly, the longest lifetime (1087 cycles) is associated with EECS and the stability region in EECS (910 cycles). EECS achieves the maximum lifetime (1101 cycles) in terms of number of alive nodes.

This chapter shows the large advantage of using EECS versus NCAEL in terms of system lifetime for a given quality measured here as the amount of effective data whose information is

Table 4.2 Values of Different Transmission Parameters

Description	Symbol	Value
Number of nodes in the system	N	100
Initial energy of the system	E_{init}	0.5 J
Energy consumption in the amplifier for transmitting at a short distance	ε_{fs}	10 pJ/bit/m^2
Energy consumption in the amplifier for transmitting at a longer distance	ε_{mp}	0.0013 pJ/bit/m^4
Energy consumed in the electronics circuit to transmit or receive	E_{elec}	50 nJ/bit
Data aggregation energy	E_{DA}	5 nJ/bit/report

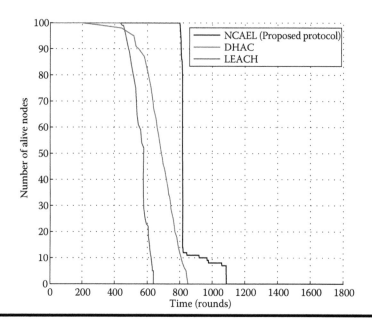

Figure 4.11   Number of alive nodes versus time.

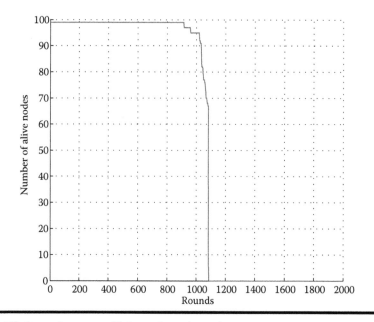

Figure 4.12   Lifetime of energy-efficient clustering schemes.

received at the sink. EECS is effective because it selects the CHs in a heterogeneous environment in which the overall energy of the system is constant. EECS is able to deliver more effective data than NCAEL, DHAC, and LEACH. Due to the nonuniform initial energy distribution in EECS, the CHs are selected from those nodes which have higher energy. This results in longer network lifetime in EECS. One of the main advantages with EECS is that this protocol adapts to the new conditions such as nodes being added to the network whereas LEACH does not fit into this criteria. Therefore, LEACH does not meet the robustness requirements for WSNs. The performance of the proposed system is better in terms of reliability and lifetime. Overall, EECS in a heterogeneous environment works better than NCAEL in a homogeneous environment in terms of network lifetime.

4.3.7 Maximum Residual Energy-Based Clustering Scheme

The present methodology is based on selecting the optimum number of CHs, forming the clusters, and then transmitting the data sensed from the environment. The sink divides the network into clusters and selects the optimum number of CHs. This is done initially by a well-known FCM clustering technique, whereby the centers of each cluster become the CH. Further, after few rounds, the CHs are selected based on their residual energy and the clusters are formed based on the shortest distance between CHs and nodes. This algorithm consists of set-up and steady-state phases. In the set-up phase, clusters are formed using FCM clustering for a few initial cycles, and then a new method is adopted for the formation of clusters based on node residual energy. In the steady state, CHs are responsible for aggregating and sending the data to the sink. Thus, static clusters are formed with the CH being located nearest to the centroid of the cluster assigned from the FCM clustering algorithm. After a few cycles of data transmission (about 100 cycles), the clusters formation adopts a new methodology in which the position of CH is optimally scheduled among the nodes present in the cluster based on the residual energy discarding the traditional FCM method. This method is hereafter referred to as static maximum residual energy-based clustering (s-MREC) in which the clusters remain static throughout the lifetime of the network, but the role of CH is rotated among other nodes of the cluster. In maximum residual energy-based clustering (MREC) (Azad and Sharma 2013b), an optimum number of CHs are selected in each cycle of data transmission based on the maximum residual energy in the whole network. Based on the selected CH, the nodes join a particular cluster based on the nearest distance (Euclidean distance). The performance of these two techniques (s-MREC and MREC) is compared with that of the results obtained from other reported protocols in similar sets of conditions and environments. It is assumed that the nodes are placed randomly in a 100 × 100 region in uniform distribution and the sink is also static and is located in the center of the field. The optimum number of clusters is taken in the range of $8 > k_{opt} > 10$. These analytical results are verified using simulations on a 100-node network by considering the average energy dissipation per round.

4.3.7.1 Static Maximum Residual Energy-Based Clustering Algorithm

The s-MREC algorithm is an improvement in the traditional approach of selecting the CHs randomly. Here, the CHs are selected using maximum residual energy in a node and their positions are rotated inside the clusters. The following are the steps of the algorithm:

Step 1: Obtain an input data set

An input data set for s-MREC is a data matrix that consists of nodes placed randomly in an area following a uniform distribution method in a 100 × 100 m² area. The following messages are used in the formation of clusters during the set-up phase:

1. Hello Message: This message is sent by all the nodes to the BS in the beginning. It contains the location information obtained through global positioning system (GPS). Each node sends its location information to the sink through a "Hello" message containing the node ID and physical location.

2. Broadcast Message: Each elected CH broadcasts its node ID to all the nodes present in the network in such a way that they should reach the farthest node in the network.
3. CH_Join Message: Each non-CH node then measures the distance with the available list of elected CHs and chooses the one at minimum distance. Then, a CH_Join message is sent from each node to the respective CH to be joined.

Step 2: Initialize the number of clusters and assign CHs

The basic purpose of the s-MREC is to generate an optimum number of clusters using the FCM technique. The CHs are chosen as the node nearest to the center as determined by this technique.

In this method, each data point has a degree of belonging to clusters rather than belonging completely to just one cluster. It was developed in 1973 (Dunn 1973) and improved in 1981 (Bezdek 1981). This method obtains the best location of clusters in an optimum manner by minimizing an objective function as given by

$$J = \sum_{i=1}^{N} \sum_{j=1}^{C} u_{ij}^{m} \| x_i - c_j \|^{2}, \tag{4.14}$$

where x_i is the set of data points, N is the number of data set, c_j is the center of the clusters, C is the number of clusters, u_{ij} is the membership degree of x_i in the cluster j, $\|*\|$ is the norm, which shows the similarity and the center (which is the distance here), and m is the weighing exponent on each fuzzy membership. The fuzzy partitioning is achieved through an iterative method of optimization by updating the membership function given by

$$u_{ij} = \frac{1}{\sum_{k=1}^{C} \left(\frac{\| x_i - c_j \|}{\| x_i - c_k \|} \right)^{\frac{2}{m-1}}}, \text{ where } c_j = \frac{\sum_{i=1}^{N} u_{ij}^{m} . x_i}{\sum_{i=1}^{N} u_{ij}^{m}} \tag{4.15}$$

where c_j are the cluster centers. These iterations stop when $\max_{ij}\{| u_{ij}^{(k+1)} - u_{ij}^{(k)} |\} < \varepsilon$, where ε lies between 0 and 1, whereas k signifies the iteration steps. This entire process converges to a local minimum by updating the cluster centers in each iteration. In the present study, sensor nodes are grouped into 10 clusters (Figure 4.13a), which is the optimum value in a scenario of 100 nodes. Each cluster contains a CH (dark gray square in Figure 4.13a) nearest to the centroid of the cluster. Each selected CH broadcasts its node id in its respective cluster in such a way that it should reach the farthest node in the cluster. The nodes then send the join-request message to their CH.

Step 3: Data transmission and CH reassignment

After the CH is allocated and clusters are formed, all the nodes are assigned a TDMA schedule in each cycle of data transmission. There is no collision between neighboring clusters, as they do not have a specified boundary; instead, the nodes present in a cluster are connected to their respective CHs as shown in Figure 4.13a. The nodes continuously monitor the environment and send the data to their respective CHs in the assigned slot—subsequently, the data can be transferred to the sink from all the CHs. After few cycles of data transmission, it is found that CHs deplete their energy drastically and lose their efficiency to act as a CH.

In this situation, new CHs are selected based on maximum residual energy among all the nodes in respective clusters (Figure 4.13b). It is to be noted that initially the number of clusters

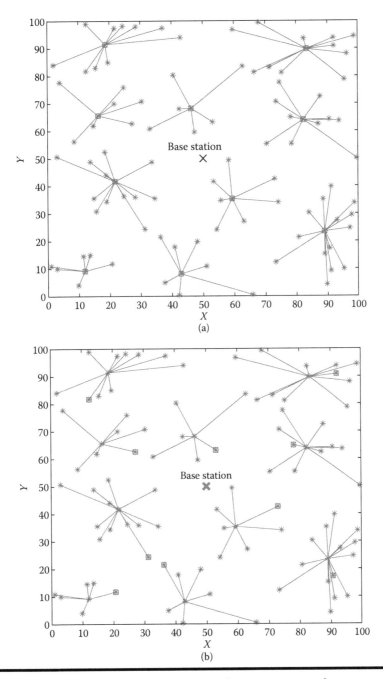

Figure 4.13   Cluster formation with (a) cluster head (CH) nearest to the center using FCM and (b) maximum residual energy node (in the same cluster) as a CH.

remains same during the data transmission. If any cluster contains fewer nodes than defined by a threshold value, then it merges into the neighboring cluster. Also, the two CHs in different clusters may be very close to each other and the nodes present in the corresponding clusters have to choose a farther distance for the transmission resulting in higher energy expenditure.

4.3.7.2 Maximum Residual Energy-Based Clustering Algorithm

In this section, the above-mentioned protocol (s-MREC) is modified; however, the methodology for CH selection is same as used in s-MREC. The difference lies in the formation of clusters, which are reformed after each cycle of data transmission. It is to be noted that the MREC method is investigated in a heterogeneous mode, where fewer nodes are assigned more initial energy (in a predefined ratio) than other nodes. The explicit methodology of MREC is as follows:

Step 1: Initial cluster formation and selection of CH

The input data set is obtained in the same manner as discussed in step 1 of Section 4.3.7.1. The methodology for electing the CH is the same as followed in s-MREC (using the FCM technique and designating the node nearest to the center of the cluster as CH).

Step 2: Reselection of CH and data transmission

In MREC, the clusters are formed dynamically after the selection of CH in each cycle. After receiving the data from the nodes for a few cycles, the node with the maximum residual energy in its respective clusters is elected as a CH at the end of each cycle. Subsequently, new clusters are formed around all elected CHs using Euclidean distance. Thus, nodes lying in the vicinity of any CH form a new cluster in each cycle. The nodes are reclustered based on the distance with the selected CH using a distance matrix, DM ($m \times n$), which is given as follows:

$$
DM = \begin{bmatrix}
d_{CH1,x_1} & d_{CH1,x_2} & \cdots & d_{CH1,x_n} \\
d_{CH2,x_1} & d_{CH2,x_2} & \cdots & d_{CH2,x_n} \\
\vdots & \vdots & \vdots & \vdots \\
d_{CHm,x_1} & d_{CHm,x_2} & d_{CHm,x_3} & d_{CHm,x_n}
\end{bmatrix}, \tag{4.16}
$$

where d is the Euclidean distance between CH and a node based on its location information. If y and z represent the location of two nodes p and q, then the Euclidean distance is

$$
d_{p,q} = [(p_x - q_x)^2 + (p_y - q_y)^2]^{1/2}. \tag{4.17}
$$

Each element $d_{i,j}$ in the distance matrix represents the distance between the ith CH and the jth node. The column containing the minimum value represents the cluster number to be joined by the corresponding node. For example, if d_{CH2,x_1} is the minimum value in the first column, in this situation, the node x_1 gets associated with the second cluster where CH2 is the CH.

The operation of reclustering and data transmission continues for many cycles (as discussed above) until the death of the first node. If the size of the cluster is smaller than the predefined threshold, the cluster merges with the neighboring clusters shown in dotted lines in Figure 4.14. The typical schematics of MREC and s-MREC are shown in a flow chart in Figure 4.15 in which both the methods are explained step-by-step. The technique for data transmission is similar to that used in the s-MREC technique.

4.3.7.3 Comparative Analysis

The results for proposed MREC and s-MREC for an average cycle within the 99% confidence bound are compared with EEHC and DEEC (for similar environment) as shown in Figure 4.16, where the total initial energy of the system is $1.5nE_{init}$ (n is the total number of nodes, E_{init} is the

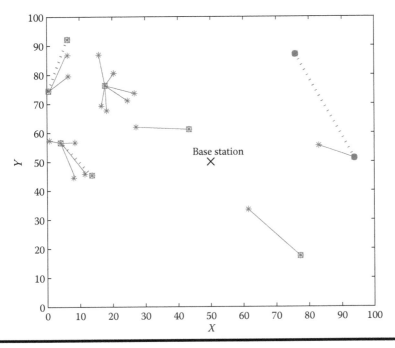

Figure 4.14 **Merging of small clusters (shown in dotted lines) and reduction in clusters with 24 nodes left.**

initial energy of each node and Eini$_t$ is the initial energy of each advanced node). The stability region, in which all the sensor nodes are sensing the environment, is found to be significantly higher for MREC and s-MREC than for other protocols (DEEC and EEHC). It is observed that almost 88% of the nodes in MREC die quickly within a time span of 26 cycles, which shows that the dissipation of energy in the majority of nodes is uniform. Particularly, the death of the first node occurs after 3561 cycles in MREC, 3239 cycles in s-MREC, 1500 cycles in DEEC, and 1008 cycles in EEHC. In EEHC, the death of the node starts very early in the 1000th cycle, due to which there is a less number of nodes collecting the information most of the time as compared to other protocols (MREC, s-MREC, and DEEC). Interestingly, the longest lifetime (5264 cycles) is associated with MREC protocol, which is almost 22% higher than s-MREC (4325 cycles), 32% than EEHC (4000 cycles), and 42% higher than DEEC (3714 cycles). The main reason behind the exceptionally good performance in the network lifetime of MREC is the selection of CHs based on highest residual energy.

Since the position of CHs is rotated in a cycle of data transmission, nodes with higher energy are selected, thereby maintaining the uniformity of energy consumption. Figure 4.17 shows the total number of data packets received at the sink from different CHs, which is found to be higher for MREC and s-MREC protocols as compared to DHAC in a homogeneous environment. More specifically, the number of data packets received (till the death of all nodes) at the sink is 149,309 in MREC, 148,242 for s-MREC, and 26,500 in DHAC protocol. Figure 4.18 shows the variation in the number of clusters for the entire life of the network. It is found that the number of clusters is invariant up to 1000 cycles using the MREC protocol. Also, the stability region is significantly higher in comparison to EEHC, which indicates a higher lifetime in MREC.

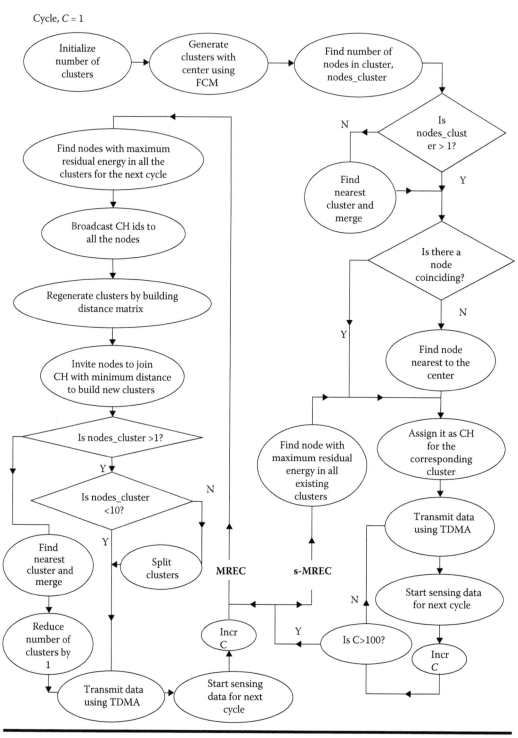

Figure 4.15 Flowchart for static maximum residual energy-based clustering and maximum residual energy-based clustering (MREC) algorithms.

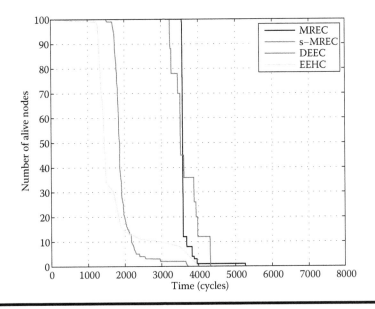

Figure 4.16   Network lifetime for a heterogeneous environment.

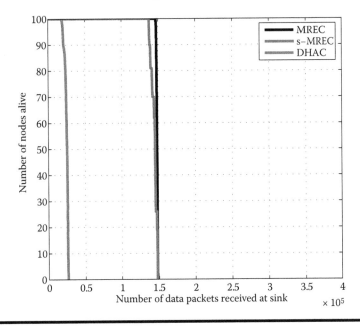

Figure 4.17   Number of nodes alive versus number of data packets received at sink.

The performance of MREC proves to be better than all the other protocols compared to s-MREC, DHAC, EEHC, and DEEC in terms of network lifetime, average energy dissipated per cycle, stability region, and number of data packets received. The main reason behind such a performance is the selection of CHs based on maximum residual energy selected in each cycle. It is also found that the MREC has the highest stability region and a 32% longer lifetime than EEHC

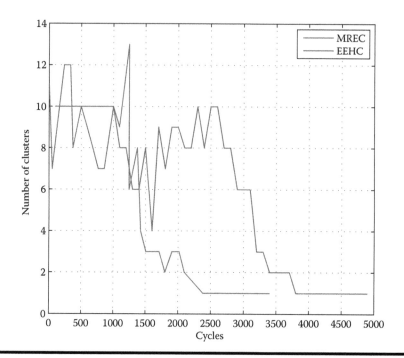

Figure 4.18 **Comparison between EEHC and MREC: Number of clusters per round.**

and 42% more than DEEC. One of the major shortcomings of the present study is to include node mobility, which creates routing difficulties as nodes move in and out of communication range with each other. The clustering works in static and dynamic modes in terms of the selection of CH, not in terms of mobility forming a dynamic environment.

4.4 Multiple Criteria-Based Algorithms

In industries as well as academia, the process of decision-making is used for selecting a particular course of action in the given set of alternatives. Thus, it is considered as the study of identification and selection of alternatives on the basis of values and preferences of the decision makers. Decision-making based on logical thinking is an important aspect in the area of science in which specialists apply their knowledge to make the decisions in a given area. For choosing the right alternative, multiple criteria should be taken into account for making any decision. There is a requirement for simple, systematic, logical, and mathematical tools to guide decision makers for the consideration of a number of selection criteria and their interrelations. Thus, identifying the appropriate selection criteria and finding the most suitable combination of criteria is always needed. To identify different criteria, efforts are needed to influence an alternative selection for a given problem using different methods and eliminate nonsuitable alternatives. Multiple criteria decision making (MCDM) techniques find their application in a number of areas in the field of science and technology. MCDM techniques are categorized into two parts: multiobjective decision-making (MODM) (Chauhan and Vaish 2013a) and multiattribute decision-making (MADM) (Chauhan and Vaish 2012) techniques. The first technique is based on the selection

Table 4.3 Decision Matrix in MADM Models

Alternative	Attributes			
	X_1	X_2	...	X_j
M_1	r_{11}	r_{12}	...	r_{1j}
M_2	r_{21}	r_{22}	...	r_{2j}
.	.	.	...	
.	.	.	...	
Mi	r_{i1}	r_{i2}	...	rij
MADM, multiattribute decision-making.				

of nondominating alternatives for all the given criteria, while the other involves the quantitative comparison of alternatives and then rank on the basis of desirability degree of different attributes. In the present work, Pareto-optimal theory as an MODM technique and technique for order preference by similarity to ideal solution (TOPSIS) as an MADM technique are employed for the selection of CHs. MADM provides different solutions to the material selection problems involving multiple objectives and conflicts. Such methods involve comparison at the level of intra- and interlevel, and explicit trade-offs are considered for the problem. The decision matrix in MADM methods has four different parts, including (1) attributes, (2) alternatives, (3) weight of each attribute, and (4) measures of performance of alternatives with respect to attributes. Of the many MADM methods, six methods are commonly used: (1) the weighted product method (WPM), (2) weighted sum method (WSM), (3) revised analytic hierarchy process (AHP), (4) AHP, (5) TOPSIS, and (6) compromise ranking method (VIKOR). The MADM problem under consideration is depicted by a decision matrix which is given in Table 4.3.

The MADM problems can be solved in two different ways: compensatory and noncompensatory. In compensatory models, attributes permit explicit trade-offs and they are based on the multiattribute utility theory (MAUT). The noncompensatory-based models depend on the pairwise comparisons of alternatives on the basis of individual criteria.

4.4.1 Pareto-Optimal Theory

Vilfredo Pareto described the Pareto-optimal theory (Kasprzak and Lewis 2001) in which the nondominated solution is discussed in a given solution space (Figure 4.19). The solutions space includes a region consisting of all possible solutions. The solutions space is divided into three categories: (1) neither dominated nor dominating, (2) completely dominated, and (3) nondominated. In the first category, few of the properties are dominated, while others are dominating. In the second method, at least one (real) alternative exists, which overshadows all the alternatives in a desirable manner. The best trade-off out of the three is nondominated solutions, which is considered as the best alternative and is not dominated by others in the solution space.

Using the Pareto-optimal theory, the selection of CHs is done using three criteria, which includes density of the cluster, node's residual energy, and number of neighbor nodes. The best solutions will exist for minimum distance between nodes and sink, maximum residual energy,

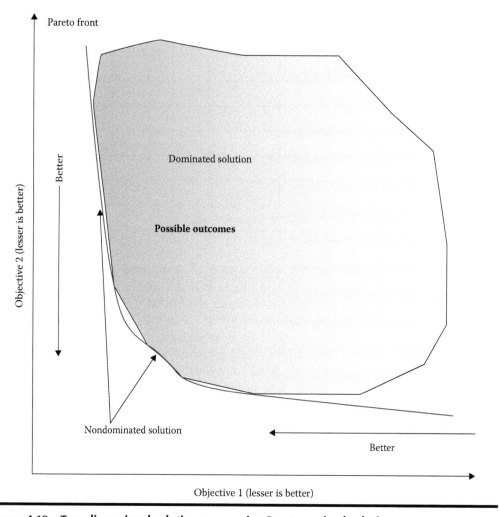

Figure 4.19 Two-dimensional solution space using Pareto-optimal solution.

and maximum number of neighbor nodes for the CH selection. If any one of the criteria is used for the selection of CH, then the performance in terms of network lifetime or complete coverage may degrade. For better results, the three criteria should be considered in an optimum percentage.

4.4.2 TOPSIS Approach

TOPSIS is one of the MADM approaches in which a matrix of n attributes and m alternatives is considered to be a problem of n dimensional hyperplane. It has m points, which is given by the value of the attributes (Hwang and Yoon 1981). The alternative will have the best possible solution in case of the shortest distance from the positive ideal solution (PIS) and the worst possible solution in case of the farthest distance from the negative ideal solution (NIS). Various engineering applications have used this technique (Chauhan and Vaish 2013b). The work is based on an aggregating function to measure the closeness with the reference point(s). Thus, the selected alternative must have the minimum distance from the PIS and the maximum distance from the NIS. The Euclidean measure is used to find the distance from the ideal solution. A situation is presented graphically in Zanakis et al. (1998). The modified version of TOPSIS uses "city block distance" (Hwang and Yoon 1981) instead of Euclidean distance to find the distance. The TOPSIS method involves in the following steps:

Step 1: Find the normalized decision matrix

$$a_{ij} = \frac{X_{ij}}{\sqrt{\left(\sum_{i=1}^{m} X_{ij}^2\right)}}; \forall j, \tag{4.18}$$

where a_{ij} is the normalized value of alternative A_i with respect to criterion C_j.

Step 2: Find the normalized decision matrix (weighted)

$$Y_{ij} = [a_{ij}]_{m \times n} * [W_j], \tag{4.19}$$

where W_j is the weight of the jth criterion and $\sum_{j=1}^{n} W_j = 1$.

Step 3: The alternatives are found using the minimum/maximum distance from the PIS (PS⁺)/NIS (NS⁻), which are defined as

$$PS^+ = \left[(\max(Y_{ij}), k \in K_1), (\min(Y_{ij}), k \in K_2), i = 1, 2, 3, ..., m\right]; \forall k \tag{4.20}$$

$$NS^- = \left[(\min(Y_{ij}), k \in K_1), (\max(Y_{ij}), k \in K_2), i = 1, 2, 3, ..., m\right]; \forall k, \tag{4.21}$$

where K_1 belongs to benefit criteria and K_2 belongs to cost criteria.

Step 4: Calculate separation measures using Euclidean distance from the PIS/NIS:

$$SM_i^+ = \left[\sum_{j=1}^{m} (Y_{ij} - Y_j^+)^2\right]^{0.5}, i = 1, 2, 3, .. \tag{4.22}$$

$$SM_i^- = \left[\sum_{j=1}^{m} (Y_{ij} - Y_j^-)^2\right]^{0.5}, i = 1, 2, 3 .. \tag{4.23}$$

Step 5: Find relative closeness

$$R_i^+ = \frac{SM_i^-}{SM_i^- + SM_i^+} \tag{4.24}$$

Sensor nodes with the highest rank index (R_i^+) are preferred and selected for CHs.

Table 4.4 shows Pareto-optimal nodes, their properties, and TOPSIS ranks in the 50th cycle. The best-ranked nodes are chosen as CHs in each round of data transmission. Table 4.5 lists 10 CHs selected in the 50th cycle. Similarly, ranking is performed in each cycle until all sensors die.

The ranks of these Pareto-selected CHs using TOPSIS for the 50th cycle are shown in Figure 4.20 as a bar chart. It is clear from the bar chart that the ranks given by TOPSIS is for the whole set of

Table 4.4 Pareto-Optimal Cluster Heads (CHs), Criteria, and Their TOPSIS Ranks

Cluster Head No.	Energy, E (Joules)	Distance from Sink, d	Cluster Density, c (Number of Neighbors)	$R_i^+ = \dfrac{SM_i^-}{SM_i^- + SM_i^+}$
CH1	0.8485	21.5830	7	0.6122
CH2	0.9313	25.5811	9	0.6388
CH3	0.8951	24.2745	9	0.6409
CH4	0.9893	34.6179	7	0.5215
CH5	0.8095	20.9972	8	0.6117
CH6	0.9812	39.3231	10	0.5243
CH7	0.9894	34.9321	8	0.5383
CH8	0.9773	23.2092	6	0.6172
CH9	0.9887	28.3985	6	0.5555
CH10	0.9888	39.8408	10	0.5222
CH11	0.9893	7.6944	3	0.5143
CH12	0.9744	4.5873	4	0.6480
CH13	0.8542	19.2526	6	0.6106
CH14	0.7921	10.6745	6	0.6459
CH15	0.9640	9.8442	5	0.7128
CH16	0.9893	28.9398	9	0.6130
CH17	0.9893	16.6420	4	0.5557
CH18	0.8396	24.2008	9	0.6112
CH19	0.9894	23.8414	5	0.5591
CH20	0.9852	26.7251	7	0.6059
CH21	0.8384	17.6095	6	0.6190
CH22	0.9360	25.8021	9	0.6376

CHs obtained using the Pareto theory. As shown, the 15th CH has the best rank while the 11th CH has the lowest rank. Similar ranking of CHs has to be done in each cycle of the entire lifetime. This consumption of extra energy may be an overhead but it is very small in quantity as compared to the energy consumption used by all the nodes in each cycle. Hence, it can be neglected. The number of CHs selected using the Pareto-optimal theory is different for each cycle, hence the need for optimum selection of number of CHs is compulsory. Using Table 4.5, selected CHs are plotted using a bar chart with their respective ranks given by TOPSIS as shown in Figure 4.21. A noticeable difference between Figures 4.20 and 4.21 is that there is a difference between the number of

Table 4.5   Ten CHs in 50th Cycle Selected Using TOPSIS Ranks

Rank	Cluster Head No.
1	15
2	12
3	14
4	3
5	2
6	22
7	21
8	8
9	16
10	1

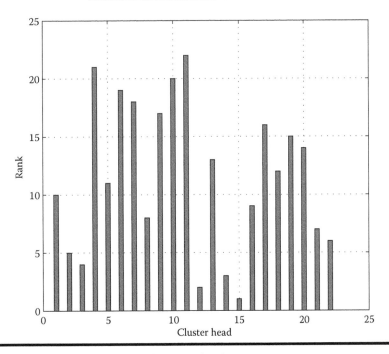

Figure 4.20   Pareto-optimal bar chart for CH selection.

selected CHs in both the cases. The reason behind this is that the optimum number of CHs never matches the number given by the Pareto-optimal theory in each round.

If the number of CHs given by the Pareto theory is more than the optimum value, the topmost CHs obtained from TOPSIS are used in each cycle. While if the number of CHs is less than the optimum value, the same number is considered without TOPSIS ranking, which may happen

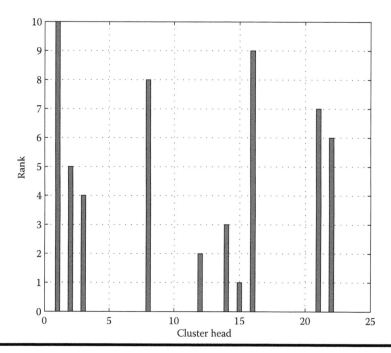

Figure 4.21 Top 10 CHs using TOPSIS.

during the last few cycles of the network lifetime. It is clear from the figure that few CHs with lower rank are not present in the bar chart and hence will act normal nodes.

For efficiently utilizing the energy, it is necessary to decide the number of clusters to be formed in the entire network in each cycle. The criterion is based on the average energy dissipation of WSNs with respect to the number of clusters as shown in Figure 4.22. It is found that with the increase in the number of clusters, the energy dissipation decreases as shown in Figure 4.22. The optimum number of clusters, k_{opt} (Comeau et al. 2006) can be estimated as

$$k_{opt} = \sqrt{\frac{\varepsilon_{fs}}{\pi(\varepsilon_{mp} d_{to\ BS}^4 - E_{elec})}} M \sqrt{N}\ , \tag{4.25}$$

where k_{opt} is the optimum number of clusters, $M \times M$ is the total area of the network, N is the total number of nodes, d_{toBS} is the distance between the CHs and the BS, and ε_{fs}, ε_{mp} is the energy consumption in the amplifier. Thus, the value of k_{opt} is found as $9 > k_{opt} > 11$ using Equation 4.25. The stability region is evaluated for a different number of clusters as shown in Table 4.6.

It is observed that the stability region is maximum when number of clusters is between 9 and 11. Thus, we have considered 10 clusters to study the effectiveness of data transmission in WSNs. Each cycle of data transmission consists of the formation of clusters and data transmission. In the first phase, the CHs are selected and clusters are formed using the Euclidean distance. The network lifetime is calculated in terms of the number of cycles.

Figure 4.23 shows the results of the experiment in which nodes are deployed randomly in an area of $100 \times 100\ m^2$ and the network lifetime is plotted. The lifetime presents the number of alive

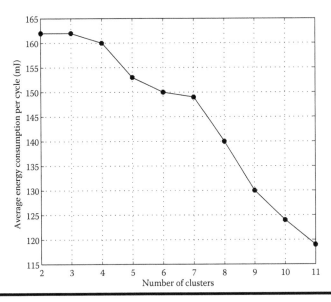

Figure 4.22 **Average energy dissipation per cycle versus number of clusters (selected using TOPSIS).**

Table 4.6 Variation of Number Clusters with Stability Region

S No.	Stability Region in Terms of Time in Cycles (No. of Cycles When First Nodes Die)	Number of Clusters
1	1756	06
2	1780	07
3	1792	08
4	1807	09
5	1818	10
6	1823	11
7	1790	12
8	1775	13
9	1780	14
10	1780	15
11	1799	16

nodes versus time in cycles in which the results are finally compared with DHAC, a well-known clustering protocol. DHAC groups similar nodes together on the basis of certain attributes including connectivity and nodes location before the CHs are selected.

In this technique, three methods including SLINK, CLINK, and UPGMA are used for the formation of a Dendrogram. Based on some threshold value, the Dendrogram is cut on the basis

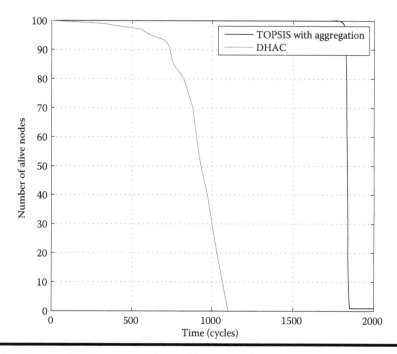

Figure 4.23 Network lifetime comparison with aggregation.

of value number of clusters, transmission radius, or cluster density. The CHs are selected on the basis of either bottom level nodes in the Dendrogram or the nodes having lower ID. In the present approach, CHs are selected first and then the clusters are formed, which results in higher lifetime in comparison to DHAC. It is observed that the number of alive nodes after 1800 cycles suddenly falls because of the uniform consumption of energy. The TOPSIS is different from other MADM techniques because the ranking of nodes is highly sensitive to weights. Sometimes, these weights are inaccurate, which leads to the wrong decision. Also, the effect of all the attributes cannot be considered. It is observed that the lifetime of the network of TOPSIS with 50% aggregation is 81% higher than that of DHAC on the total runtime of the entire network (Azad and Sharma 2015). Also, the first node dies at the 1818th cycle in TOPSIS and the 320th cycle in DHAC, which is a considered as a significant improvement as far as overall stability is concerned.

4.4.3 Ranking Using Fuzzy TOPSIS

In MADM, the data is first compared and then ranked on the basis of the attribute's degree of desirability. In the present study, MODM (Pareto-optimal technique) and MADM (fuzzy-TOPSIS) approaches are used for the selection of CHs.

4.4.3.1 Fuzzy TOPSIS Approach

As explained in Section 4.4.2, TOPSIS is widely used in the field of science and engineering; but sometimes it is not possible to assign a precise rating to an alternative for the attributes under consideration. In such cases, the fuzzy approach is used for the assignment of the attribute's relative importance using fuzzy numbers. This section presents the TOPSIS method in fuzzy environment

(Yang and Hung 2007; Rathod and Kanzaria 2011; Torfi et al. 2010), which works under fuzzy environment. The rationale of fuzzy theory is discussed as follows:

Definition 1. A fuzzy set a of discourse X in a universe is defined by a membership function $\mu_a(x)$, which is associated with each element x in X. Here a is a real number in [0, 1]. The function value $\mu_a(x)$ is termed the grade of membership of x in a (Chen 2000). A triangular fuzzy number a can be defined by a triplet (a_1, a_2, a_3). Its conceptual schema and mathematical form are shown as

$$\mu(x) = \begin{cases} 0, & \ldots x \leq a_1 \\ \dfrac{x - a_1}{a_2 - a_1}, & \ldots a_1 < x \leq a_2 \\ \dfrac{a_3 - x}{a_3 - a_2}, & \ldots a_2 < x \leq a_3 \\ 0, & \ldots x > a_3. \end{cases} \tag{4.26}$$

Definition 2. Let $a = (a_1, a_2, a_3)$ and $b = (b_1, b_2, b_3)$ be two triangular fuzzy numbers. Then the vertex method is defined to calculate the distance between them in Equation 4.31 as

$$d(a,b) = \sqrt{\frac{1}{3}\left[(a_1 - b_1)^2 + (a_2 - b_2)^2 + (a_3 - b_3)^2\right]} \tag{4.27}$$

Property 1. Assuming that both a and b are real numbers, then d is the Euclidean distance between a and b.

Property 2. If a, b, and c are three triangular fuzzy numbers, then d(a,b) > d(a,c) only when b is closer to a than c.

Also,

$$a \times b = (a_1 \times b_1, a_2 \times b_2, a_3 \times b_3) \text{ for multiplication} \tag{4.28}$$

$$a + b = (a_1 + b_1, a_2 + b_2, a_3 + b_3) \text{ for addition} \tag{4.29}$$

The fuzzy TOPSIS can be applied on a decision matrix as

$$D = \begin{array}{c} \\ A1 \\ A2 \\ A3 \\ A4 \\ A5 \end{array} \begin{array}{ccccc} C_1 & C_2 & C_3 & C_4 & C_5 \\ \left[\begin{array}{ccccc} \tilde{x}_{11} & \tilde{x}_{12} & \tilde{x}_{13} & \cdots & \tilde{x}_{1n} \\ \tilde{x}_{21} & \tilde{x}_{22} & \tilde{x}_{23} & \cdots & \tilde{x}_{2n} \\ \vdots & \vdots & \vdots & \vdots & \vdots \\ \vdots & \vdots & \vdots & \vdots & \vdots \\ \tilde{x}_{m1} & \tilde{x}_{m2} & \tilde{x}_{m3} & \cdots & \tilde{x}_{mn} \end{array}\right] \end{array} \tag{4.30}$$

$$\tilde{W} = [w_1, w_2, \ldots, w_n],$$

where $\tilde{x}_{ij}$, $i = 1, 2, ..., m$, $j = 1, 2, ..., n$ and $\tilde{w}_j$, $j = 1, 2, ..., n$ are linguistic triangular Fuzzy numbers, $x_{ij} = (a_{ij}, b_{ij}, c_{ij})$ and $\tilde{w}_j = (w_{j1}, w_{j2}, w_{j3})$. x_{ij} is the rating of the ith alternative. $\tilde{w}_j$ represents the weight of the jth criterion, C_j. The normalized Fuzzy decision matrix denoted by $\tilde{R}$ is given as

$$\tilde{R} = [\tilde{r}_{ij}]_{mxn} \tag{4.31}$$

The weighted fuzzy normalized decision matrix is

$$\tilde{V} = \begin{bmatrix} w_1\tilde{r}_{11} & w_2\tilde{r}_{12} & ..... & w_n\tilde{r}_{1n} \\ w_1\tilde{r}_{21} & w_2\tilde{r}_{22} & ..... & w_n\tilde{r}_{2n} \\ w_1\tilde{r}_{31} & \vdots & ..... & \vdots \\ \vdots & \vdots & ..... & \vdots \\ w_1\tilde{r}_{ml} & w_2\tilde{r}_{n2} & ..... & w_n\tilde{r}_{mn} \end{bmatrix}$$

$$\tilde{V} = \begin{bmatrix} \tilde{v}_{11} & \tilde{v}_{12} & ..... & \tilde{v}_{1n} \\ \tilde{v}_{21} & \tilde{v}_{22} & ..... & \tilde{v}_{2n} \\ \tilde{v}_{31} & \vdots & ..... & \vdots \\ \vdots & \vdots & ..... & \vdots \\ \tilde{v}_{ml} & \tilde{v}_{n2} & ..... & \tilde{v}_{mn} \end{bmatrix} \tag{4.32}$$

Steps for the fuzzy TOPSIS procedure are follows:

Step 1: Choose the linguistic ratings $(\tilde{x}_{ij})$, $i = 1, 2, 3, ..., m$; $j = 1, 2, 3, ..., n$ for alternatives with respect to criteria and the appropriate linguistic variables $(\tilde{w}_j, j = 1, 2, ..., n)$ for the weight of the criteria. If the range of triangular fuzzy numbers belongs to $[0,1]$, then there is no need for a normalization.

Step 2: Obtain the weighted normalized fuzzy decision matrix given in Equation 4.32.

Step 3: The selection of an alternative is based on the shortest distance from the PIS (A^+) and the furthest from the NIS (A^-), which are defined as

$$A^+ = \{\tilde{v}_1^+, \tilde{v}_2^+, ..., \tilde{v}_n^+\} = \{(\max_i \tilde{v}_{ij} \mid i = 1, ..., m), j = 1, 2, ..., n\} \tag{4.33}$$

$$A^- = \{\tilde{v}_1^-, \tilde{v}_2^-, ..., \tilde{v}_n^-\} = \{(\min_i \tilde{v}_{ij} \mid i = 1, ..., m), j = 1, 2, ..., n\}. \tag{4.34}$$

Step 4: The separation measures are the distances of each alternative from A^+ and A^- given as

$$d_i^+ = \sum_{j=1}^{n} d(\tilde{v}_{ij}, v_j^+), i = 1, 2, ..., m \tag{4.35}$$

$$d_i^- = \sum_{j=1}^{n} d(\tilde{v}_{ij}, v_j^-), i = 1, 2, ..., m. \tag{4.36}$$

Step 5: TOPSIS rank indices can be estimated as

$$CC_i = \frac{d_i^-}{d_i^+ + d_i^{-.}}$$ (4.37)

CHs are those nodes that have higher TOPSIS index. Further, clusters are formed using the minimum Euclidean distance criterion as done in the previous section for TOPSIS approach.

4.4.3.2 Performance Evaluation of Fuzzy TOPSIS

In each cycle, it is important to decide the number of clusters/CHs that exist in the WSN for maximizing the energy efficiency. We have estimated the optimum number of clusters, k_{opt} (Comeau et al. 2006) as

$$k_{opt} = \sqrt{\frac{\varepsilon_{fs}}{\pi(\varepsilon_{mp}d_{toBS}^4 - E_{elec})}} M \sqrt{N}.$$ (4.38)

The value of k_{opt} is estimated in the range of $9 > k_{opt} > 11$ similar to the TOPSIS method. The Pareto-optimal nodes are the optimum selection in view of the above-mentioned three criteria whose values are shown in Table 4.7 for the second cycle of simulation. Similar calculations are performed in each cycle for short listing the Pareto-optimal sensor nodes. These attributes of the criteria are further normalized in the range [0,1] given in Table 4.8. We have assigned 0.5, 0.25, and 0.25

Table 4.7 Decision Matrix for Fuzzy TOPSIS Analysis in Second Cycle

Cluster Head No.	Number of Neighbors, n C_2	Residual Energy, $E_o(J)$, C_1	Distance from Sink, d C_3
CH1	7	0.9998	21.583
CH2	9	0.9998	24.2745
CH3	8	0.9887	20.9972
CH4	10	0.9894	39.3231
CH5	6	0.9998	23.2092
CH6	10	0.9998	39.8408
CH7	3	0.9998	7.6944
CH8	4	0.9988	4.5873
CH9	9	0.9998	25.5698
CH10	6	0.9947	10.6745
CH11	5	0.9964	9.8442
CH12	4	0.9998	16.642
CH13	9	0.9919	24.2008
CH14	6	0.9998	17.6095

Table 4.8 Normalized Decision Matrix for Fuzzy TOPSIS

Cluster Head No.	C_1	C_2	C_3
CH1	0.5714	0.9996	0.5179
CH2	0.8571	0.9988	0.4416
CH3	0.7143	0.0000	0.5345
CH4	1.0000	0.0590	0.0147
CH5	0.4286	0.9998	0.4718
CH6	1.0000	0.9998	0.0000
CH7	0.0000	0.9995	0.9119
CH8	0.1429	0.9096	1.0000
CH9	0.8571	1.0000	0.4048
CH10	0.4286	0.5433	0.8273
CH11	0.2857	0.6900	0.8509
CH12	0.1429	0.9996	0.6581
CH13	0.8571	0.2917	0.4436
CH14	0.4286	0.9998	0.6306
Weight	0.25	0.5	0.25

subjective weights to residual energy, number of neighbors, and distance from sink, respectively. Membership function is used to convert the values (in Table 4.8) into linguistic variables as shown in Table 4.9. Further fuzzy linguistic variables are transformed into fuzzy triangular membership functions shown in Table 4.10 and a fuzzy weighted decision matrix using Equation 4.32 as shown in Table 4.11. We define the fuzzy PIS and NIS (step 3) and computed separation measures (step 4) and rank indices (step 5) for the Pareto-optimal sensor nodes. Table 4.12 shows the Pareto-optimal nodes, their properties, and the Fuzzy TOPSIS indices in the second cycle. Table 4.13 lists the top 10 CHs (from Table 4.12) selected in the second cycle.

Figure 4.24 shows the results of the experiment in which nodes are placed randomly in an area of 100×100 m² and the network lifetime is plotted, which shows the number of alive nodes over the time in cycles (Azad and Sharma 2013a). The results are finally compared with a well-known protocol DHAC. It is reported that DHAC is more energy-efficient than that of other methods including LEACH and LEACH-C.

The network lifetime is higher for the fuzzy-TOPSIS approach than that of DHAC (Figure 4.24). It shows that the present approach is more effective in WSNs.

4.4.4 Ranking Using VIKOR

Pareto-optimal CHs are next ranked using another method called VIKOR (Azad and Sharma 2014), another MADM technique. With the initial weights, it determines the compromise solution,

Table 4.9 Decision Matrix Using Fuzzy Linguistic Variables

Cluster Head No.	C_1	C_2	C_3
CH1	M	VH	M
CH2	VH	VH	M
CH3	H	VL	M
CH4	VL	VL	VL
CH5	M	VH	M
CH6	VL	VH	VL
CH7	VL	VH	VH
CH8	VL	VH	VL
CH9	VH	VL	M
CH10	M	M	VH
CH11	L	H	H
CH12	VL	VH	H
CH13	VH	L	M
CH14	M	VH	H
Weight	M	VH	M

compromise ranking-list and the weight stability intervals for preference stability of the compromise solution. This method involves ranking and selection using a given set of alternatives. This method introduces the ranking index based on the particular measure of "closeness" to the "ideal" solution. The use of utility weight in VIKOR allows the users to apply expert opinion. Similar to TOPSIS, the solutions depend on the PIS and NIS in which the values of the decision matrix are easily accessed. The closeness is measured on the basis of an aggregating function, which represents the distance from the ideal solution. The highest ranked candidate is the nearest to the ideal solution (Zanakis et al. 1998). The Lp-metric, which is derived from compromising programming method, finds the multicriteria merit for compromise ranking. The following steps define the CH selection procedure using VIKOR method:

Step 1: Find the PIS (A^+) and NIS (A^-) given as

$$A^+ = \left[(\max(f_{ij}), j \in J) \; or \; (\min(f_{ij}), j \in J') \right] = \{ f_1^*, f_2^*, ..., f_n^* \} \tag{4.39}$$

$$A^- = \left[(\min(f_{ij}), j \in J) \; or \; (\max(f_{ij}), j \in J') \right] = \{ f_1^-, f_2^-, ..., f_n^- \}, \tag{4.40}$$

where f_{ij} is the jth attribute of ith alternative and J is benefit criteria and J' is cost criteria.

Table 4.10 Fuzzy Decision Matrix and Fuzzy Attribute Weights

Cluster Head No.	C_1	C_2	C_3
CH1	(0.35, 0.5, 0.65)	(0.75, 0.9, 1)	(0.35, 0.5, 0.65)
CH2	(0.75, 0.9, 1)	(0.75, 0.9, 1)	(0.35, 0.5, 0.65)
CH3	(0.55, 0.7, 0.85)	(0, 0.1, 0.25)	(0.35, 0.5, 0.65)
CH4	(0.75, 0.9, 1)	(0, 0.1, 0.25)	(0, 0.1, 0.25)
CH5	(0.35, 0.5, 0.65)	(0.75, 0.9, 1)	(0.35, 0.5, 0.65)
CH6	(0.75, 0.9, 1)	(0.75, 0.9, 1)	(0, 0.1, 0.25)
CH7	(0, 0.1, 0.25)	(0.75, 0.9, 1)	(0.75, 0.9, 1)
CH8	(0, 0.1, 0.25)	(0.75, 0.9, 1)	(0.75, 0.9, 1)
CH9	(0.75, 0.9, 1)	(0.75, 0.9, 1)	(0.35, 0.5, 0.65)
CH10	(0.35, 0.5, 0.65)	(0.35, 0.5, 0.65)	(0.75, 0.9, 1)
CH11	(0.15, 0.3, 0.45)	(0.55, 0.7, 0.85)	(0.75, 0.9, 1)
CH12	(0, 0.1, 0.25)	(0.75, 0.9, 1)	(0.55, 0.7, 0.85)
CH13	(0.75, 0.9, 1)	(0.15, 0.3, 0.45)	(0.35, 0.5, 0.65)
CH14	(0.35, 0.5, 0.65)	(0.75, 0.9, 1)	(0.55, 0.7, 0.85)
Weight	(0.35, 0.5, 0.65)	(0.75, 0.9, 1)	(0.35, 0.5, 0.65)

Step 2: Find the regret and utility measures given as

$$S_i = \sum_{j=1}^{n} W_j \frac{(f_j^* - f_{ij})}{(f_i^* - f_i^-)}; \forall i \tag{4.41}$$

$$R_i = \text{Max}_j \left[W_j \frac{(f_i^* - f_{ij})}{(f_i^* - f_j^-)} \right]; \forall i, \tag{4.42}$$

where S_i and R_i are the utility and regret measures and W_j is the relative weight given to the jth attribute.

Step 3: Calculate VIKOR index:

$$Q_i = v \left[\frac{S_i - S^*}{S^- - S^*} \right] + (1 - v) \left[\frac{R_i - R^*}{R^- - R^*} \right]; \forall i, \tag{4.43}$$

where Q_i represents the ith alternative VIKOR value, v is the group utility weight, which is generally taken as 0.5 (unsupervised):

$$S^* = \text{Min}_i(S_i) \tag{4.44}$$

Table 4.11   Fuzzy-Weighted Decision Matrix

Cluster Head No.	C_1	C_2	C_3
CH1	(0.1225, 0.2500, 0.4225)	(0.5625, 0.8100, 1.0000)	(0.1225, 0.2500, 0.4225)
CH2	(0.2625, 0.4500, 0.6500)	(0.5625, 0.8100, 1.0000)	(0.1225, 0.2500, 0.4225)
CH3	(0.1925, 0.3500, 0.5525)	(0, 0.0900, 0.2500)	(0.1225, 0.2500, 0.4225)
CH4	(0.2625, 0.4500, 0.6500)	(0, 0.0900, 0.2500)	(0, 0.0500, 0.1625)
CH5	(0.1225, 0.2500, 0.4225)	(0.5625, 0.8100, 1.0000)	(0.1225, 0.2500, 0.4225)
CH6	(0.2625, 0.4500, 0.6500)	(0.5625, 0.8100, 1.0000)	(0, 0.0500, 0.1625)
CH7	(0, 0.0500, 0.1625)	(0.5625, 0.8100, 1.0000)	(0.2625, 0.4500, 0.6500)
CH8	(0, 0.0500, 0.1625)	(0.5625, 0.8100, 1.0000)	(0.2625, 0.4500, 0.6500)
CH9	(0.2625, 0.4500, 0.6500)	(0.5625, 0.8100, 1.0000)	(0.1225, 0.2500, 0.4225)
CH10	(0.1225, 0.2500, 0.4225)	(0.2625, 0.4500, 0.6500)	(0.2625, 0.4500, 0.6500)
CH11	(0.0525, 0.1500, 0.2925)	(0.4125, 0.6300, 0.8500)	(0.2625, 0.4500, 0.6500)
CH12	(0, 0.0500, 0.1625)	(0.5625, 0.8100, 1.0000)	(0.1925, 0.3500, 0.5525)
CH13	(0.2625, 0.4500, 0.6500)	(0.1125, 0.2700, 0.4500)	(0.1225, 0.2500, 0.4225)
CH14	(0.1225, 0.2500, 0.4225)	(0.5625, 0.8100, 1.0000)	(0.1925, 0.3500, 0.5525)

$$S^- = \text{Max}_i(S_i) \tag{4.45}$$

$$R^* = \text{Min}_i(R_i) \tag{4.46}$$

$$R^- = \text{Max}_i(R_i). \tag{4.47}$$

The alternatives with lower values of VIKOR index (Qi) are preferred.

4.4.4.1 Performance Evaluation of VIKOR

Similar to TOPSIS, CHs selected using the Pareto-optimal theory are selected using three different criteria including minimum distance from the sink, residual energy of the node, and node density. Figure 4.25 shows Pareto-optimal nodes in dark gray color in three-dimensional space. These nodes are selected optimally taking care of the above-mentioned three criteria. These criteria are assigned equal weights as they are considered equally important. Different steps of the VIKOR method are evaluated for all the selected CHs in each cycle as shown in Table 4.14.

It shows the positive (A^+) and negative (A^-) ideal solution, utility (S_i) and regret (R_i) measures, and finally, the VIKOR index (Q_i) and ranks as discussed in Section 4.2 for the second cycle. The ranks of these Pareto-selected CHs using VIKOR for the second cycle are observed and it is noticed that the fourth CH has the best rank while 12th CH has the lowest rank. Using

Table 4.12 Fuzzy TOPSIS Analysis

Cluster Head No.	C_1	C_2	C_3	d_i^+	d_i^-	cc_i
CH1	(0.1225, 0.2500, 0.4225)	(0.5625, 0.8100, 1.0000)	(0.1225, 0.2500, 0.4225)	1.0893	0.9100	0.5448
CH2	(0.2625, 0.4500, 0.6500)	(0.5625, 0.8100, 1.0000)	(0.1225, 0.2500, 0.4225)	0.9768	0.9870	0.4974
CH3	(0.1925, 0.3500, 0.5525)	(0, 0.0900, 0.2500)	(0.1225, 0.2500, 0.4225)	1.3331	0.5136	0.7219
CH4	(0.2625, 0.4500, 0.6500)	(0, 0.0900, 0.2500)	(0, 0.0500, 0.1625)	1.4099	0.5143	0.7327
CH5	(0.1225, 0.2500, 0.4225)	(0.5625, 0.8100, 1.0000)	(0.1225, 0.2500, 0.4225)	1.0893	0.9100	0.5448
CH6	(0.2625, 0.4500, 0.6500)	(0.5625, 0.8100, 1.0000)	(0, 0.0500, 0.1625)	1.1255	0.9479	0.5428
CH7	(0, 0.0500, 0.1625)	(0.5625, 0.8100, 1.0000)	(0.2625, 0.4500, 0.6500)	1.1255	0.9479	0.5428
CH8	(0, 0.0500, 0.1625)	(0.5625, 0.8100, 1.0000)	(0.2625, 0.4500, 0.6500)	1.1255	0.9479	0.5428
CH9	(0.2625, 0.4500, 0.6500)	(0.5625, 0.8100, 1.0000)	(0.1225, 0.2500, 0.4225)	0.9768	0.9870	0.4974
CH10	(0.1225, 0.2500, 0.4225)	(0.2625, 0.4500, 0.6500)	(0.2625, 0.4500, 0.6500)	1.0960	0.7402	0.5969
CH11	(0.0525, 0.1500, 0.2925)	(0.4125, 0.6300, 0.8500)	(0.2625, 0.4500, 0.6500)	1.0946	0.8355	0.5671
CH12	(0, 0.0500, 0.1625)	(0.5625, 0.8100, 1.0000)	(0.1925, 0.3500, 0.5525)	1.1699	0.9067	0.56f
CH13	(0.2625, 0.4500, 0.6500)	(0.1125, 0.2700, 0.4500)	(0.1225, 0.2500, 0.4225)	1.1914	0.6424	0.6497
CH14	(0.1225, 0.2500, 0.4225)	(0.5625, 0.8100, 1.0000)	(0.1925, 0.3500, 0.5525)	1.0277	0.9475	0.5203
A^+	$\tilde{V}_1^+ = (1,1,1)$	$\tilde{V}_2^+ = (1,1,1)$	$\tilde{V}_3^+ = (1,1,1)$			
A^-	$\tilde{V}_1^- = (1,1,1)$	$\tilde{V}_2^- = (0,0,0)$	$\tilde{V}_3^- = (0,0,0)$			
Weight		(0.75, 0.9, 1)	(0.35, 0.5, 0.65)			

Table 4.13 Best CHs in the Second Cycle Selected Based on Fuzzy TOPSIS Ranks

Rank	Cluster Head
1	4
2	3
3	13
4	10
5	11
6	12
7	1
8	5
9	6
10	7

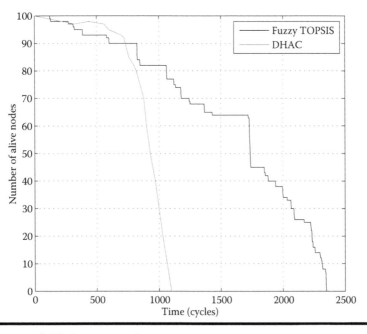

Figure 4.24 Network lifetime comparison.

the VIKOR method, the CHs are ranked as given in Table 4.15 in which 11 CHs are selected in each cycle. Average energy dissipation is estimated for different numbers of clusters to investigate the optimum number of CHs as shown in Figure 4.26. It is observed that the energy dissipation decreases significantly for number of clusters equal to or higher than 11, which can be set as an optimum value. The overall network lifetime will be higher for lower energy dissipation.

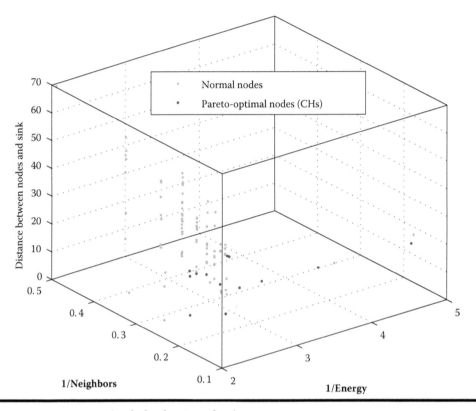

Figure 4.25 Pareto-optimal plot for CH selection.

The method of selecting the optimum number of clusters depends upon the maximum network lifetime and stability region as shown in Figure 4.27. In this section, we consider the number of clusters to be around 10 for an acceptable lifetime and stability.

The second technique involves the optimum number of clusters, k_{opt} (Comeau et al. 2006) as

$$k_{opt} = \sqrt{\frac{\varepsilon_{fs}}{\pi(\varepsilon_{mp}d^4_{to\ BS} - E_{elec})}}.M\sqrt{N}, \tag{4.48}$$

where n is the number of nodes, M^2 is the network area, and d^4 is the fourth power distance to the sink. k_{opt} is estimated between $11 > k_{opt} > 18$ in case the sink is away from the field. These two techniques are in good agreement with our results. Thus, we have considered 11 number of clusters as the initial value for building clusters. Table 4.16 provides the simulation parameters used in our experiments. Figure 4.28 shows the network lifetime using the VIKOR method, which is compared with the fuzzy analytic hierarchy process (FAHP). It is shown that the network lifetime of VIKOR increases as much as 72% more than FAHP. Also, the time at which the first node dies is at the 917th second in VIKOR and the 390th second in FAHP, which presents a good stability region. In addition, the energy dissipation analysis is also plotted as shown in Figure 4.29.

Table 4.14 CHs and Their VIKOR Ranks

Cluster Head	Number of Neighbors, n	Residual Energy, E_o (Joules) (C_1)	Distance from Sink, d	S_i	R_i	Q_i	VIKOR Rank
1	8	1.9998	79.4768	0.1939	0.0073	0.0839	2
2	18	1.9998	94.3348	0.1022	0.0489	0.1222	12
3	17	1.9998	136.7213	0.2343	0.1677	0.5138	22
4	7	1.9981	76.8822	0.2061	0	0.0744	1
5	14	1.9998	94.9085	0.1572	0.0505	0.1656	14
6	17	1.97	94.2645	0.2213	0.0487	0.2071	19
7	19	1.9376	136.9486	0.4295	0.1683	0.6552	24
8	8	1.9998	79.7971	0.1948	0.0082	0.0868	3
9	20	1.963	141.1936	0.3377	0.1802	0.6192	23
10	9	1.9998	82.2292	0.1883	0.015	0.0991	7
11	12	1.9998	86.2287	0.1595	0.0262	0.1065	9
12	22	1.831	148.2669	0.8	0.2	1	25
13	17	1.9998	97.6696	0.1249	0.0582	0.1618	13
14	13	1.9993	86.5914	0.149	0.0272	0.1015	8
15	15	1.9998	96.0275	0.147	0.0536	0.1662	15
16	16	1.9998	98.503	0.1406	0.0606	0.1789	16
17	15	1.9782	94.0414	0.2181	0.0481	0.2032	18
18	16	1.9998	133.5827	0.2389	0.1589	0.495	20
19	18	1.9998	135.7117	0.2182	0.1648	0.4951	21
20	12	1.9944	85.9306	0.1777	0.0254	0.1175	10
21	14	1.995	88.2301	0.1555	0.0318	0.1177	11
22	10	1.9998	82.1805	0.1749	0.0148	0.0891	4
23	14	1.9998	97.4749	0.1644	0.0577	0.1888	17
24	10	1.996	81.384	0.1861	0.0126	0.0916	5
25	10	1.9998	82.8507	0.1767	0.0167	0.0952	6
A^+	22.0000	1.9998	76.8822				
A^-	7.0000	1.8310	148.2669				

Table 4.15 CHs Ranks in Second Cycle Using VIKOR Method

Ranks	1	2	3	4	5	6	7	8	9	10	11
Cluster Heads	CH4	CH1	CH8	CH18	CH24	CH25	CH10	CH14	CH11	CH20	CH21

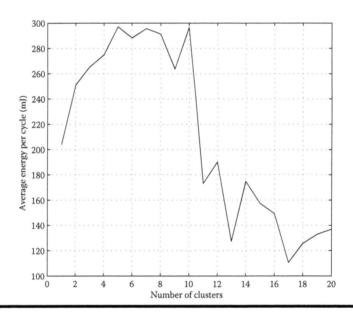

Figure 4.26 Average energy dissipation per cycle versus number of clusters.

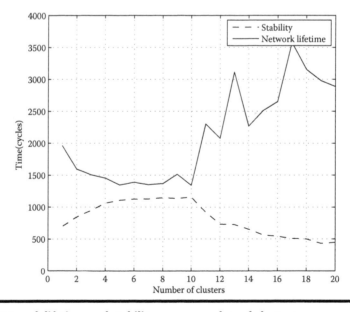

Figure 4.27 Network lifetime and stability versus number of clusters.

Table 4.16   Simulation Parameters for Data Transmission for VIKOR

Description	Symbol	Value
Total no. of nodes	N	200
BS location	–	(50,175)
Data packet	–	500 bytes
Hello/broadcast/join message	–	25 bytes
Energy consumption by the amplifier to transmit at a longer distance	ε_{mp}	0.0013 pJ/bit/m^4
Energy consumption in the amplifier to transmit at a short distance	ε_{fs}	10 pJ/bit/m^2
Energy consumption in the electronics circuit to transmit or receive the signal	E_{elec}	50 nJ/bit

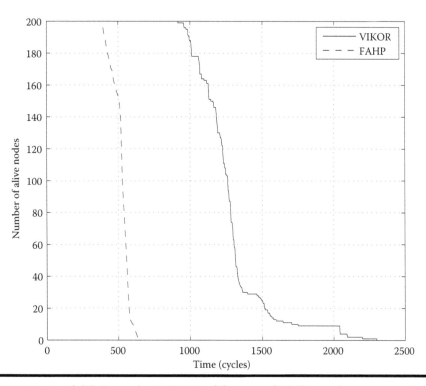

Figure 4.28   Network lifetime using VIKOR and fuzzy analytic hierarchy process schemes.

VIKOR is used in various multigoal decision-making as it helps in finding the final decision index and also considers the side effects. It also uses different computing modes, relative distances, and weights to find and reduce the overall target minority injury when considering the side effect in situation.

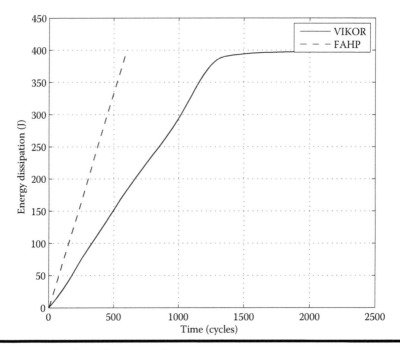

Figure 4.29 **Energy analysis for 200 nodes.**

4.5 Conclusion

Various single- and multicriteria-based EECSs in WSNs are proposed in this chapter. Residual energy is the prime criterion while working with single-criterion-based algorithms for building of clusters in a WSN. However, three criteria including residual energy distance of the nodes from the BS and the number of neighbor nodes are considered in order to optimize the number of clusters/CHs in multicriteria-based algorithms. Single-criteria algorithms include localization estimation algorithm (LEA), EEHC, DEEC, NCAEL, EECS, and MREC and multicriteria algorithms use different approaches including the Pareto-optimal theory, TOPSIS, fuzzy TOPSIS, and VIKOR. The present algorithms have been compared with some standard algorithms and it is found that the algorithms with a single criterion for selection are not as efficient and have overall poor network lifetime except MREC. However, the algorithms with multiple criteria are better in terms of network lifetime and energy savings. It is also observed that if the number of clusters is between 9 and 11, the stability region attains the maximum values in multiple criteria-based algorithms. In the future, we would consider a realistic model for the location of nodes without considering GPS, which has certain limitations of power in our future work. We also anticipate that further improvements in energy efficiency can be achieved by considering the concept of limited mobility in sensor nodes.

References

Abbasi, A.A. and Younis, M. 2007. A survey on clustering algorithms for wireless sensor networks. *Comput Commun*. 30:2826–2841.

Akylidiz, I.F. and Wang, X. 2005. A survey on wireless mesh networks. *IEEE Commun Mag*. 43(9):23–40.

Azad, P. and Sharma, V. 2013a. Cluster head selection in wireless sensor networks under fuzzy environment. *ISRN Sensor Netw.* 2013(Article ID 909086):8p. doi:10.1155/2013/909086

Azad, P. and Sharma, V. 2013b. Maximum residual energy based clustering scheme for wireless sensor networks. *Adv Sci Focus.* 1(2):111–119.

Azad, P. and Sharma, V. 2014. Energy efficient clustered scheme for wireless sensor networks using multi-criteria decision making approach. *Int J Comput Aided Eng Technol.* 6(3). doi:10.1504/IJCAET.2014.0631130

Azad, P. and Sharma, V. 2015. Pareto-optimal clustering scheme using data aggregation for wireless sensor networks. *Int J Electron.* 102:1165–1176.

Azad, P., Singh, B., and Sharma, V. 2011. A novel clustering approach for extending the lifetime for wireless sensor networks. *Int J Adv Eng Tech.* 1(5):441–446.

Bezdek, J.C. 1981. *Pattern Recognition with Fuzzy Objective Function Algorithms.* New York, NY: Plenum Press.

Chauhan, A. and Vaish, R. 2012. Magnetic material selection using multiple attribute decision making approach. *Mater Des.* 36:1–5.

Chauhan, A. and Vaish, R. 2013a. Hard coating material selection using multi-criteria decision making. *Mater Des.* 44:240–245.

Chauhan, A. and Vaish, R. 2013b. Pareto optimal microwave dielectric materials. *Adv Sci Eng Med.* 5:149–155.

Chen, C.T. 2000. Extensions of the TOPSIS for group decision-making under fuzzy environment. *Fuzzy Set Syst.* 114:1–9.

Comeau, F., Sivakumar, S.C., Robertson, W., and Phillips, W.J. 2006. Energy conserving architectures and algorithms for wireless sensor networks. In *Proceedings of the 39th Annual Hawaii International Conference on System Sciences*, Kauai, HI, USA, 4–7 January, 2006, 9(1).

Dunn, J. 1973. A fuzzy relative of the ISODATA process and its use in detecting compact well-separated clusters. *Cybern Syst.* 3:32–57.

Heinzelman, W.B., Chandrakasan, A.P., and Balakrishnan, H. 2002. An application-specific protocol architecture for wireless microsensor networks. *IEEE Trans Wireless Commun.* 1(4):660–670.

Hwang, C.L. and Yoon, K. 1981. *Multiple Attribute Decision Making-Methods and Applications, A State-of-the-Art Survey.* New York, NY: Springer-Verlag.

Kasprzak, E.M. and Lewis, K.E. 2001. Pareto analysis in multiobjective optimization using the colinearity theorem and scaling method. *Struct Multidiscip Optim.* 22:208–218.

Kumar, D., Aseri, T.C., and Patel, R.B. 2009. EEHC: Energy efficient heterogeneous clustered scheme for wireless sensor networks. *Comput Commun.* 32:662–667.

Lung, C.H. and Zhou, C. 2009. Using hierarchical agglomerative clustering in wireless sensor networks: An energy-efficient and flexible approach *Adhoc Netw.* 8:328–344.

Qing, L., Zhu, Q., and Wang, M. 2006. Design of a distributed energy-efficient clustering algorithm for heterogeneous wireless sensor networks. *Comput Commun.* 29:2230–2237.

Rathod, M.K. and Kanzaria, H.V. 2011. A methodological concept for phase change material selection based on multiple criteria decision analysis with and without fuzzy environment. *Mater Des.* 32(6):3578–3585.

Romesburg, H.C. 1990. *Cluster Analysis for Researchers.* Belmont, CA: Lifetime Learning Publications.

Smaragdakis, G., Matta, I., and Bestavros, A. 2004. SEP: A stable election protocol for clustered heterogeneous wireless sensor networks. In *Proceedings of Second International Workshop on Sensor and Actor Network Protocols and Applications*, SANPA.

Torfi, F., Farahani, R.Z., and Rezapour, S. 2010. Fuzzy AHP to determine the relative weights of evaluation criteria and fuzzy TOPSIS to rank the alternatives. *Appl Soft Comput.* 10:520–528.

Yang, T. and Hung, C.C. 2007. Multiple-attribute decision making methods for plant layout design problem. *Robot Comput Integr Manuf.* 23:126–137.

Ye, M., Li, C., Chen, G., and Wu, J. 2006. An energy efficient clustering scheme in wireless sensor networks. *Ad Hoc Sensor Wireless Netw.* 3:99–119.

Yu, J.Y. and Chong, P.H.J. 2005. A survey of clustering schemes for mobile ad hoc networks. *IEEE Commun Surveys Tuts.* 7(1):32–48.

Zanakis, S.H., Solomon, A., Wishart, N., and Dublish, S. 1998. Multi-attribute decision making: A simulation comparison of select methods. *Eur J Oper Res.* 107:507–529.

Chapter 5

Power Management in Sensor Node

Anuradha Pughat and Vidushi Sharma

Contents

5.1 Introduction..116
5.2 Sources of Power Consumption..117
    5.2.1 Microcontroller Unit..118
    5.2.2 Radio Unit...118
    5.2.3 Interfacing Unit...119
    5.2.4 Storage Unit...119
    5.2.5 Power Unit...119
    5.2.6 Sensing Unit..120
5.3 Design Techniques for Low Power..120
    5.3.1 Static Power Management...120
    5.3.2 Dynamic Power Management...120
5.4 Standard OSs for WSNs..121
5.5 Power-Aware WSNs..121
    5.5.1 Power-Aware Hardware..122
    5.5.2 Power-Aware Software..123
    5.5.3 Power-Aware Sensing...123
    5.5.4 Power-Aware Computing..124
    5.5.5 Power-Aware Communication..125
5.6 Power Management...125
    5.6.1 Operational-Level PM Techniques...126
        5.6.1.1 Greedy Scheme..126
        5.6.1.2 Time-Out Scheme..127
        5.6.1.3 Predictive Scheme..127
        5.6.1.4 Stochastic Scheme..127
    5.6.2 DVS Techniques...129
    5.6.3 Task Scheduling Techniques...130

5.6.4 Game-Theoretic DPM 130
5.6.5 Coordinated Power Management 130
5.7 Challenges in Power Management 132
Acknowledgment 132
References 132

5.1 Introduction

The cost of the wireless sensor network (WSN) depends on the cost of an individual node because of the infrastructure-less feature of the network. Each sensor node in WSNs works on limited energy and the lifetime of the network depends on the number of dead sensor nodes or the number of alive nodes, which are able to communicate to serve the purpose of the sensor network. In the state of the art, the relation between the death of a node and the sensor network has been found. Some authors state that a network is alive until all its nodes are dead, while some declare the death of a network on a single dead node (Bhardwaj and Chandrakasan 2002). However, the lifetime of an individual sensor node generally depends on the battery depletion rate and time. These nodes such as TelosB, Tmote Sky, Mica2, MicaZ, and IRIS motes use two AA batteries, which operate in a voltage range of 2.1–3.6 V DC. Other sensor nodes such as SunSPOT and SHIMMER use rechargeable lithium-ion batteries of 750 and 250 mAh, respectively. There have been several standard energy-scavenging or battery-recharging techniques, which help in recharging a node's battery up to a great extent. The deployed sensor nodes require some energy-efficient power management (PM) techniques to handle the workload even if energy scavenging helps in recharging the battery of a resource-constrained node. In an energy-scavenged sensor node, the power conditioning unit does the operation of dynamic power management (DPM) according to the workload. Thus, the workload characteristics should be known for the energy-efficient WSN application. Now, the power requirement of a sensor node within a specified time depends on the application requirement. Some monitoring applications like health monitoring need faster sensing and consume more power, while environment monitoring applications require slower sensing and low power in processing and communicating information from one node to another. It has been observed that data transmission consumes a huge amount of power than processing it (Raghunathan et al. 2002).

Thus, the power consumption of a sensor node/network depends on many factors and requires efficient power saving techniques/algorithms at the node and network levels. An energy management-based architecture for WSNs was given by Jiang et al. (2007).

This architecture suggests the following concepts in energy management:

- ◼ The energy of the components in an operating system (OS) should be classified (Zeng et al. 2005).
- ◼ The system should identify the priority level for the resource requests (Banga et al. 1999).
- ◼ The OS should be responsible for accountability and energy pricing (Neugebauer and Mcauley 2001).
- ◼ The resource allocation should be based on the dynamic constraint of the system (Flinn and Satyanarayanan 2004).
- ◼ The feedback in the system should guarantee the quality of service (Leslie et al. 1996).
- ◼ Besides the energy-efficient networking protocols and PM techniques on the sensor node, hardware should be energy efficient (Mplemenos and Papaefstathiou 2012).

The OS provides schedule and organization of the task execution between node hardware and application layer based on their priority level. The main design challenge with the sensor node designer is the trade-off between power consumption and performance when the DPM technique is implemented. The hardware/software codesign issues also need to be considered for making it a more energy-efficient node.

5.2 Sources of Power Consumption

A sensor node's components such as microcontroller unit, a sensor unit, transceiver unit, memory unit, interfacing unit, and power unit are the main sources of power consumption. Their power consumption depends on their use within certain operational conditions of a node such as duty cycle, workload pattern at input, application requirement, and algorithmic implementation. The total power consumed in a circuit is represented in terms of the dynamic power consumption and static power consumption as shown in

$$P_t = P_d + P_s \tag{5.1}$$

Here, the terms P_t, P_d, and P_s represent the total power consumption, dynamic power consumption, and static power consumption of the digital circuit. The dynamic power is related to the average transition and switching capacitance of wireless sensor node components from one power state to another (C), their operating/supply voltages (V_{dd}), frequency of operation (f), and how much time a component is active. On the other hand, static power consumption (P_s) reflects the leakage behavior of the circuit at a sensor node (Abdollahi et al. 2004):

$$P_t = CV_{dd}^2 \times f + P_s \tag{5.2}$$

The low-voltage requirement of DPM reduces static power consumption, but increases the latency overhead. Therefore, a balance between both is highly essential for total power consumption reduction. The leakage or static current reduction requires transistor threshold voltage to be increased but this further restricts the supply voltage (Roy et al. 2003). Thus, the minimum supply voltage should not be greater than the threshold voltage of the transistor to minimize static and dynamic power consumption on a wireless sensor node.

Apart from the basic sensor node components (i.e., sensor, processor, memory, transceiver), an additional unit, i.e., a PM unit, is required for reducing the power consumption during system idle time (Pughat and Sharma 2015a). The PM unit utilizes the system and its components' inactivity by operating them in a low power or making them completely off. Here, the point to be noted is that the transition or switching from one power mode to another should be such that the total power saving dominates the switching overheads. Thus, the transition or switching frequency is a critical choice in the designing of dynamic power-managed wireless sensor node.

Another important factor is that, in most of the WSN applications, a sensor node follows low-duty cycle and low-event arrival rate at the input of sensor node. Is there really any need for PM on the system that already has much less activity, or will the switching activities make the system more complex and energy consuming?

The answer to this question is trivial and we will give its solution after analysis and evaluation of different existing DPM techniques throughout this chapter. The essential and applicable condition of energy saving > energy overhead must be fulfilled for any type of DPM technique on a sensor node.

The main power-consuming sources of a sensor node are microcontroller unit, radio unit, interfacing unit, storage unit, and power unit. Here, we explain how these components consume power during functioning and communicating the desired information.

5.2.1 Microcontroller Unit

The microcontroller unit of a sensor node connects to the sensors/actuators, transceiver, serial port, LEDs, and the coprocessors. It helps in sensing, collecting, and processing useful information from neighboring nodes. It provides control to communication and signal processing algorithms/protocols. All these above-mentioned tasks and protocols require huge power from the sensor node. An energy-efficient microcontroller commands the radio unit for a desired event only and saves a large amount of power consumption during communicating the nodes. Depending on the workload for any application, OS schedules the system activities and on/off of the sensor node components to achieve low power consumption. The most popular sensor nodes use the microcontrollers such as TI MSP430, Atmel ATmega128L, Atmel ARM920T, Intel PXA271 Xscale, Cortex®M3, Intel StrongARM SA-1100, Atmel ATmega1281, and many more. A detailed power analysis of Rockwell's Wins Nodes and Medusa II Nodes has been performed (Raghunathan et al. 2002). The SA1100 processor of a µAMPS node can operate on dynamic voltage scaling (DVS) (Klaiber 2000). It can control power consumption by adjusting the operating voltage and frequency at different power levels. The MSP430 microcontroller consumes approximately 2 mA and 760 pJ energy/instruction at a frequency of 8 MHz and a voltage of 3 V while few µA in a sleep state. Today, some of the PM-based sensor nodes use two processors: one for high workload and another for low workload processing and communication.

5.2.2 Radio Unit

The radio unit consumes power during transmission, reception, overhearing, idle listening, and even idle mode of the information through a wireless medium. The radio unit can work in transmitting, receiving, idle, and sleep modes in a node. Levis et al. (2005) designed and developed an event-driven, tiny, and faster OS for modular and concurrent operations. It has been observed that a node takes more power in transmitting than receiving a bit of information. For example, an RFM TR1000 transceiver consumes 1 µJ for transmitting while 0.5 µJ for receiving a bit of useful information (Hill et al. 2000). It becomes quite interesting when the receiving power is more than the transmitting power in the network scenario where the transmission distance is small. Then, the protocols, which are designed assuming very small/negligible receiving power, fail in the system of small transmission distance. Moreover, the preobservation of the transmission and reception power at a different distances is required for developing local-level protocol design. Otherwise, the dynamic protocols are required, which details the transmission distance, transmitting power, receiving power, CPU utilization, and energy use by an individual sensor node. The details of % CPU utilization and % energy use by each node component during transmission and reception of a packet have been tabulated.

Table 5.1 illustrates the power consumption of the radio unit in µAMPS, Rockwell's Wins, and Medusa II nodes at different modes of operation. It shows the maximum power consumption when a node is active and receiving the packets. Minimum power is consumed during idle

Table 5.1 µAMPS, Rockwell's Wins, and Medusa II Nodes (Power Consumption of Radio Unit)

Operating Mode	Power Consumption µAMPS Node (mW)	Power Consumption Rockwell's Wins Node (mW)	Power Consumption Medusa II Node (mW)
Active (Transmission)	157	771.1	24.58
Active (Reception)	276	751.6	22.20
Idle (Transmission)	4	727.5	22.06

transmission mode (Shih et al. 2004). One can notice in the table that idle power consumption is close to receiving power consumption in the Rockwell's Wins node. Instead, complete shut-off/sleep on the radio unit is desirable during the no activity period. This implies that standard protocols designed such as XMAC (Buettner et al. 2006) and RI-MAC (Sun et al. 2008) do not implement universally on the sensor nodes. Moreover, operating the radio unit in different modes and their transition from one mode to another consume a significant amount of power. In addition, the intermediate amplifiers, local oscillator, and phase-locked loop (PLL) are the power consumers when in use. Thus, they require more flexibility, robustness, and durability before implementation.

5.2.3 Interfacing Unit

The interfacing unit provides a connection between the sensor node components. The subsystem consists of serial/parallel buses for use of high-speed and low-speed data transfer. Most often, the serial peripheral interface bus is used to communicate to controller and radio units. The frequency of interaction between processor and other devices via bus interface consumes power. Consumption also depends on communication bandwidth. On the other hand, inter-ICs buses can be used for data transfer between the sensor, memory, analog-to-digital converter, and coprocessor (Dargie 2012).

5.2.4 Storage Unit

Power consumption in memory depends on the speed and amount of data transfer between microprocessor and memory. The choice of memory on a sensor node, which can store useful data and routing table, etc., is also a very critical issue for the designer. The nonvolatile memory does not fulfill the low-cost node requirement, and flash memory has its design issues. Ideally, dynamic RAM has been used in commercially available sensor nodes, which requires refreshing techniques and this consumes power. Other factors such as RAM timing, decoding rate, multiplexing, and latency in memory access are the good power consumers (Weissel and Bellosa 2002). Thus, the memory to store a DPM technique needs to be considered. Power saving is possible if the OS is capable for dynamic memory allocation.

5.2.5 Power Unit

The DC–DC converter provides varying DC supply to hardware components. It is a major contributor in the implementation of DPM schemes on sensor nodes. This supplies different levels of voltages to sensor components and adopts the low power. The conversion efficiency should be higher but it consumes more power. Here, the scope of research lays in the conversion schemes that should be accurate and faster. The battery is another component of a power unit and the lifetime of a sensor node depends on the discharge rate and temperature of the battery itself. There are thousands of commercially available batteries (such as NiCd, alkaline, lithium ion, NiMH, etc.) but the choice of battery for a sensor node depends on the type of application, amount of current drawn, and voltage need of sensor components (Hausmann and Depcik 2013). The relaxation effect in batteries improves the lifetime of a battery, sensor node, and, thus, the entire network.

5.2.6 Sensing Unit

Power consumption in a sensing unit is due to the operation in the sensor itself, analog-to-digital converter, and signal conditioning. The sampling frequency is set to fulfill the application requirement and it is then decided with the amount of power consumption. The higher the sampling rate, the greater the power consumed. Actually, active sensors consume more power than passive sensors. Other than varying sampling frequency, there is no dynamic method that can be implemented on a sensing unit of the sensor. This area does not have much potential, but still scope for PM in ADC and signal conditioning exist.

5.3 Design Techniques for Low Power

5.3.1 Static Power Management

Static power management (SPM) includes compilation and synthesis at design time. This technique can be implemented at the procedure level, register-transfer level (RTL), instruction level, circuit level, process level, and system simulation level. Useful simulators for SPM are *Simplescalar* simulator, *Wattch* simulator, *SimplePower* simulator, *JouleTrack* simulator, and *ARMulator* simulators. In SPM, the sensor node can suspend and resume its operations for saving power. SPM provides power efficiency in the idle state of the system. Sometimes, this technique avoids sensor node damage from overheating and low battery strength indication (Ortega et al. 2010).

5.3.2 Dynamic Power Management

DPM traces the runtime behavior and workload of the device and utilizes its low workload and idle state for further power reduction. This technique can be implemented at the component level, processor level, and system hardware and software levels. The next section details the existing DPM techniques; the first technique is the system-level DPM, i.e., changing the operating state of system components when the workload varies and making components *off* or *idle* when not in use. The second technique is processor level, i.e., dynamic voltage and frequency scaling which require the system processor to be capable of scale-down frequency and voltage. The third technique uses ordering and scheduling of tasks and processes in such a way as to enhance the energy efficiency and execution within a specified time. Finally, the cooperative PM technique uses a combination of two or more specified DPM techniques or a modulation technique with them. Figure 5.1 illustrates input and output factors, which play an important role in decision-making on the DPM-based sensor node.

The DPM-implemented sensor node keeps on tracking the input workload, arrival rate, specific event arrival/interruption, and a resource used. Based on the energy use of the available resources in the current cycle, it takes a decision on the power mode for next cycle. The observation on the battery discharge rate prevents loss of useful information before the node dies due to complete battery drain. The node passes on information to the neighboring node when it is about to die. The details are explained in subsequent sections.

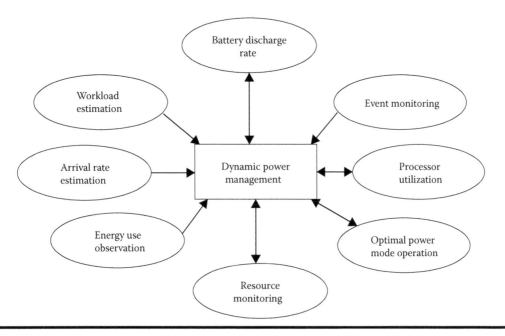

Figure 5.1   Input and output factors for dynamic power management.

5.4 Standard OSs for WSNs

The widespread WSN applications generally require a real-time operating system (RTOS) for real-time task scheduling and operations. However, the use of an OS increases the power consumption of a sensor node but makes the system more flexible for developing several applications. Depending on the criticality of an application, different levels of power consumption and complexity are tolerated. A list of most popular OSs for resource-constrained WSNs has been discussed in Chapter 1, which details the operating requirements and the type of sensor node to be used. The selection of a sensor node for an application lies under the choice of microcontroller unit, radio unit, OS, data rate, memory, sensor, actuator, protocol development interface, energy source, programming interface, and cost. The commercially available sensor nodes are categorized such as "SPEC" for specialized sensing, "Berkeley" for generic sensing, "iMote" for high bandwidth sensing, and the Gateway "Stargate" for connecting low-level nodes with Internet/ user. Although their hardware structure is almost the same, the OS provides and controls the hardware resources to application software. The energy-efficient OS for WSN is not similar to the traditional OS. They do not provide separate memory space for process protection and context switching from one process to another.

An earliest-deadline-first (EDF) scheduler for WSN is used for real-time applications. This limits the number of processes and increases time complexity with an increase in a number of processes.

5.5 Power-Aware WSNs

The power-aware system is energy consumption adaptive to the changes in operating conditions: environment, user requests, available resources, and quality of service. Power-aware WSNs scale the energy consumption depending on the sensor's sampling frequency, workload pattern, available resources, required performance, and quality of the output of the node. A power-aware component has the capability to work on different power and performance levels. This is then utilized for energy-efficient computing at software levels such as handling the varying duty cycle and workload conditions. Thus, the coordination between power-aware hardware and software is necessary for the development of an energy-efficient application on the sensor node. It gives the flexibility to an individual for taking a decision between the lifetime and performance trade-offs. Sometimes, the network is deployed for mission-critical applications where performance is preferred over battery depletion rate, while a few applications require a longer lifetime of the entire network at the cost of the signal-to-noise ratio (SNR). The first prototype of a micropower sensor node was developed with the name μAMPS (μ-adaptive multidomain power-aware sensors). Calhoun et al. (2005) gave instantaneous power scaling and measurement during data collection, line-of-bearing calculation, data transmission, and sleep times. They presented the challenges faced in designing energy-aware architecture by adding hardware components to work on a range of application scenarios. The radio module has been optimized for high data rate, faster start-up time, and low power.

The research on the optimization at the physical, data-link, medium access, transport, and routing layers gives solutions to the distributed microsensor networks in power-consuming WSN applications.

5.5.1 Power-Aware Hardware

The microprocessor of the microcontroller unit performs task execution and monitors timers/counters for interrupt generation at time intervals. A power-aware microprocessor works on different power modes such as active, idle, power down, etc. There can be more than one power-down mode in a processor, which can be modeled according to the workload on sensor input. The active and idle modes make the processor ON and OFF, respectively, while power-down modes make the other components OFF, except asynchronous interrupts for waking up the processor. The multiple power nodes not only save power but also consume during switching operations. An experimental cost is given as (Benini et al. 2000):

StrongARM1100:

■ Active state power consumption: 400 mW
■ Idle state power consumption: 50 mW
■ Sleep state power consumption: 160 μW
■ Active state to idle state transition time: 10 μs
■ Active state to sleep state transition time: 90 μs
■ Idle state to active state transition time: 10 μs
■ Sleep state to active state transition time: 160 μs

Dynamic random access memory (DRAM):

■ Active state power consumption: 300 mW
■ Nap state power consumption: 3 mW
■ Nap state to active state transition time: 120 ns
■ Nap state to active state power consumption: 165 mW

Transceiver CC2420:

■ Lock-up time required for the PLL: 192 μs

Transition from one power state to another is fruitful when energy saving becomes greater than the cost of energy consumption. For a two-state system, energy saving due to transition from upper power state (P_u) to lower power state (P_l) is given as

$$E_{saved,l} = P_u \times (t_l + t_{u,l} + t_{l,u}) - (P_{u,l} \times t_{u,l} + P_{l,u} \times t_{l,u} + P_l \times t_l) \tag{5.3}$$

where t_l and t_u represent the time the system stays in lower and upper power states. The transition times are represented by $t_{l,u}$ and $t_{u,l}$. The transition to a lower power state is successful only when energy is saved ($E_{saved,l} > 0$).

Another aspect of power-aware hardware is DVS in the microprocessor of the sensor node. This requires a DC–DC regulator and algorithms at the OS, which provides power awareness from workload to the resources/hardware used.

5.5.2 Power-Aware Software

The power-aware software plays a critical role in the power saving of WSNs as well as the sensor node. The control and application software can save up to 70% of energy even if low power hardware is used. Based on the duty cycle and sampling frequency of the desired application, the power-aware software controls the operation of available resources. The power-aware OS tracks the system activities and commands the node components to operate in different power-down modes. For example, the available active, sleep, deep-sleep, and idle modes of power-aware components are controlled with software/algorithm. The open source TinyOS provides support to Atmel ATmega 128L and TI MSP430 microcontrollers, which have been very commonly used in event-driven sensor nodes. The DVS scheme is also controlled within the OS and minimizes unnecessary power loss.

5.5.3 Power-Aware Sensing

Power awareness in sensing provides quality in sensing along with low power. Event detection can be classified as trigger driven, prediction-based, and significant event. A trigger-driven event is not necessarily significant. It does not require immediate attention. For example, in temperature monitoring, if the temperature exceeds a specified limit, any control action may be taken automatically. The higher the sampling frequency, the more the power consumption. Actually, we cannot limit the sampling frequency for our convenience to low power. The sampling frequency depends on the application requirement. The following list of applications gives a range of sampling frequencies:

- Monitoring
  - Atmospheric temperature, barometric pressure: 0.015–1 Hz
- Medical
  - Heart rate monitoring: 0.8–3.3 Hz
  - ECG monitoring: 100–250 Hz
  - Hearing

- Natural disaster
  - Volcanic infrasound, seismic vibration: up to 100 Hz
  - Earthquake vibrations: 100–150 Hz
- Industrial
  - Instrument vibrations: 40 kHz

Medical equipment requires the lifetime of sensor nodes in few number of days and months sometimes. Applications such as military and security applications require the desired lifetime of the nodes to be years/decades. Thus, power-aware sensing requires the technique of workload prediction to obtain future power use. The predictive event helps in determining the future situation. It is based on the knowledge gained due to past events. This requires a system with memory. Sometimes the predicted event may be inaccurate. On the other hand, the significant event indicates the potential change. These events require special attention and need to be processed within a short and specified time. Otherwise it may create serious harm. In this type of sensing, high detection probability and poor false event rate are desirable. Adaptive sampling is another type of power-aware sampling, which provides fine-grained sampling in the time duration where a change in the signal value is faster. Multiscale sensing divides the sensing region into low-resolution and high-resolution zones but requires the sensor to be always in the ON state. Another more generalized sampling scheme is model-based adaptive sampling in which the sensing energy is optimized according to the spatiotemporal relationships between the measured values at the node input. However, this requires the effective and energy-efficient predictive techniques for resource-limited sensor nodes.

5.5.4 Power-Aware Computing

Figure 5.2 shows the pyramid to represent the factors affecting low power in a power-managed system. Generally, three types of management, PM, task management (TM), and event management, provide the optimal operating point of node components.

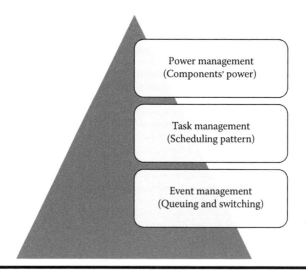

Figure 5.2 Components of power-aware computing.

First, event management includes collection, observation, and queuing of the incoming event. Here, energy-efficient event management techniques are required. After that, switching of events is required for processing purposes. This may include priority-based or nonpriority-based switching (Pughat and Sharma 2015c; Huang et al. 2012).

Second, the TM takes a decision on the scheduling pattern of the tasks. There can be first-in-first-out (FIFO), last-in-first-out (LIFO), EDF, rate monotonic (RM), and priority-based scheduling patterns for energy-constrained WSNs. The third type of management for low power is PM, which observes the activity of sensor node components and is based on the present and anticipated workload; it commands the components to operate in a suitable power mode. Thus, power-aware computing is essential for minimizing unnecessary use of power on the sensor node.

5.5.5 Power-Aware Communication

Power-aware communication includes auto-adjustment of transmission distance, choice of modulation technique, and PM in radio frequency (RF) circuitry. A sensor node can transmit useful information to other nodes in a schedule-driven or trigger-driven manner. It has been found that operating the radio unit at lower power mode can increase overall power consumption. So, an effective and adaptive PM policy is required to mitigate low-performance effects on the node. PM without affecting its performance is a nontrivial problem. Intelligent techniques for packet forwarding and modulation scheme selection can solve the problem to an extent. A power-aware radio unit provides energy-aware packet scheduling and modulation scaling, which reduces the power of the major power hog subsystem, i.e., communication system in WSNs.

5.6 Power Management

Power-aware components provide efficient PM techniques with hardware/software codesign issues. The detailed taxonomy for PM approaches is presented in Figure 5.3. The diagram depicts the main PM schemes in three vertical levels.

One level is for the operational-level PM, DVS, task scheduling, game theoretic, and coordinated PM schemes. The operational-level PM is then categorized into greedy, time-out, predictive, and stochastic schemes. The third level of PM presents the subclasses of stochastic schemes, i.e.,

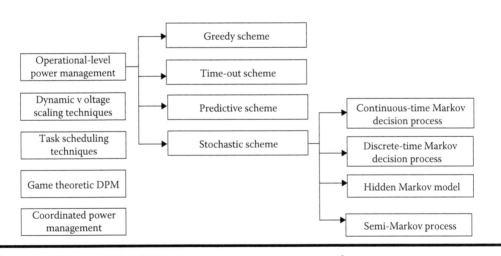

Figure 5.3   Taxonomy for different power management approaches.

continuous-time Markov decision process, discrete-time Markov decision process, hidden Markov models (HMMs), and semi-Markov process (SMP). The details about each scheme are given below.

5.6.1 Operational-Level PM Techniques

In operational-level PM techniques, a sensor node observes the event arrival at its input and state-of-the-node components. Then the power consumption and transition cost of each power-aware component are estimated. Based on the estimated cost, the system takes a decision on the transition to optimal power state or remains in the same state. These techniques are implemented at OS and observe the operations of the available components on the sensor node.

5.6.1.1 Greedy Scheme

A sensor node can be either schedule driven or trigger/event driven. So far, we have discussed event-driven sensor nodes. The main objectives for such a type of sensing are listed below:

- To monitor the sensor field in monitoring applications. This can be used by the field engineers to observe the environmental situations.
- To detect field anomalies and observations to take necessary corrective action. For applications such as raising factory alarm for fire detection, controlling the thermostat, and managing any disaster.

Greedy schemes manage power consumption, comparing threshold value with the actual value of the input. They wake up the sensor node when input crosses the set threshold value. The system processes and communicates for every value which crosses the threshold or simply stating for every event whether it is significant or not. Thus, the scheme is not adaptive to the input workload. The models for these applications can be used with so many amendments. One idea is to use timers for sleep and wake periods, i.e., *time-out scheme*. This will increase flexibility and reduce consumption. However, this may increase the missed events during the sleep time of the sensor. The application design and modeling for low power in WSNs are another tedious task. For beginners, an example for fire monitoring is given below.

Fire monitoring scenario:

The low-power application can be designed according to the choice of the programmer. An adaptive fire monitoring scenario is given as a reference that will help in understanding the design parameters for real-time WSN applications.

- Event arrival rate should be high, medium, or low. For example, one event/hour for high rate, one event/8 hours for medium, and one event/24 hours for low event arrival rate.
- The sampling frequency of sensors can be high for critical applications and low for normal operations.
- Any change in temperature is an indication of an event occurrence. The threshold level is set to compare the change.
- Instead of taking action on every changed value in temperature, a trend of historical data is a better choice for future predictions.
- Set a time to redeploy and reconfigure the system. For example, one time unit.

There is no hard-and-fast rule for scenario design, but it affects the power consumption a lot. An expert should keep in mind higher performance as well as low power consumption in the designed system.

5.6.1.2 Time-Out Scheme

This is a heuristic approach to PM and nonadaptive too. In the component-based scheme, certain components can be made off/sleep at the OS level while a watchdog will only be active. In *time-out* scheme, the timers are used to provide on and off timings to the sensor node. The scheme is effective only when the idle period of the system is larger and can accommodate the switching overheads. This reduces the chances of missed events and increases the flexibility to designed system.

5.6.1.3 Predictive Scheme

Based on past values, the *predictive scheme* predicts the future behavior of the input at the sensor node. It uses regression equation to predict the next transition time for the system component. Predictive schemes for DPM are used to shut down the event-driven system (Hwang and Wu 2000; Srivastava et al. 1996). In WSNs, the sensor nodes use prediction techniques for event arrival and power mode selection. A parameter, i.e., breakeven time, is used to decide the idle time in which the node can be powered down. Idle time is compared with breakeven time and the system goes to a power-down state if the idle time is greater than the breakeven time. However, these predictive schemes are slower and prone to error in the application of the higher sampling rate. It requires high correlation between past and present events. Heuristic approaches such as time-out and predictive are complex for components with more than two states and exhibit power-performance trade-offs.

5.6.1.4 Stochastic Scheme

PM can also be done using a stochastic control scheme such as a Markov decision process. These schemes can save up to 20% more power consumption than greedy and time-out schemes. Input workload distribution and state transitions of a power-managed sensor node can be modeled using stochastic techniques. An example of this in the queuing system is represented in Figure 5.4.

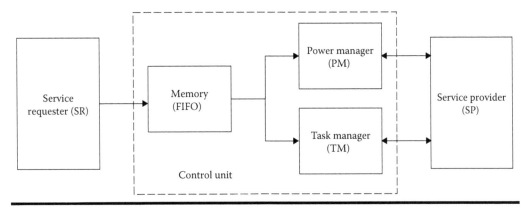

Figure 5.4   An abstract model for stochastically power-managed system.

As analogous to the sensor node behavior, the service requester (SR) is presented with event arrival and service provider (SP) with power mode of operation. Here, the power control unit consists of a queue, power manager, and task manager. PM observes the present power state of the node components, event arrival rate, and type, then takes a decision on the next power state of the components (Pughat and Sharma 2016a). TM schedules the task execution in energy-efficient way.

5.6.1.4.1 Continuous-Time Markov Decision Process

In continuous-time Markov decision process (CTMDP), service requests are generated in continuous time and correspond to the behavior of a real-time system. The DPM system can be stochastic DVS based, which models the processor voltage according to the workload (Kargahi and Movaghar 2008). The system consists of a single processor and the event arrival follows a Poisson distribution. An energy-aware stochastic model for WSNs gives the sensor nodes to be placed in active or sleep mode. Then, power consumption, delay, and throughput are analyzed (Zhang and Li 2011; Xu and Wang 2014). A steady-state probability distribution is derived for obtaining power consumption in the sensor states.

5.6.1.4.2 Discrete-Time Markov Decision Process

In a discrete-time Markov decision process (DTMDP), service requests and commands for state transitions are generated at discrete time. This increases power saving as compared to CTMDP. However, a wrong decision on the arrival of a new event can take the system to a higher power state and a resulting higher power consumption. The stochastic model checking of continuous- and discrete-time DPM systems has been presented using PRISM (probabilistic model checker) (Kwiatkowska et al. 2007).

5.6.1.4.3 Hidden Markov Model

The HMM maximizes the likelihood of an observed sequence of events. According to the observed sequence, HMMs are trained and optimized for low power (Tan and Qiu 2008). As far as the WSNs are concerned, the implementation of these models is complex and gives better results if designed properly. The selection of hidden layers in the model is a challenging task for optimization.

5.6.1.4.4 Semi-Markov Process

The SMP is generalized CTMDP. An analyzer-based model is applicable to the events which are significant or not. Today, researchers are trying to model a DPM scheme that can be implemented on trigger-driven as well as the schedule-driven sensor nodes. One solution to this is an "analyzer-based model" which can be applicable to both event-driven and schedule-driven sensor node applications with some amendments (Pughat and Sharma 2015b).

In an event-driven sensor node, the preprocessor sends a filtered signal to the analyzer. Then this filtered signal is passed to the analyzer, where a change in the detected signal is compared with the previous filtered value. If any change is detected, only the information is further passed for processing and communication.

For schedule driven, it becomes "on" and "off" alternately and a wait for an event during its awake period. It requires a real-time clock and wakes up the node if any significant event occurs that needs to be processed and communicated within a specified time interval. The schedule-driven approach can be used for target tracking applications (He et al. 2006). The change difference goes very high all of a sudden. This change decreases in every hour delay of response and drops to zero at another specified time. This indicates that the desired event was not processed because node was sleeping. This increases the event miss rate. An analyzer and preprocessor-based node model decreases the event miss rate by making low power watchdog/analyzer "on" and other components of the node "off." When the desired event occurs, the analyzer makes the rest of the node components "on" to process further and communicate. A suitable combination of hardware and software use is represented in Jung et al. (2009). This paper gives the derived and compared semi-Markov models for schedule-driven and event-driven sensor nodes. This explores the implementation and power analysis of these models at the networked environment.

5.6.2 DVS Techniques

A sensor node does not work continuously at its peak performance (refer Section 5.5.3). For example, an event-driven sensor node schedules the processing and communication on the occurrence of an event at its input. Otherwise, it remains inactive for a long time. We know the voltage scaling and frequency scaling give quadratic and linear power saving, respectively. Here, DVS in the microprocessor provides flexibility in the node to operate at different voltage levels. The power consumed during lower voltage levels will be lower when a node is in an active state. Thus, DVS prevents unnecessary power consumption when the node does not need to be operated at its peak performance. It should be noted that voltage scaling alone does not effectively work without the frequency scaling in the microprocessor. Hence, the term "dynamic voltage scaling" reveals frequency scaling too. Here, voltage as well as frequency of the sensor node is scaled down according to the workload at its input. The relation between normalized energy, operating voltage, and frequency is given in Equation 5.4 (Sinha and Chandrakasan 2001). The relation shows that the energy cost increases with the sampling time (T_s), normalized frequency (f_{norm}), and the supply voltage (V_{dd}).

$$E_{norm} = \frac{k(V_{dd} - V_t)^2 T_s f_{ref} f_{norm} \left[\dfrac{V_t V_{dd}}{(V_{dd} - V_t)^2} + \dfrac{f_{norm}}{2} + \sqrt{\dfrac{V_t V_{dd} f_{norm}}{(V_{dd} - V_t)^2} + \left(\dfrac{f_{norm}}{2} \right)^2} \right]}{V_{dd}} \quad (5.4)$$

where the term k represents the constant for switching capacitance and V_t as supply voltage.

However, the scaling has technology limitations and cannot be done endlessly. Further, the requirement of hardware such as stable clock generator and voltage converter makes the system expensive. We have presented the optimal power and performance trade-offs for DVS on a sensor node (Pughat and Sharma 2016b). Work includes a faster gradient descent optimization technique in which the adaptive step size determines the optimal voltage and frequency. The technique has been implemented on the MSP430 microcontroller, which is used in various commercially available sensor motes. Results show that the DVS is effective within the boundary of performance constraints.

Intel's iMote2 is a commercially available mote, which includes dynamic voltage and frequency scaling (DVFS) and an integrated circuit (IC) for PM. This is compatible with TinyOS and controls the component power by operating them in different power modes (Nachman et al. 2008). The power-aware "XYZ" motes provide low power for several motions-enabled applications (Lymberopoulos and Savvides 2005). The tiny, reliable, and low power "Hitachi watch" is application specific. They have an external real-time clock for waking up the sensor node from sleep power modes (Yamashita et al. 2006).

The DVFS minimizes the power consumption at the cost of processing speed. This requires the scheduler to be energy aware, reliable, and faster in scheduling. Another requirement of voltage scaling includes the faster speed of a DC–DC converter, which provides different voltage levels to system components. Thus, slower optimization and higher chances of synchronization error are two limitation factors of DVS. Moreover, two bus lines—one for high speed and high voltage and another for low speed and low voltage—can also be used for providing static voltage levels to the node components.

5.6.3 Task Scheduling Techniques

The task scheduling is another implementation of the DPM technique. This can be done at the scheduler of OS. Generally, the simplest FIFO scheduler is used for the WSN applications. The energy-efficient weighted fair queuing (E^2 WFQ) is a rate-adaptive packet scheduling technique that gives up to 10× energy consumption reduction during the transmission of packets. The scheme does not affect the throughput allocation but increases the packet latency within small bounds. Other scheduling techniques (e.g., EDF, RM, etc.) are less suitable for resource-constrained WSNs. However, ongoing research may fill the gap in the future. Workload estimation can be done with the filtering scheme or by predictive techniques. The observation of task activities for a period and its future prediction can be useful but requires accurate estimation in mission-critical applications. On the other hand, filters such as exponential weighted average (EWA) and least mean square (LMS) filters are used preferably, but at the cost of hardware. Sometimes, the event counter at the scheduler itself helps in estimating the number and duration of hardware accesses. This way the workload is observed and the decision on the activity of hardware component during next cycle is taken (Merkel and Bellosa 2008).

5.6.4 Game-Theoretic DPM

In a game-theoretic PM scheme, the game is played between the PM committee and its opponent committee. The PM committee consists of power-aware components, which save power workings on different power modes. On the other hand, the opponent committee is responsible for generating events for consuming power. They consume a common resource, i.e., battery, in the case of WSNs. Here, the objective is to optimize the system power such that power saving could never be lesser than the power consumed in policy overheads. A system has been designed as a finite state machine (Shukla and Gupta 2001). Up to 60% power savings on the device can be achieved using this approach. Further, the reward and penalties can be introduced in the above-mentioned model to increase the closer look of the policy optimization. Based on the interest of both the committees or for maximum profit, the game rules can be altered but should be known to both. The game-theoretic policies are more suitable for power saving on larger systems. Still, obtaining policy for optimal power and performance is an open challenge due to the indefinite workload behavior of the sensor node (Ren and Meng 2009; Ganeshpure and Kundu 2013). In this work, the authors designed a game-theoretic DPM for the sensor node. Recent advancements, open issues, and solutions for game-theoretic policy on sensor networks need to be explored in depth (Shi et al. 2012).

5.6.5 Coordinated Power Management

The newer and advanced research area for the low power today comes with a combination of the different PM techniques. This preferably would be called coordinated PM. A combination of DPM and DVFS are implemented at the sensor node level. Techniques such as dynamic modulation scaling (DMS) and signal processing scaling (SPS) are implemented at the radio module and sensor module of the node and this ensures the energy efficiency in communication. Actually, modulation scaling belongs to the communication subsystem and voltage scaling to the digital subsystem. The modulation scaling technique scales packet schedules before transmission. It observes the idle time after packet transmission and if found free then schedules and stretches the transmission time before next transmission of the packet. The number of bits per symbol can be varied to optimize the transmission power (Madan et al. 2007). In other words, packets are transmitted at the slowest speed of transmission with the lowest transmission power and without missing deadlines. Detailed scaling is presented in Figure 5.5. This represents how and when the tasks are scheduled within the deadline limits. The degradation of system performance is tolerable up to some extent, beyond that the quality of service degrades and thus violates the actual application need.

Execution deadlines for each task and % utilization have been given. Task duration with and without scaling shows how scaling affects task execution time and therefore the system performance and resulting power saving. For real-time applications, the cost of scaling or delay in task execution must be below the limits of task execution deadlines.

Schurgers et al. (2001) developed modulation scaling to EDF-based communication subsystem, which scales the transmission rate.

SPS (Wang and Chandrakasan 2002) entails changes in signal statistics/features and changes in the data rate at node/network. Out of various methods of SPS, fixed-point arithmetic is one of the most frequently used methods and found implemented on real-time applications. Thus, power-aware sensing, processing, and communication are provided to distributed sensor networks. This reduces processing and communication complexity at the cost of operational delays. Therefore, power fidelity and performance trade-offs require effective optimization techniques on the node and network levels, too. Cost overhead is another research challenge requiring immediate attention.

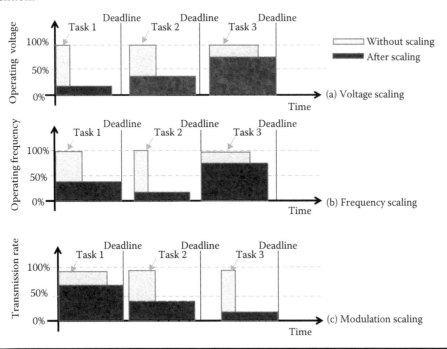

Figure 5.5 **Scaling techniques for low power consumption (a) voltage scaling, (b) frequency scaling, and (c) modulation scaling.**

5.7 Challenges in Power Management

PM at hardware and software of a sensor node requires immediate attention of the researcher, when it becomes a trivial issue from power, performance, and complexity points of view. The following points mention the research highlights to create a spark in a thirsty mind:

- Use of a high-performance processor and a low-performance processor on a single sensor board remains an architectural issue. Is there really any need of a coprocessor in saving the energy of an event-based, power-aware sensor node?
- The performance requirement along with the exact prediction and calculation of workload is a challenging task for real-time applications.
- The technique that extrapolates the speed requirement of the processor for mapping voltage and operating frequency should be explored.
- Which implementation results in lesser overheads and more power saving (globally implemented DPM or locally implemented DPM).
- Hardware support for event management and data transmission management is required. This can save a significant amount of power. Further, this requires the subsequent changes at the MAC layer.
- Hardware and software codesign issues are still under research and development.
- On-chip reconfigurable computing will be available in future nodes. This will increase flexibility in the functionality of the components.
- Policy optimization under a performance constraint is a big designing issue.
- To reduce complexity in stochastically power-managed systems when components have more than two power states is the thirst area for WSNs.

Acknowledgment

The authors would like to acknowledge DST, Govt. of India, for granting financial support (SR/WOS-A/ET-1043/2014 (G&C)) for this research work.

References

Abdollahi, A., Fallah, F., and Pedram, M. 2004. Leakage current reduction in CMOS VLSI circuits by input vector control. *IEEE Trans Very Large Scale Integr (VLSI) Syst.* 12(2):140–154. doi:10.1109/TVLSI.2003.821546.

Banga, G., Druschel, P., and Mogul, JC. 1999. Resource containers: A new facility for resource management in server systems. *Computer.* 33:45–58. doi:10.1145/224057.225831.

Benini, L., Bogliolo, A., and De Micheli, G. 2000. A survey of design techniques for system-level dynamic power\ management. *IEEE Trans Very Large Scale Integr (VLSI) Syst.* 8(3: 299–316. doi:10.1109/92.845896.

Bhardwaj, M., and Chandrakasan, A.P. 2002. Bounding the lifetime of sensor networks via optimal role assignments. In *Proceedings of the 21st Annual Joint Conference of the IEEE Computer and Communications Societies.* Hilton New York Hotel, New York, USA, 25–27 June 2002. doi:10.1109/INFCOM.2002.1019410.

Buettner, M., Yee, G.V., Anderson, E., and Han, R. 2006. X-MAC: A short preamble MAC protocol for duty-cycled wireless sensor networks. In *Proceedings of the 4th International Conference on Embedded Networked Sensor Systems (SenSys 2006).* Boulder, Colorado, USA, October 31–November 03, 2006, 307–320. doi:10.1145/1182807.1182838.

Calhoun, B.H., Daly, D.C., Verma, N., 2005. Design considerations for ultra-low energy wireless microsensor nodes. *IEEE Trans Comput.* 54(6):727–740. doi:10.1109/TC.2005.98.

Dargie, W. 2012. Dynamic power management in wireless sensor networks: State-of-the-art. *IEEE Sens J.* 12(5):1518–1528. doi:10.1109/JSEN.2011.2174149.

Flinn, J. and Satyanarayanan, M. 2004. Managing battery lifetime with energy-aware adaptation. *ACM Trans Comput Syst.* 22(2):137–179. doi:10.1145/986533.986534.

Ganeshpure, K. and Kundu, S. 2013. Game theoretic approach for run-time task scheduling on an multi-processor system on chip. *IET CircDev Syst.* 7(5):243–252. doi:10.1049/iet-cds.2013.0091.

Hausmann, A. and Depcik, C. 2013. Expanding the Peukert equation for battery capacity modeling through inclusion of a temperature dependency. *J Power Sources.* 235:148–158. doi:10.1016/j.jpowsour.2013.01.174.

He, T., Krishnamurthy, S., Luo, L. et al. 2006. VigilNet: An integrated sensor network system for energy-efficient surveillance. *ACM Trans Sen Netw.* 2(1):1–38. doi:10.1145/1138127.1138128.

Hill, J., Szewczyk, R., Woo, A., Hollar, S., Culler, D., and Pister, K. 2000. System architecture directions for networked sensors. *ACM SIGOPS Oper Syst Rev.* 34(5):93–104. doi:10.1145/384264.379006.

Huang, D., Chien Tseng, H., Jiunn Deng, D., and Chieh Chao, H. 2012. A queue-based prolong lifetime methods for wireless sensor node. *Comput Commun.* 35(9):1098–1106. doi:10.1016/j.comcom.2011.11.006.

Hwang, C.-H. and Wu, A.C.-H. 2000. A predictive system shutdown method for energy saving of event-driven computation. *ACM Trans Des Autom Electron Syst.* 5(2):226–241. doi:10.1145/335043.335046.

Jiang, X., Taneja, J., Ortiz, J. et al. 2007. An architecture for energy management in wireless sensor networks. *ACM SIGBED Rev.* 4(3):31–36. doi:10.1145/1317103.1317109.

Jung, D., Teixeira, T., and Savvides, A. 2009. Sensor node lifetime analysis. *ACM Trans Sens Netw.* 5(1):1–33. doi:10.1145/1464420.1464423.

Kargahi, M. and Movaghar, A. 2008. Stochastic DVS-based dynamic power management for soft real-time systems. *Microprocessors Microsyst.* 32(3):121–144. doi:10.1016/j.micpro.2007.06.001.

Klaiber, A. 2000. The Technology behind Crusoe Processors. Transmeta Technical Brief, no. January. Available at: http://scholar.google.com/scholar?hl=en&btnG=Search&q=intitle:The+Technology+Behind+Crusoe?+Processors#0.

Kwiatkowska, M., Norman, G., and Parker, D. 2007. Stochastic model checking. *Formal Methods Perform Eval SE–6.* 4486:220–270. doi:10.1007/978-3-540-72522-0_6.

Leslie, I.M., McAuley, D.R., Roscoe, T. et al. 1996. The design and implementation of an operating system to support distributed multimedia applications. *IEEE J Sel Area Comm.* 14(7):1280–1297. doi:10.1109/49.536480.

Levis, P., Madden, S., Polastre, J. et al. 2005. TinyOS: An operating system for sensor networks. In *Ambient Intelligence,* Weber, W., Rabaey, J.M., Aarts, E. (Eds.), Springer Berlin Heidelberg, 115–148. doi:10.1007/3-540-27139-2_7.

Lymberopoulos, D. and Savvides, A. 2005. XYZ: A motion-enabled, power aware sensor node platform for distributed sensor network applications. In *2005 4th International Symposium on Information Processing in Sensor Networks, IPSN.* Boise, ID, USA, 15 April 2005, 449–454. doi:10.1109/IPSN.2005.1440970.

Madan, R., Cui, S., Lall, S., and Goldsmith, A.J. 2007. Modeling and optimization of transmission schemes in energy-constrained wireless sensor networks. *IEEE/ACM Trans Netw.* 15(6):1359–1372. doi:10.1109/TNET.2007.897945.

Merkel, A. and Bellosa, F. 2008. Task activity vectors: A new metric for temperature-aware scheduling. In *ACM SIGOPS/EuroSys European Conference on Computer Systems (Eurosys).* Glasgow, Scotland, 31st March–4 April 2008, 1. doi:10.1145/1352592.1352594.

Mplemenos, G. and Papaefstathiou, I. 2012. Fast and power-efficient hardware implementation of a routing scheme for WSNs. In *IEEE Wireless Communications and Networking Conference, WCNC.* Paris, France, 1–4 April 2012, 1710–1714. doi:10.1109/WCNC.2012.6214059.

Nachman, L., Huang, J., Shahabdeen, J., Adler, R., and Kling, R. 2008. IMOTE2: Serious computation at the edge. In *IWCMC 2008—International Wireless Communications and Mobile Computing Conference.* Crete, Island, 6–8 August 2008, 1118–1123. doi:10.1109/IWCMC.2008.194.

Neugebauer, R. and Mcauley, D. 2001. Energy is just another resource: Energy accounting and energy pricing in the nemesis OS. In *Proceedings Eighth Workshop on Hot Topics in Operating Systems.* Elmau, Germany, 20–22 May 2001, 67–72. doi:10.1109/HOTOS.2001.990063.

Ortega, C., Tse, J., and Manohar, R. 2010. Static power reduction techniques for asynchronous circuits. In *Asynchronous Circuits and Systems (ASYNC), 2010 IEEE Symposium on* 3–6 May 2010, Grenoble, France, 52–61. doi:10.1109/ASYNC.2010.18.

Pughat, A. and Sharma, V. 2015a. A review on stochastic approach for dynamic power management in wireless sensor networks. *Hum Centric Comput Info Sci.* 5(4): 1–14. doi:10.1186/s13673-015-0021-6.

Pughat, A. and Sharma, V. 2015b. Stochastic model for lifetime improvement of wireless sensor node. In *2015 8th International Conference on Contemporary Computing, IC3 2015.* Noida, India, 20–22 August 2015. doi:10.1109/IC3.2015.7346718.

Pughat, A. and Sharma, V. 2015c. Queue discipline analysis for dynamic power management in wireless sensor node. In *2015 Annual IEEE India Conference (INDICON).* New Delhi, India, 17–20 December 2015, 1–5. IEEE. doi:10.1109/INDICON.2015.7443670.

Pughat, A. and Sharma, V. 2016a. Performance analysis of an improved dynamic power management model in wireless sensor node. *Digit Commun Netw.* 3(1): 19–29. . doi:10.1016/j.dcan.2016.10.008.

Pughat, A. and Sharma, V. 2016b. Optimal power and performance trade-offs for dynamic voltage scaling in power management based wireless sensor node. *Perspect Sci.* 8(September):536–539. doi:10.1016/j.pisc.2016.06.013.

Raghunathan, V., Schurgers, C., Park, S., and Srivastava, M.B. 2002. Energy-aware wireless microsensor networks. *IEEE Signal Process Mag.* 19(2):40–50. doi:10.1109/79.985679.

Ren, H. and Meng, M.Q H. 2009. Game-theoretic modeling of joint topology control and power scheduling for wireless heterogeneous sensor networks. *IEEE Trans Autom Sci Eng.* 6(4):610–625. doi:10.1109/TASE.2009.2021321.

Roy, K., Mukhopadhyay, S., and Mahmoodi-Meimand, H. 2003. Leakage current mechanisms and leakage reduction techniques in deep-submicrometer CMOS circuits. *Proc IEEE.* 91(2):305–327. doi:10.1109/JPROC.2002.808156.

Schurgers, C., Aberthorne, O., and Srivastava, M.B. 2001. Modulation scaling for energy aware communication systems. In ISLPED'01: *Proceedings of the 2001 International Symposium on Low Power Electronics and Design (IEEE Cat. No.01TH8581),* Huntington Beach, CA, USA, August 06–07, 2001, 96–99. doi:10.1109/LPE.2001.945382.

Shi, H.-Y., Wang, W.-L., Kwok, N.-M., and Chen, S.-Y. 2012. Game theory for wireless sensor networks: A survey. *Sensors.* 12(7):9055–9097. doi:10.3390/s120709055.

Shih, E., Cho, S., Lee, F.S., Calhoun, B.H., and Chandrakasan, A. 2004. Design considerations for energy-efficient radios in wireless microsensor networks. *J VLSI Signal Process Sys Signal Image Video Technol.* 37(1):77–94. doi:10.1023/B:VLSI.0000017004.57230.91.

Shukla, S.K. and Gupta, R.K. 2001. A model checking approach to evaluating system level dynamic power management policies for embedded systems. In *Sixth IEEE International High-Level Design Validation and Test Workshop.* 7–9 November 2001, Monterey, CA, USA, 53–57. doi:10.1109/HLDVT.2001.972807.

Sinha, A. and Chandrakasan, A. 2001. Dynamic power management in wireless sensor networks. *IEEE Des Test Comput.* 18(2):62–74. doi:10.1109/54.914626.

Srivastava, M.B., Chandrakasan, A.P., and Brodersen, R.W. 1996. Predictive system shutdown and other architectural techniques for energy efficient programmable computation. *IEEE Trans Very Large Scale Integr (VLSI) Syst,* 4(1):42–55. doi:10.1109/92.486080.

Sun, Y., Gurewitz, O., and Johnson, D.B. 2008. {RI-MAC:} A receiver initiated asynchronous duty cycle {MAC} protocol for dynamic traffic loads in wireless sensor networks. In *The 6th ACM Conference on Embedded Networked Sensor Systems.* Raleigh, NC, USA, November 04–07 2008.

Tan, Y. and Qiu, Q. 2008. A framework of stochastic power management using hidden Markov model. In *Design, Automation and Test in Europe, 2008. DATE '08.* Munich, Germany, 10–14 March 2008, 92–97. doi:10.1109/DATE.2008.4484668.

Wang, A. and Chandrakasan, A. 2002. Energy-efficient DSPs for wireless sensor networks. *IEEE Signal Process Mag.* 19(4):68–78. doi:10.1109/MSP.2002.1012351.

Weissel, A. and Bellosa, F. 2002. Process cruise control: Event-driven clock scaling for dynamic power management. In *Proceedings of the 2002 International Conference on Compilers, Architecture, and Synthesis for Embedded Systems.* Grenoble, France, 8–11 October 2002, 238–246. doi:10.1145/581630.581668.

Xu, D. and Wang, K. 2014. Stochastic modeling and analysis with energy optimization for wireless sensor networks. *Int J Distrib Sens Netw.* 10(5): 1–5. doi:10.1155/2014/672494.

Yamashita, S., Shimura, T., Aiki, K. et al. 2006. A 15 × 15 mm, 1 µA, reliable sensor-net module : Enabling application-specific nodes. In *Proceedings of the 5th International Symposium on Information Processing in Sensor Networks.* Nashville, TN, 383–390. doi:10.1145/1127777.1127835.

Zeng, H., Ellis, C.S., and Lebeck, A.R. 2005. Experiences in managing energy with ECOSystem. *IEEE Pervasive Comput.* 4(1): 62–68. doi:10.1109/MPRV.2005.10.

Zhang, Y. and Li, W. 2011. An energy-based stochastic model for wireless sensor networks. *Wireless Sens Netw.* 03(09):322–328. doi:10.4236/wsn.2011.39035.

Chapter 6

Energy Harvesting Issues in Wireless Sensor Networks

Gourav Verma, Vidushi Sharma, and Anuradha Pughat

Contents

6.1 Introduction 138
    6.1.1 Lifetime Issues in WSNs...138
    6.1.2 Storage Capacity 139
6.2 Types of Energy Harvesting Techniques 141
    6.2.1 Solar Energy Harvesting...141
        6.2.1.1 PV Cell...142
        6.2.1.2 Research Issues in Solar Energy Harvesting............................142
    6.2.2 Wind Energy Harvesting...145
        6.2.2.1 Wind Speed...146
        6.2.2.2 Wind Power...146
        6.2.2.3 Research Issues 146
    6.2.3 Thermal Energy Harvesting..147
        6.2.3.1 Thermoelectric Generator 148
        6.2.3.2 Research Challenges on Thermal Energy Harvesting...............149
    6.2.4 Energy Harvesting through Flow of Liquid 149
        6.2.4.1 Basic Principle 150
        6.2.4.2 Research Challenges in Energy through Flow of Liquid 152
    6.2.5 Acoustic Energy or Mechanical Energy Harvesting................................152
        6.2.5.1 Piezoelectric Crystal 152
        6.2.5.2 Research Challenges of Acoustic Energy Harvesting 153
    6.2.6 Ambient RF Energy Harvesting 156
        6.2.6.1 Basic Principle 156
6.3 Observations on the Reviewed Energy Harvesting Mechanisms 159
References 159

6.1 Introduction

A wireless sensor network (WSN) consists of battery-operated sensor nodes. These small sized nodes have limited, onboard battery capacity, memory, and space. The lifetime of WSNs can be improved by implementing energy efficient protocol/hardware at either the network or operating level of the sensor node. Various low-power design techniques have been proposed and implemented on the sensor node and its components, but they face limitations because any software/hardware change itself consumes energy. Thus, the idea is to introduce a technique that has the potential to generate power. Energy harvesting is a technique in which power is extracted and stored from the ambient environment, e.g., radio frequency (RF) energy, solar energy, thermal energy, wind energy, acoustic energy, etc. The extracted power is then converted to its desired form and used in the operation of the sensor nodes. This replaces manual battery recharging or replacement from the on-site nodes. The onetime additional cost of the energy harvesting circuitry at the node is compensated by saving the redeployment cost of the battery-depleted or dead sensor nodes.

Apart from the interesting features of different harvesting techniques, they have some limitations such as—the solar energy harvesting is effective only in the daytime when energy is available and the thermoelectric generator (TEG) works only when a temperature difference is present in the environment. As far as the availability of the RF energy harvesting is concerned, it can be found from easily available resources, e.g., mobile towers, TV towers, radio towers, etc. However, the path loss between the transmitter and receiver provides the least power in RF energy harvesting than the other harvesting techniques.

Efficient energy harvesting architecture, efficient energy transferring, good mechanical design for the transducers, e.g., TEG, and an efficient framework are a few of the research challenges for energy harvesting techniques in WSNs.

One needs to understand first the lifetime issues in WSN systems and storage capacity of the sensor node, before actually understanding energy harvesting.

6.1.1 Lifetime Issues in WSNs

The lifetime of any battery-operated device or WSN depends on the battery capacity and the load across which the battery is connected. The lifetime of any WSN system can be shown as follows:

$$\text{Lifetime} = \frac{\text{Battery capacity (mAhr)}}{\text{Current consumtion of load (mA)}} \tag{6.1}$$

Another important factor is the duty cycle of the WSN node on which the average current consumption depends. The duty cycle is the fraction of the time for which the node is active. On the basis of the average current consumption, the lifetime may increase or decrease. The facts of a lifetime also depend on the type of sensor node (e.g., Xbee, CC2420, Telosb, Heliomote, etc.), or software used such as media access control (MAC) layer type, routing algorithm, etc. Mostly, 99% of the lifetime of a WSN is spent in sleep mode (RF Monolithics 2010). Figure 6.1 shows the status of the sensor node for the time frame of 5 minutes. The node stays active for 500 ms, captures the data from the sensor within this period, a process in the microcontroller, and transmits the data to the network head.

So, the average current consumption using Figure 6.1 can be derived as shown in Equations 6.2 through 6.4.

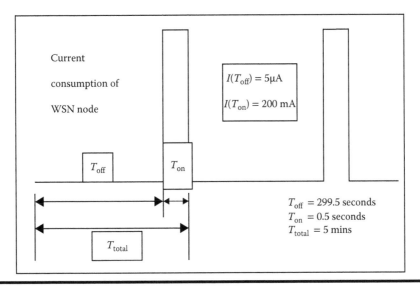

Figure 6.1 Width of periodic, active mode of sensor node.

$$I_{av} = 5 \times 10^{-6} \left(\frac{299.5}{300} \right) + 200 \times 10^{-3} \left(\frac{0.5}{300} \right) \tag{6.2}$$

$$I_{av} = 4.99 \times 10^{-6} + 334 \times 10^{-6} = 339 \, \mu A \tag{6.3}$$

The required current for 24 hours:

$$I_{av} = 339 \, \mu A \times 24 \, hr = 8.14 \, mAhr \tag{6.4}$$

6.1.2 Storage Capacity

For a Li-ion battery of 3.6 V, the required energy for 24 hours is

$$E_{\text{Total for 24 hr}} = 8.14 \, mAhr \times 3600 \, \frac{s}{hr} = 105.5 \, J \tag{6.5}$$

So, the energy required to be accumulated is 105.5 J. This is an important aspect, which provides us the target for a design specification of any kind of energy harvester circuit. For example, if someone is designing a solar energy harvester, then the time for the dark period can be assumed and the required energy can be calculated. This estimation can further help in choosing the specific solar panel, voltage booster, etc. From Equations 6.4 and 6.5, we come to know that the battery capacity for a day must be a minimum of 8.14 mAhr for satisfactory operation. Assuming the number of days for which the harvester is unable to harvest energy from the source (i.e., dark days in case of solar energy harvesting) to be 15 days, the battery capacity should be

$$\text{Battery capacity} = 8.14 \, \frac{mAhr}{day} \times 15 \, day = 122.1 \, mAhr \tag{6.6}$$

Thus, the node needs 122.1 mAhr battery capacity for 15 days of work. The designing of any harvester circuit mainly depends on these predictable equations. Figure 6.2 depicts the 1-day average power consumption of a sensor node, microcontroller, and transceiver.

The WSN node does not operate all the time due to which a lot of power is saved; hence, duty cycle is an important parameter of energy requirement for which the energy harvesting is to be done. Figures 6.3 and 6.4 show the current consumptions for different possible duty cycles (2%, 10%, 25%, and 50%) for an Xbee sensor mote and transceiver CC2420, respectively.

The $I_{\_avg}$, $I_{\_sleep}$, $I_{\_awake}$, and $I_{\_node}$ represent the average current consumption of node, current consumption in sleep state, transition current consumption from sleep to awake state, and current consumption during transmission and reception, respectively. This results show that the CC2420 outperforms Xbee.

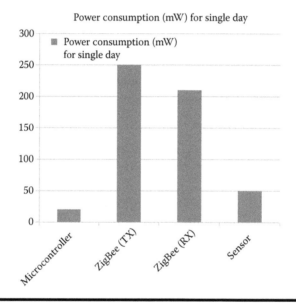

Figure 6.2 Average power consumption of WSN node

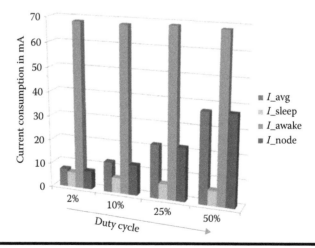

Figure 6.3 Xbee current consumption profile to size up the harvester circuit.

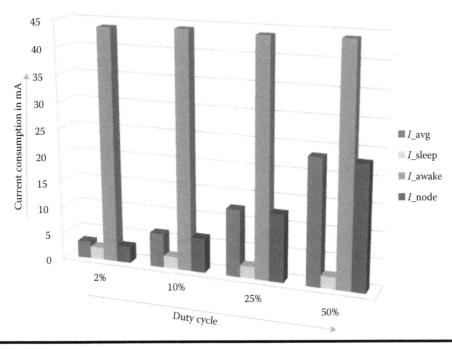

Figure 6.4   CC2420 current consumption profile to size up the harvester circuit.

6.2 Types of Energy Harvesting Techniques

Researchers have been developing various promising energy harvesting techniques such as solar energy harvesting, wind energy harvesting, thermal energy harvesting, RF energy harvesting, and acoustic or mechanical energy harvesting. In WSNs, this energy is captured and stored in an electrical form, which is then used for the operation of sensor nodes. This requires rechargeable batteries and additional circuitry to harvest energy. Next, we discuss different aspects of various harvesting techniques and their research challenges.

6.2.1 Solar Energy Harvesting

The solar energy is harvested from solar light with the use of a photovoltaic (PV) cell. The PV cell/ solar cell works on the PV effect in which light energy is converted to electrical energy. Today, we have available solar cells, which work with a maximum efficiency of 21%. The wavelength of solar radiation, which could be converted into electrical energy using a PV cell is 380–750 nm. The radiation at the surface of the earth depends on a number of atmospheric parameters such as the presence of ozone ions, earth atmosphere, the distance of the earth from the sun, longitude and latitude, time and date, cloudy or clear weather conditions, etc. In the sun's radiation, only 7% of the visible spectrum is available, while in ultraviolet (UV) 46% (with wavelength = 10–380 nm) and in infrared (IR) 47% (with a wavelength 750–1,000,000 nm) are available. The maximum and minimum energy densities of the visible spectrum are 1413 and 1321 W/m^2, respectively. Actually, the energy density of the visible spectrum, which could be converted into the electrical energy, is 1120 W/m^2.

6.2.1.1 PV Cell

A solar panel of indium, gallium, and nitrogen can convert virtually the spectrum of light into electrical energy. Biohybrid cells can provide the best efficiency among all but they are under research. Table 6.1 provides a brief comparative chart of all the solar cells. Among monocrystalline, polycrystalline, amorphous, cadmium telluride (CdTe), or copper indium (gallium) selenide (CIS/CIGS) PV cell technology, the monocrystalline provides the maximum efficiency. The amorphous silicon, CdTe, and CIS/CIGS are classified under "thin-film solar cells." The voltage across the PV cell is directly related to the solar light intensity I_{dn} as

$$V'_{OC} = \frac{nKT}{q} \ln\left(\frac{I_{sc}I_{dn}}{I_O}\right) \tag{6.7}$$

where I_{sc} represents the short circuit current, the constant $n = 1$ for Ge and 2 for Si, the Boltzmann constant $(K) = 1.38 \times 10^{-23}\,\mathrm{m^2 kg\,s^{-2}K^{-1}}$, electronic charge $(q) = 1.6 \times 10^{-19}\mathrm{C}$, open circuit voltage of PV cell (V_{OC}), and temperature (T) in Kelvin. Figure 6.5 depicts a comparison between the characteristics of the real and ideal PV cell. The real cell includes the effects of R_s and R_{sh} on the overall performance of the cell. The resistive parameters $(R_s$ and $R_{sh})$ depend on the type of solar cell. The open circuit voltage (V_{oc}) and short circuit current (I_{sc}) along with form factor give the smaller difference between two cells.

6.2.1.2 Research Issues in Solar Energy Harvesting

Raghunathan et al. (2005) indicated the important considerations needed for the designing of a solar energy harvesting system. They designed and implemented a solar power energy harvesting device and analyzed the device using the Heliomote. The features of the solar cell and its utilization for WSNs as an energy harvesting component were presented. The energy harvesting circuit was used to manage the power received from the solar panel, maintain the record of battery level, and accumulation of current in energy storage. The efficiency achieved by the complete assembly is 80%–84%. Three design matrices have been analyzed:

Table 6.1 Comparison of Efficiency and Area for 1 kW Power for Different Available Solar Cell

Parameters	Monocrystalline	Poly Crystalline	Amorphous	CdTe	CIS/CIGS
Maximum typical module efficiency (%)	21	16	8	11	12
Best research cell (%)	25	20.4	13.4	18.7	20.4
Area required for 1 kW power (m²)	9–8	7	16	12	11

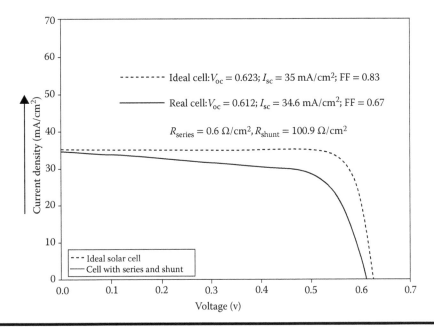

Figure 6.5 **Voltage and current characteristics of ideal and real solar cells.**

- Solar cell characteristics—The behavior of a PV cell is like a current source and due to the dependence of supply voltage over time varying load impedance, it is hard to directly power the target application. The short circuit current (I_{SC}) is directly proportional to solar radiation. Solar radiation increases I_{SC} and if solar radiation decreases, I_{SC} decreases, but the open circuit voltage remains constant. After analysis of solar cell characteristics, it is important to use some storage element like a battery to provide stable voltage.
- Energy storage technologies—Batteries and supercapacitors can be used as energy storage devices. Both have their own pros and cons. The energy density is higher in the battery, while power density is higher in supercapacitors. Supercapacitors have the problem of leakage due to parasitic paths in the external circuits. The batteries can be Ni–Cd, sealed lead acid (SLA), lithium-ion-based (Li$^+$), nickel–metal hydride, etc.
- Harvesting circuit design—The radiation intensity changes all through the day so it is required to deploy maximum power point tracking (MPPT) in the system, to track maximum radiation. The circuit has a DC–DC converter to obtain the constant output voltage. Presently, a buck–boost converter is commonly used for these purposes.

The output power of a PV cell also depends on the connection with energy storage. If there is some stored energy in the battery, then there is some voltage level, e.g., V_{BAT}, and this battery level forces the open circuit voltage of a PV cell to lower than the optimal value. In Park and Chou (2004), a switch matrix was used to power individual components of the system using a solar panel or the battery. To design an energy harvesting system, it is important to find the current status of available energy and estimate the future consumption of power (Kansal et al. 2006; Moser et al. 2007). In Ward et al. (2006) and Stanley-Marbell and Marculescu (2007), PV harvesters were used for ultralow-power devices designed for wearable devices. They did not use the MPPT in such harvesting device. In Simjee and Chou (2006), the author designed an MPPT system along with a supercapacitor. The system included the tracking algorithm, which controlled the power converter using pulse width modulation (PWM). The MPPT-based system required a microcontroller and some other changes in the firmware. The improvement in this work was done by Park et al. (2006). They designed an algorithm, which was more optimized and less power consuming than their previously designed system (Simjee and Chou 2006).

Brunelli et al. (2008) designed a maximal power point (MPP) system for solar energy. The system consumed less than 10 mW power. Figure 6.6 shows the three major blocks: the MPP regulator, MPP tracker, and MPP power supply. The MPP regulator generally contains a buck–boost converter circuit, which is used to provide sufficient output voltage. The MPP tracker uses a small solar cell, namely the *Pilot Cell*, to track the solar radiation present in the environment and control the output of the comparator. Another power supply unit is required to provide power to the comparator. Ultralow-power devices have been used so that the power consumption of the device is considerably reduced. For verification, the author powered a Tmote Sky sensor node using the designed system. One 50 F supercapacitor was used to power the mote and as an energy storage device. The amount of energy was 693 J by means of which the sensor node continuously worked for 64 seconds. Moreover, the paper showed how the solar-energy efficiency was optimized under nonstationary light conditions.

In González et al. (2012), the author implemented the solar energy harvesting system on an open wireless sensor (Wi-Se) node. The Advanced RISC Machine (ARM) cortex was used for the purpose of processing the data. Xbee and CC2420 were analyzed and used with the gas detection sensors. In this work, a solar energy harvesting system was designed. The authors investigated sources of available designs on different platforms such as Tmote sky and Fox board, UC-Berkeley, SENTIO, XYZ, etc. The authors fabricated the system and maintained the duty cycle to enhance the lifetime of the system. It can be concluded from the experimental results that using different sleep and wake up strategies, the lifetime of the WSN can be increased. A system was designed, which used a fluorescent lamp of 34 W as an energy source (Hande et al. 2006). The power management circuit was employed to manage the power and store the energy in ultracapacitors. The author analyzed the system on the basis of its current consumption and this showed that a large amount of energy was wasted in the OFF state.

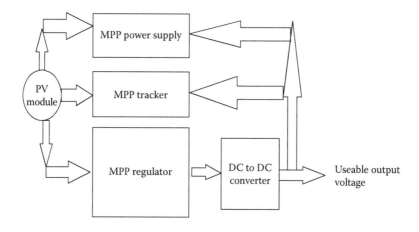

FIGURE 6.6 Solar energy harvesting system architecture.

Carvalho and Paulino (2013) provided an indoor solution for energy harvesting by means of a solar cell. The authors analyzed in detail about indoor light availability. Experimentally, it was shown that a power density of 0.13 W/m^2 was achieved in the worst case when the irradiance was 15.6 lux and 0.7 W/m^2 was achieved when an irradiance of 84 lux was available. After analyzing different solar panels, it was inferred that amorphous silicon provided better power than others to harvest energy indoors.

In Abbas et al. (2014), the author discussed the models of energy harvesting. Three types of energy harvesting models were discussed: first was the ambient energy model, the second a two-storage device-based model, and the last one an ambient energy and two-storage device model. The first model received its energy from the surroundings and stored it. The node started working when the threshold was achieved. When this value became lower than the threshold value, the node stopped working and started to accumulate energy again. The disadvantage of this model was that the node had to stop its working while accumulating energy from the surroundings. This model is described in Seah et al. (2009).

Another work showed a model with an implemented storage battery and ultracapacitor (Jiang et al. 2005). First, the supercapacitor delivered the energy and then the battery started to deliver the energy. In the meanwhile, the supercapacitor started to charge. Their improved model is presented in Saggini et al. (2010). The authors utilized the facility of DC–DC converter and power manager to implement the empowered version of above two models.

Solar energy harvesting is the best choice as a power management mechanism as it is a reliable power source among other kind of energy sources such as wind, thermal, vibration, etc. (Khan et al. 2015). In Gao et al. (2009), the authors adapted a low-power design to save energy and solar energy was utilized by the device to improve the lifetime of the system. In Naveen and Manjunath (2011), the system was implemented and located in a building. The designed system is a low-power solar energy harvester. The primary energy storage, in this case, was an ultracapacitor, while the secondary was an alkaline battery. Researchers were also able to achieve low charging time using small capacitors (Chuang et al. 2014).

A hybrid storage solar energy harvesting system was also developed (Xiang and Pasricha 2015; Taneja et al. 2008; Alberola et al. 2008) using a combination of a supercapacitor and battery. The supercapacitor charged the battery when the voltage dropped to a threshold value and the microprocessor was programmed to control the charging of the battery. The performance analysis of the hybrid systems was carried out in Ecker et al. (2012), Takahashi et al. (2004), and Liu et al. (2009). It was proved that the capacity of the Li-battery may discharge when it is in a high self-state of charge (SOC) for a long duration. Researchers have improved the system that recharges the battery when the voltage level is lower than the threshold value (Jiang et al. 2005; Li et al. 2010). The hybrid system has been made architecturally intelligent (Li and Shi 2015) to check on the threshold level and trigger the charging.

The solar energy harvesting system has yet to achieve a milestone in terms of reducing the size to improve their applicability in sensor networks. Further research and improvement are required to manage harvesting with intermittent supply of solar energy.

6.2.2 Wind Energy Harvesting

Wind is one of the green energy sources and also an erratic one. An uneven heating condition in the atmosphere is responsible for the wind. Wind energy is represented in terms of wind speed and wind power.

6.2.2.1 Wind Speed

The rate of flow of air through a point above the earth's surface is known as wind speed. It is varying in nature and depends on many environmental parameters such as temperature (T) and pressure (P).

6.2.2.2 Wind Power

The wind power is a function of the cube of air velocity (v in m/s):

$$\text{Wind power} = F(v^3) \tag{6.8}$$

This relationship shows that the wind power directly relates to cube of wind velocity (v). The wind power can also be termed as

$$w = \frac{1}{2} pAv^3 \tag{6.9}$$

In this equation, p is air density and A is the area of cross section swept out by the wind turbine. The turbine is used for large-scale energy generation and the circular shaped rotors in turbine have the area $(A) = \pi r^2$, where r is the radius of the turbine. The air density (p) is a factor of elevation, temperature, and weather fronts:

$$p = \frac{1.325 \times P}{T} \tag{6.10}$$

where P is the atmospheric pressure in inches of mercury adjusted for elevation and T is the temperature in Fahrenheit + 459.69. Now, the wind power density (W/m²) can also be formulated as

$$w = (0.652v^3) \frac{W}{m^2} \tag{6.11}$$

So, the area of the turbine plays an important role when designing wind energy-based harvesting system.

6.2.2.3 Research Issues

- The large-scale wind energy harvesting technique has already captured a considerable area among other natural renewable energy sources. A study of the wind in an urban environment has demonstrated that the wind speed is below 5 m/s most of the time. Today, researchers are attracted to wind energy harvesting for small battery-operated devices such as WSNs. The choice of the storage technology is a great challenge for the designing of wind energy-based harvesting systems.
- Research on the design of a fast charging circuit, the maximum peak power tracking, and use of the super capacitors as storage are required.
- The efficiency of a wind generator is the ratio of maximum output power to the power of wind. Research on the techniques to improve energy efficiency is required.

Table 6.2 Comparative Study of Different Wind Energy Systems

Turbine Characteristics	Turbine Radius (cm)	Maximum Power (mW)	Wind Speed (m/s)	Efficiency (%)
3-Blade (KMAC 2012)	16	200	5.4	2.5
Piezoelectric windmill (ISHM 2012)	6.5	5	4.5	0.7
Vertical-axis wind turbine (VAWT)six blades (Rice et al. 2010)	6	45	5	0.5
Wind turbine (Hankuk 2012)	3	24	4.5	14.9
Wind turbine with maximum power point tracking (MPPT) (Hankuk 2012)	3	7.86	3.62	9.4
Horizontal-axis wind turbine (HAWT) (Tan and Panda 2011)	7.5	243	5	16.3
VAWT (Tan and Panda 2011)	19	83	5	8.2

Harris et al. (2012) developed a wireless network in which the airflow energy harvester was used. The network consists of 24 nodes to work at least for 1 year. They installed three microwind turbines to verify its actual power-charging capability (Park et al. 2012, D'Souza et al. 2007). They tested it for a wind speed of 7 m/s, which recovers the power loss within 33 minutes. A node without a harvesting system drains its energy by 0.1 V. For structural health monitoring, microwind turbines have received good attraction (Park et al. 2012). Many analysts are targeting the design of the microwind turbine for WSN applications (Xu et al. 2010; Federspiel and Chen 2003; Priya et al. 2005; Rancourt et al. 2007). In Federspiel and Chen (2003), a maximum power of 8 mW with a power density of 0.1 mW/cm^2 was achieved using four number of blades, 10 cm rotor diameter, and 2.5 m/s airspeed. The 12 number of blades of 10.2 cm rotor diameter and 4.4 m/s wind speed gave a maximum power of 5 nW (Priya et al. 2005). A very good amount of power, i.e., 130 mW, was achieved in Rancourt et al. (2007) using three number of the blade, 4.2 cm rotor tip, and 11.8 m/s air speed. The power density, in this case, is 9.38 mW/cm^2. Table 6.2 gives a comparative study of different works associated with the wind energy harvesting techniques.

An implementation of the wind system to power the Wi-Se node for remote sensing was done in Azevedo and Mendonça (2015). This gave 10%–30% efficiency for turbines with a radius of 5–15 cm. This work concluded that the horizontal-axis turbines were more power efficient than the vertical-axis turbines.

Similarly, another work (Zhang et al. 2015) presented a wireless temperature node that used the wind energy harvester. The nodes communicated the temperature to neighboring nodes. The results showed that the output power was approximately 1.59 mW and resistor value of 20 kΩ, for a wind speed of 11.2 m/s. The node temperature was updated for transmission with an interval of 13 seconds and the total energy consumption was found to be about 0.3566 mJ for an operation.

6.2.3 Thermal Energy Harvesting

The technique works on the thermoelectric effect/Seebeck effect in which a temperature difference $(T_{hot}-T_{cold})$ is created between two thermoelectric semiconductors. The "n"-type and "p"-type semiconductors are electrically connected in series through a conducting plate on the top (hot plate) and bottom (cold plate). Electrons and holes are accumulated in the cold side and an electric field starts to develop due to the concentration gradient or temperature difference.

6.2.3.1 Thermoelectric Generator

The output power of a TEG can be calculated using the product of potential difference and resultant current flow from the circuit. The n-type and p-type semiconductors have high electric conductivity (σ) and low thermal conductivity (κ). The low thermal conductivity produces a temperature gradient between the hot and cold sides. Figure 6.7 shows the working of the TEG.

The power generated in the thermoelectric converter can be calculated using

$$P = Q \times A \tag{6.12}$$

where A = area (in m²) and Q = radiation flux. The efficiency of the converter and figure of merit (Z) depends on the type of material:

$$\eta_{max} = \frac{T_H - T_C}{T_H} \times \frac{\sqrt{1 + Z \times T} - 1}{\sqrt{1 + Z \times T} + T_H / T_C} \tag{6.13}$$

$$Z = \frac{\sigma S^2}{k} \tag{6.14}$$

where Z = figure of merit, S = Seebeck coefficient, k = thermal conductivity, σ = electrical conductivity, T_H = hot-side temperature, T_C = cold-side temperature, and T = 27°C or 300 K. The combination of materials such as bismuth telluride (Bi_2Te_3), lead telluride (PbTe), etc., is used to increase the converter efficiency.

Many designs have been analyzed for thermo-electric generator energy harvesting (TEG-EH) in Hudak and Amatucci (2008) and Carmo et al. (2010). Micropelt (Bottner 2005) and Sieko thermic wristwatch (Kishi et al. 1999) are the commercialized products available in the market based on the TEG-EH. In Venkatasubramanian et al. (2001), telleride super lattice

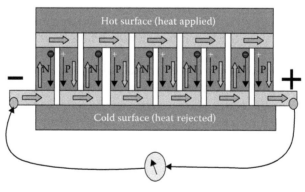

Thermoelectric generator (TEG)

Figure 6.7   Thermoelectric generator internal diagram.

thermocouples were used in design with 2% efficiency at $\Delta T = 20°C$, which is close to room temperature. In Bailly et al. (2008) and Samson et al. (2010, 2011), the authors used a heat storage unit (HSU) to generate temperature variation. For structural health monitoring of aircraft, phase change material (PCM) was used (Bailly et al. 2008), which can also be used as an HSU. Using an HSU, 13 J of energy is generated. In Samson et al. (2011), 27 J of energy was captured at a temperature sweep of +20°C to –20°C. In Kiziroglou et al. (2011), an analysis was done on the geometry of HSUs and the conductivity (k) of TEGs. It was shown that for the given parameters of TEG (device dimension and temperature), the output energy can be maximized by controlling the conductivity.

In Sukesha and Kaur (2014), the control of room temperature and appliances such as AC, cooler, and heater were powered by an artificial lighting-based energy harvester. They used the MPPT algorithm to get the power from the PV panel. A power management unit controls the use of harvested energy. The illuminance and irradiance of the environment depend on the weather conditions. Therefore, the maximum consumption occurs during the afternoon and between 9 AM and 5 PM the consumption is less. The experimental work was done and found a minimum of 0.6°C ΔT_{TEG} across the thermoelectric module, which generates 23 mV of input voltage to activate the harvester in approximately 3 minutes (Mullen et al. 2015). Thus, we can say that the energy harvester can provide a current up to 30 mA at a voltage of 2.2 V in 3 ms pulses for over a second. This is sufficient to power the sensor components and communication system. A large number of the pellets (nearly 2300 couples) in Wang et al. (2009) were used to find the solution for the low voltage generated due to limited cooling. In Su et al. (2010), nearly 1000 couples were used in order to activate a pulse oximeter. The size of the device was of the order of a wrist watch. Research also implemented 254 couples that produced 85 mV at $\Delta T = 8.7°C$ (Settaluri et al. 2011). In Roy et al. (2013), the author studied the mathematical relation between the geometry of thermoelectric pellets and the output power received. According to the researchers, after deriving the relation and extensive experimentation (Moser et al. 2012), it was observed that $\Delta T = 1.2°C$ was required for data transmission.

A buck–boost converter has been used to handle micropower (Ramadass and Chandrakasan 2011). A boost topology-based DC–DC converter was also used by the author to boost the voltage level of 35 mV to a level on which the sensor node can work properly.

6.2.3.2 Research Challenges on Thermal Energy Harvesting

- The main challenge when designing a TEG-EH is to create the temperature gradient (ΔT).
- The essential part of thermoelectric energy harvesting techniques is a substantial temperature difference ΔT, which is not available in the applications, e.g., heat engine structural monitoring, etc. (Kiziroglou et al. 2013).
- Instead of artificial sources, piezoelectric patches can be used to harvest energy in future sensor networks.

6.2.4 Energy Harvesting through Flow of Liquid

Any matter in the form of liquid (e.g., water, oil, sea water, etc.) containing movement with some velocity, pressure, or force can be utilized to generate electrical energy. The energy of flow of a liquid can rotate the dynamo or can provide pressure over a piezoelectric crystal. Generating energy using a dynamo is a good technique provided that the mechanical structure is designed carefully so that the turbines run properly.

6.2.4.1 Basic Principle

A moving conductor in a magnetic field induces a voltage in the conductor and it generates current when connected to the load. The generator has a prime mover, which moves with the help of renewable energy of resources. The energy due to the flow of liquid is considered for harvesting and converting to electrical energy. The average flux density and the current through each conductor are

$$B_{av} = \frac{\phi p}{\pi DL} \tag{6.15}$$

$$\text{Current flowing through each conductor} = \frac{I}{A} \tag{6.16}$$

where I = current produced in the conductor, A = number of parallel paths, z = number of conductors, and N = speed of the rotating part. If D and L are the rotor diameter and the length of the machine in meters, respectively, then the torque on a single conductor and the total torque are represented as

$$\text{Torque on a single conductor} = B_{av} \times \frac{I}{A} \times L \times \frac{D}{2} \tag{6.17}$$

$$\text{Total torque}, T = z \times B_{av} \times \frac{I}{A} \times L \times \frac{D}{2} \tag{6.18}$$

Putting the value of B_{av}, we get motor torque (Equation 6.19) and generator current (Equation 6.20).

$$T = \frac{pz\phi I}{2\pi A} \text{ (for motor condition)} \tag{6.19}$$

$$I = \frac{2\pi A}{pz\phi} \times T \text{ (for generator condition)} \tag{6.20}$$

The direction of the electromagnetic torque, T, will be along the direction of rotation. It is opposite to the direction of rotation in case of generator operation.

Let us assume an imaginary circular surface from which the water is passed due to which the turbine on the rotor side starts moving with an unbalanced force, which can be in any direction. The unbalanced force is applied on the conductor due to the tides of the sea; therefore, an unbalanced torque is also applied on the rotating part which is

$$T_{net} = \frac{dL}{dt} = \frac{d(I_m \omega)}{dt} \tag{6.21}$$

where L = angular momentum, I_m = moment of inertia, and ω = angular velocity. Putting the value of $\omega = \dfrac{v\,(\text{linear velocity})}{r\,(\text{radius})}$ in Equation 6.21, we get

$$T_{\text{net}} = I_m \times \frac{d(v/r)}{dt} = \frac{I_m}{r} \times \frac{dv}{dt} \tag{6.22}$$

The velocity in Equation 6.22 can be calculated by Bernoulli's pressure equation. The total pressure combines the static, dynamic, and hydrostatic pressures. Then, the Bernoulli equation states that the total pressure along a streamline is constant. The stagnation pressure results due to the sum of the static and dynamic pressures. It is expressed as

$$P_{\text{stag}} = P + \frac{V^2}{2} \tag{6.23}$$

By using Bernoulli's equation, we can calculate the velocity of the water at any pressure, which is given as

$$V = \frac{\sqrt{2(P_s - P)}}{\rho} \tag{6.24}$$

where P_s = stagnation pressure, P = static pressure, and ρ = density of water.

The average flux density is $B_{\text{av}} = \dfrac{\Phi}{(\pi D / p)L}$. Since the induced voltage in a single conductor is $(B_{\text{av}} \times Lv)$ and the number of conductors in each parallel path is $\left(\frac{Z}{A}\right)$, if v is the tangential velocity, then

$$v = \pi DN \tag{6.25}$$

$$\text{Total voltages appearing across the brushes} = \frac{Z}{A} \times B_{\text{av}} \times Lv \tag{6.26}$$

Thus, the voltage induced across the conductor, $V_c = \dfrac{pZ}{A} \times \Phi N$ \qquad (6.27)

As we know that the power $P = V_c \times I$,

$$P = \frac{pZ}{A} \times \Phi N \times \frac{2\pi A}{pz\phi} \times \frac{I_m}{r} \times \frac{d(\sqrt{2(P_s - P)}/\rho)}{dt}$$

$$P = 2\pi N \times \frac{I_m}{r} \times \frac{d(\sqrt{2(Ps - P)}/\rho)}{dt} \tag{6.28}$$

Equation 6.28 tells us that the output power depends on some factors such as speed, moment of inertia, and the radius of the prime mover. It also depends on the static, dynamic pressure, and density of the fluid.

6.2.4.2 Research Challenges in Energy through Flow of Liquid

Researchers are trying to harvest energy from fluids using different kind of methods, e.g., by enhancing the impeller structures, using hydraulic pressure and vortex shedding, and utilizing the flapping motion. Vortex shedding can produce the kinetic energy in the liquid and this kinetic energy could be harvested using the piezoelectric materials (Koyvanich et al. 2015). The output power of 0.18 μW can be achieved with 4.4% of efficiency and at the liquid velocity of 6.8 m/s.

Some energy has been harvested with simple water droplets of 4 mm diameter and very low wind flow velocity of 0.5 m/s (Hoffmann et al. 2013). The kinetic energy of the droplet has been converted into electrical energy. The amount of voltage from energy harvesting circuit is 0.7–1 V AC. The basic principle used for energy harvesting is electromagnetic induction and cantilever.

In a small scale, the energy can be harvested from domestic pipelines (Lee et al. 2015). With a flow rate of 20 L/min, 720 mW is harvested through a domestic pipeline and 2 mW of power is harvested with 3 L/min flow of water. Still, there is scope for research in optimization of the mechanical structure of the impeller and design of the magnetic circuit to enhance the output power with lower flow rate.

By means of a cantilever beam placed in a converging–diverging flow channel, the vibrational energy of the fluid can be utilized (Morarka and Ghaisas 2016). The piezoelectric energy harvesting method can be employed for harvesting energy of fluid vibration. At 20 L/min, 20 mW of energy can be harvested using the model (Morarka and Ghaisas 2016).

6.2.5 Acoustic Energy or Mechanical Energy Harvesting

The acoustic energy/mechanical energy is captured from the vibration or mechanical waves in gases, liquids, and solids including sound, ultrasound, and infrasound. The input stress/vibration power can be generated by machines, humans, or nature. Whenever pressure is applied to the transducer or a pressure sensor, it starts to flow the electron and thus an electrical energy produced in terms of electric current across a load resistor. This current can be calculated by Thevenin's equivalent circuit.

6.2.5.1 Piezoelectric Crystal

A piezoelectric generator can harvest the electrical energy to charge the Wi-Se node. Figure 6.8 shows the piezoelectric effect in which a force is applied over piezoelectric material and the voltage starts to develop across the load resistance.

The input pressure applied on a surface is $P = \dfrac{F}{A} = \dfrac{1}{2}|F||U|$ and the output current produced is $I_l = \dfrac{V_{th}}{Z_{th} + R_L}$, where F = force exerted, which is equal to the pressure, U = displacement in the material, and A = piezoelectric area, and the root mean square (RMS) power across the load is obtained as

$$P_L = \frac{1}{2}|I_L|^2 R_L \tag{6.29}$$

The power transfer efficiency of the piezoelectric material can be calculated in terms of current across the load, the force applied on that material, and the displacement in the material due to the applied force:

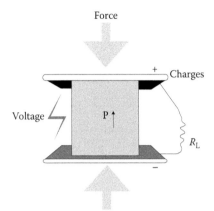

Figure 6.8 **Pressure converted into electrical energy.**

$$\eta = \frac{P_L}{P_{in}} = \frac{|I_L|^2 R_L}{|F||U|} \tag{6.30}$$

Therefore, the efficiency of harvested energy can be improved considering these parameters.

6.2.5.2 Research Challenges of Acoustic Energy Harvesting

The parameter needed to find the input physical value in the case of acoustic energy is available as sound pressure level (SPL) and frequency range. The SPL and frequency range for different sources have different values. These parameters for different ambient sources are shown in Table 6.3.

On the basis of the piezoelectricity effect, most of the acoustic system has been developed (Anton and Sodano 2007). The produced charge is directly proportional to applied compression (Cady 1964). For an application like aeroacoustic, an acoustic energy harvester (AEH) was developed Horowitz et al. (2006). A power of 0.34 μW/cm² was achieved at 149 dB of SPL in (Horowitz et al. (2006). The schematic of AEH is presented in Figure 6.9. The proposed idea depends on the conventional Helmholtz resonator (HR) as a tuned, second-order resonator for pressure amplification.

In most of the techniques, an electromechanical Helmholtz resonator (EMHR) is designed as a resonator. The sol-gel fabrication technique is used to fabricate piezoelectric membrane (Shigeki et al. 2010). Using this change, 11 pW at 100 dB of SPL is achieved at 24.02 kHz resonant frequency. The sol-gel lead zirconate titanate (PZT) process is a method for spin-on deposition and crystallization of thin-film PZT (0.53/0.47). In Kimura et al. (2011), a sol-gel PZT processed membrane was used and 140 pW of power was obtained at 100 dB of SPL with 16.7 kHz resonant frequency.

Piezoelectric material and sonic crystals have been used in Wu et al. (2009) to design the AEH. At a resonant frequency of 4.2 KHz, 3.9 kΩ load resistance, and 45 dB of SPL, a 40 nW of power was achieved. The EMHR was also used in designing of an AEH (Liu et al. 2008). The author achieved 30 mW of power reported at 161 dB of SPL. In Tomioka et al. (2011), the piezoelectric membrane was implemented with dual top electrodes. This arrangement utilized the maximum polarization of charge, which developed over the PZT diaphragm surface. With this arrangement, the author reported 42.5 pW of power at 4.92 KHz of resonance frequency with 100 dB of SPL. The output power of 42.5 pW with 30 Ω load resistance and 52.8 pW with 25 Ω load resistance was achieved at 100 dB of SPL and at 4.92 kHz resonant frequency, respectively

Table 6.3 Sources of Acoustic Energy with Their Sound Pressure Levels (SPLs) and Frequency Ranges

Source	SPL (dB)	Frequency Range
Surrounding of aircraft (Issarayangyun et al. 2005)	80	20 Hz–20 kHz
In vehicle (automobile) (Jung et al. 2009)	70–90	1–100 Hz
Turbo fan engines (Horowitz et al. 2006; Taylor et al. 2004)	150	20 Hz–20 kHz
Automobile without engine muffler (Mieszkowski 2004)	194	1–20 Hz
Car air conditioning system (Yokoyama and Hashimoto 2010)	• 71.9 (supply duct) • 65.1 (return duct) • 47.7 (exhaust duct)	20 Hz–20 kHz

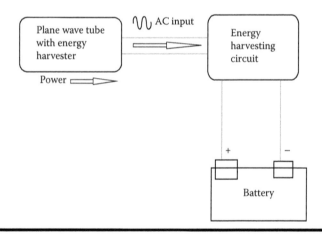

Figure 6.9 Acoustic energy harvester schematic.

(Iizumi et al. 2011). With some improved design in the same technology, 82.8 pW with 50 Ω load was achieved at 100 dB of SPL and 4.92 kHz resonant frequency (Atrah and Salleh 2013).

In Khan and Izhar (2013), a scavenging acoustic energy device was designed. Different SPLs were used for harvesting. At 120 dB SPL, the maximum power delivered to the load was 1503.4 μW/cm^2. There is a discussion about clamped structures and the authors experienced increased energy density with an increase in force on the piezoelectric (Sherrit 2008). This increased the power density too. The multilayer structure decreased the source impedance for inductive tuning.

Hassan et al. (2014) conducted experimentations on a piezoelectric generator (i.e., PZT-5A cantilever type). This extracted the sound energy from the loudspeakers placed at various locations and converted the sound energy into electrical energy. A piezoelectric generator generated the maximum voltage at its resonant frequency. This resonance frequency had to be operating near the sound frequency. A maximum voltage of 26.7 mV$_{rms}$ was generated at 62 Hz resonance frequency with the sound intensity of 78.6 dB and 1 cm distance.

In Chen et al. (2015), an acoustic microenergy harvester (AMEH) was modeled using lumped element modeling (LEM) and an Euler-Bernoulli beam (EBB). This converted the wasted acoustical energy into useful electrical energy. The mathematical modeling was validated for feasibility check and performance was optimized with the bandwidth tuning. This way, the harvester generated 0.9 V/(m/s^2) and 1.79 µW/(m^2/s^4) at 60 Hz and 400 kΩ resistive load. However, a variance of about 7% was found from the validity test, when the results were compared with the theoretical data. An averaged output power of 2.2 mW was obtained that can operate the sensor node components and communication peripherals.

In Ostaseviciusa et al. (2015), the authors proposed an energy harvesting scheme for low-power sensor nodes that could monitor the tool's conditions. This showed the increase in the vibration amplitudes. Edvinsson (2012) proposed two types of energy harvesters. One used *piezoelectricity* for absorbing vibrations and to convert these vibrations into electricity. The second harvester used a *thermoelectric effect* to convert heat flow to electricity. Then, the performance was tested for the off-the-shelf components. The thermoelectric microgenerator produced a power of 2.7 mW at 20°C temperature and 10 Ω load. The other microgenerator, i.e., piezoelectric type, produced a power of 2.3 mW at 56.1 Hz and 565 Ω load. Therefore, we can say that the power density of these generators ranged between 2 and 3 W/m^2. A comparative table of the above discussion is provided in Table 6.4.

Table 6.4 Comparative Analysis of Different Technologies

Technology	SPL (dB)	Resonance Frequency (kHz)	Output
EMHR—second order Device 1 (Horowitz et al. 2006)	149	–	0.34 µW/cm^2
EMHR—second order Device 2 (Horowitz et al. 2006)	149	–	252 µW/cm^2
Sol-gel PZT (Shigeki et al. 2010)	100	24.02	11 pW/cm^2
Sol-gel PZT processed membrane (Shigeki et al. 2011)	100	16.7	140 pW/cm^2
Load resistance = 3.9 kΩ (Wu et al. 2009)	45	4.2	40 nW/cm^2
EMHR—piezoelectric membrane is implemented (Liu et al. 2008)	161	4.2	30 mW/cm^2
a. Load resistance = 30 Ω b. Load resistance = 25 Ω c. Load resistance = 50 Ω (Tomioka et al. 2011; Iizumi et al. 2011; Atrah and Salleh 2013)	100	4.92	42.5 pW/cm^2 52.8 pW/cm^2 82.8 pW/cm^2
a. Increasing frequency sweep b. Decreasing frequency sweep (Khan and Izhar 2013)	120	–	1503.4 µW/cm^2 191.4 µW/cm^2

MHR, electromechanical Helmholtz resonator; PZT, lead zirconate titanate.

6.2.6 Ambient RF Energy Harvesting

6.2.6.1 Basic Principle

RF energy harvesting converts RF waves into DC power using a rectifying antenna, which is a combination of antenna and rectifying circuit. The amount of harvesting power depends on the incident power density of the location, efficiency of power conversion, and size constraints of the energy collection device, i.e., antenna or an induction coil.

The RF radiation is normally quantified in terms of electric field strength E, which is measured in V/m. In the far-field region, the electric field strength can be converted into incident power density using

$$P_i = \frac{E^2}{Z_0} \tag{6.31}$$

where P_i is the incident power density (in W/m²) and Z_0 is the characteristic impedance of free space (approximated to 377 Ω). The incident power density is also a function of the distance between the source and the receivers. The possible power to be harvested in the far-field region can be estimated using the Friis transmission equation:

$$P_r = P_t G_t G_r \left(\frac{\lambda}{4\pi d}\right)^2 \tag{6.32}$$

where P_r is the received power (watts), P_t is the transmitted power (watts), G_t and G_r are the gain of the transmitting and receiving antenna (with respect to an isotropic antenna), λ is the wavelength of the operating frequency (meters), and d is the distance between the source and the harvester (meters). Equation 6.32 assumes there is an unobstructed line-of-sight (LOS) path between the transmitting and receiving antennas.

In the block diagram of RF energy harvesting in Figure 6.10, the power transmitter is the dedicated RF power transmitter or ambient RF source, e.g., mobile tower, TV tower, etc. The matching network is the matching circuit used for impedance matching. A voltage multiplier is used for the voltage multiplication to attain the minimum possible working potential and finally the storage device is used, which may be a supercapacitor. Researchers have considered various energy harvesting resources and have tried to explore their different facets, but there are several issues that need to be addressed.

RF energy harvesting is the power reception either from available RF power sources, e.g., cell phone tower, TV tower, DTV, Wi-Fi, etc., or by a dedicated power transmitter, which is mainly designed for RF energy transmission. When the power is received from available or ambient RF energy sources, then this is known as energy scavenging, while dedicated power transmission and reception is termed as energy harvesting. In wireless power transmission, there are RF power transmitters and power receiver circuits, which are tuned to the same frequency. As the distance increases, the power will decrease as a function of the square of the distance between the transmitter and receiver. Here, the distance is the main constraint. A lot of research work is happening in this field to develop power efficient reception at the receiving end. This has been done in many ways, either by designing highly efficient circuits or by developing very low-threshold voltage devices such as Schottky diodes or MOSFETs.

In Figure 6.11, an RF–DC converter has been shown. The RF–DC converter is usually a voltage multiplier. Figure 6.11 depicts the internal schematic of an RF–DC converter. Internal

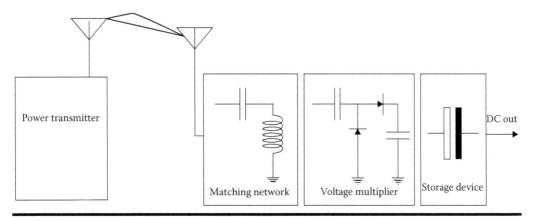

Figure 6.10 Block diagram of radio frequency (RF) energy harvester.

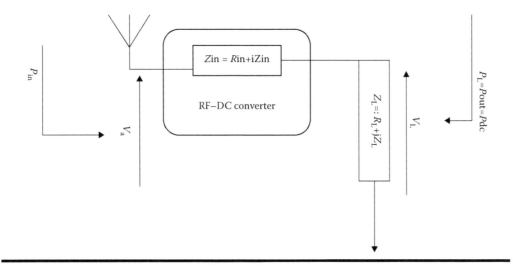

Figure 6.11 Schematic of RF–DC converter.

impedance has real and imaginary component as R_{in} and Z_{in}, respectively. Load impedance also has the components R_L and Z_L as real and imaginary components, respectively. The V_a and V_L are the input and load voltage. The P_{in} is the input power and P_L is the output power.

The power delivered to the load R_L is

$$P_L = \frac{V_L^2}{R_L} = \frac{A_V^2 V_a^2}{R_L} \tag{6.33}$$

where A_V is the voltage gain provided by the impedance transformation network. The antenna can deliver the maximum power to the impedance transformation network when the following relation is satisfied:

$$R_a = Z_{in}^* \tag{6.34}$$

where Z_{in} is the input impedance of the impedance transformation network. The superscript * denotes the complex conjugation. The current I_a can be shown as

$$I_a = \frac{V_a}{R_a + Z_{in}} = \frac{V_a}{2R_a}$$

(6.35)

Therefore, the maximum power from the antenna to the impedance transformation network is obtained:

$$P_{L\,max} = |I_a^2|\,R_a$$

(6.36)

The power efficiency of a voltage multiplier is defined as the ratio of the output power of the voltage multiplier (P_{out}) to the input power of the voltage multiplier (P_{in}):

$$\eta_V = \frac{P_{out}}{P_{in}}$$

(6.37)

The rectifiers in the voltage multiplier consume some power and this reduces the efficiency of the voltage multipliers. The power efficiency of an impedance transformation network is defined as the ratio of the power delivered to the multiplier (P_L) to the power available at the input of the impedance transformation network (P_{in}):

$$\eta = \frac{P_L}{P_{in}}$$

(6.38)

The global power efficiency of an RF power harvester is defined as the ratio of the DC output power of voltage multiplier to the incident power of the RF signal:

$$\eta = \frac{\text{DC output power}}{\text{Incident RF power}}$$

(6.39)

The incident RF power (i.e., P_{RF}) is derived from Friis equation:

$$P_{RF} = P_B G_B G_M \left(\frac{\lambda^2}{4\pi r} \right)$$

(6.40)

where P_B presents the base station power provided for antenna, G_B presents the gain of the base station antenna, G_M is the antenna gain of the passive wireless microsystem, λ represents the wavelength, and r is the distance between the passive wireless microsystem and the base station. The most effective is tropically radiated power, i.e., P_{EIRP}. It is derived in Equation 6.41.

$$P_{EIRP} = P_B G_B$$

(6.41)

Actually, P_{EIRP} is the received power, which can be converted into DC power by the voltage multiplier.

6.3 Observations on the Reviewed Energy Harvesting Mechanisms

- Power extraction for high-power devices are much easier to achieve, but energy harvesting for lower power devices is still a challenge for the researchers.
- At a small level, designing of the energy harvesting system is critical due to the size constraint of WSNs. Thus, designing energy harvesting systems that consume less power is a great challenge.
- The choice of the energy harvesting technique based on the duty cycle of an application and the available energy resource is still a research challenge for designers. Enhancing the lifetime of the WSN is a major challenge before researchers.
- There are two ways by which the lifetime of sensor network can be increased. One of the ways is by power management, by which minimum power is consumed using optimized algorithms, and another is by generating power from the available environment, which is also known as energy harvesting.
- Lifetime and performance are design goals of the Wi-Se node and these design goals are fulfilled by means of energy harvesting technology.
- In a PV effect, the solar energy is harvested from the environment. The advantage of the device is that a good amount of energy can be harvested when there is sunlight, but the major drawback is that we could not harvest energy in a dark environment.
- In mechanical energy harvesting, a piezoelectric effect is used, but for such kinds of energy harvesters, we need continuous vibration from the source. This type of harvesting could be employed where large machines are available.
- TEG needed temperature differences. Temperature difference creation is a challenge in this technology. This type of technology could be employed where a good amount of high temperature can be provided to the energy harvester arrangement. The rest of the design (cooling arrangement for temperature difference) depends on efficient heat transfer.
- Dynamic fluid energy harvesting includes the speed of the fluid, the density of the fluid, and the use of an efficient dynamo from which the fluid passes. Change of flux for the generation of electrical energy is used for magnetic energy harvesting. This kind of energy harvesting is used, where the magnetic flux changes rapidly, e.g., power transmission lines. It is near-field energy harvesting.
- Another and one of the most popular energy harvesting technique is far-field RF energy harvesting. We can transmit power over large distances wirelessly. It is the least efficient energy harvesting method, but good in every environmental situation. There is one power transmitter and a power receiver. The receiver converts the RF energy into usable electrical energy. The whole assembly is made of the antenna, matching network, and voltage multiplier. The output power directly could be fed to load (sensor node or any ultralow-power device) or could be stored in battery storage.
- Thus, we can say that energy harvesting techniques can be successfully implemented for increasing the sensor node lifetime.

References

Abbas, M.M., Tawhid, M.A., Saleem, K. et al. 2014. Solar energy harvesting and management in wireless sensor networks. *Int J Distrib Sensor Netw.* 2014: 8p. doi:10.1155/2014/436107.

Alberola, J., Pelegri, J., Lajara, R., and Perez, J.J. 2008. Solar inexhaustible power source for wireless sensor node. In *IEEE Proceeding IMTC*, Victoria, Canada, 12–15 May 2008, 657–662.

Anton, S.R. and Sodano, H.A. 2007. A review of power harvesting using piezoelectric materials. *Smart Mater Struct*. 16:1–21.

Atrah, A.B. and Salleh, H. 2013. Simulation of acoustic energy harvester using Helmholtz resonator with piezoelectric backplate. In *Proceedings of the 2nd International Congress on Sound and Vibration*. Bangkok, Thailand, 7–11 July 2013.

Azevedo, J. and Mendonça, F. 2015. Small scale wind energy harvesting with maximum power tracking. *Energy*. 3(3):297–315.

Bailly, N., Dilhac, J.-M., Escriba, C., Vanhecke, C., Mauran, N., and Bafleur, M. 2008. Energy scavenging based on transient thermal gradients: Applications to structural health monitoring of aircraft. In *PowerMEMS*. Sendai, Japan, November 2013, 205–208.

Bottner, H. 2005. Micropelt miniaturized thermoelectric devices: Small size, high cooling power densities, short response time. In *24th International Conference Thermoelectrics*. Clemson, SC, 19–23 June 2005, 1–8.

Brunelli, D., Benini L., Moser, C., and Thiele, L. 2008. An efficient solar energy harvester for wireless sensor nodes. In *Design, Automation and Test in Europe Conference and Exhibition*, Munich, Germany, March 2008, 104–109. doi:10.1.1.301.7015. www.tik.ee.ethz.ch/file/31f00ef294d8a1bfeac25f03e8526fb2/DATE08-HW.pdf

Cady, W.G. 1964. *Piezoelectricity*. pp. 1–8. New York, NY: Dover Publications.

Carmo, J.P., Goncalves, L.M., and Correia, J.H. 2010. Thermoelectric microconverter for energy harvesting systems. *IEEE Trans Ind Electron*. 57:861–867.

Carvalho, C. and Paulino, N. 2013. On the feasibility of indoor light energy harvesting for wireless sensor networks. In *Conference on Electronics, Telecommunications and Computers*, Procedia Technology, ISEL - Instituto Superior de Engenharia de Lisboa, Rua Conselheiro Emídio Navarro, 1, Lisbon, Portugal, 17:343–350. doi:10.1016/j.protcy.2014.10.206.

Chen, K.F., Ho, J.H., and Yap, E.H. 2015. Piezoelectric approach on harvesting acoustic energy. *Int J Electr Comput Energetic Electron Commun Eng*. 9:726–732.

Chuang, W.Y., Lee, C.H., Lin, C.T., Lien, Y.C., and Wu, W.J. 2014. Self-sustain wireless sensor module. In *IEEE International Conference on Internet Things and Green Computing and Communication and Cyber, Physical and Social Computing*, 1–3 September 2014, 288–291. doi:10.1109/iThings.2014.51.

D'Souza, M., Bialkowski, K., Postula, A., and Ros, M. 2007. A wireless sensor node architecture using remote power charging, for interaction applications. In *Proceedings 10th Euromicro Conference on Digital System Design Architectures, Methods and Tools DSD*, Los Alamitos, CA, 2007, 485–492.

Ecker, M., Gerschler, J.B., Vogel, J. et al. 2012. Development of a lifetime prediction model for lithium-ion batteries based on extended accelerated aging test data. *J Power Sources*. 215:248–257.

Edvinsson, N. 2012. Energy harvesting power supply for wireless sensor networks. *Thesis*. Uppsala University, Sweden.

Federspiel, C.C. and Chen, J. 2003. Air-powered sensor. In *Proceedings of the IEEE Sensors*, Toronto, Canada, 22–24 October 2003, 22–25.

Gao, Y., Sun, G., Li, W., and Yong, P. 2009. Wireless sensor node design based on solar energy supply. *IEEE 2nd International Conference on PEITS*, Shenzhen, China, 19–20 December 2009, 3:203–207. doi:10.1109/PEITS.2009.5406835.

Giorgetti, G., Cidronali, A., Gupta, S.K.S., and Manes, G., 2007. Exploiting low-cost directional antennas in 2.4 GHz IEEE 802.15.4 wireless sensor networks. In *European Conference on Wireless Technologies*, Munich, Germany, 2098–2101.

González, A., Aquino, R., Mata, W., Ochoa, A., Saldaña, P., and Edwards, A. 2012. Open-WiSe: A solar powered wireless sensor network platform. *Sensors*. 12(6):8204–8217. doi:10.3390/s120608204.

Hande, A., Polk, T., Walker, W., and Bhatia, D. 2006. Self-powered wireless sensor networks for remote patient monitoring in hospitals. *Sensors*. 6(9):1102–1117. doi:10.1.1.105.2975.

Hankuk Relay Co. Ltd. 2012. Accessed 29 February 2012. Available at: http://www.hankukrelay.co.kr/.

Harris, N.R., Grabham, N.G., Tudor, J., Beeby, S.P., and White, N.M. 2012. Practical implementation of a novel wind energy harvesting network. In *Proceedings of the 26th European Sensor*. 9–12 September 2012, Southampton: Elsevier, 961–964.

Hassan, H.F., Hassan, S.I.S., and Rahim, R.A. 2014. Acoustic energy harvesting using piezoelectric generator for low frequency sound waves energy conversion. *Int J Eng Technol.* 5:4702–4707.

Hoffmann, D., Willmann, A., Gopfert, R., Becker, P., Folkmer, B., and Manoli, Y. 2013. Energy harvesting from fluid flow in water pipelines for smart metering applications. *J Phys Conf Series.* 476(1):1–5.

Horowitz, S.B., Sheplak, M., Cattafesta, L.N., and Nishida, T. 2006. A MEMS acoustic energy harvester. *J Micromech Microeng.* 16(9):S174.

Hudak, N.S. and Amatucci, G.G. 2008. Small-scale energy harvesting through thermoelectric, vibration, and radiofrequency power conversion. *J App Phys.* 103:101–301.

Iizumi, S., Kimura, S., Tomioka, S. et al. 2011. Lead zirconate titanate acoustic energy harvesters utilizing different polarizations on diaphragm. *Res Article Procedia Eng.* 25:187–190.

Illinois Structural Health Monitoring Project (ISHM). 2012. Accessed 12 March 2012. Available at: http://shm.cs.uiuc.edu/.

Issarayangyun, T., Black, D., Black, J., and Samuels, S. 2005. Aircraft noise and methods for the study of community health and well-being. *J Eastern Asia Soc Transp Stud.* 6:3293–3308.

Jiang, X., Polastre, J., and Culler, D.E. 2005. Perpetual environmentally powered sensor networks. In *Proceedings of the Fourth International Symposium on Information Processing in Sensor Networks, IPSN 2005.* UCLA, Los Angeles, CA, 24–27 April 2005, 65:463–468. doi:10.1109/IPSN.2005.1440974.

Joung-Hu, P., Ahn, J.-Y., Cho B.-H., and Yu G.-J. 2006. Dual-module-based maximum power point tracking control of photovoltaic systems. *Indust. Elec. IEEE Trans.* 53(4): 1036–1047. doi:10.1109/TIE.2006.878330.

Jung, S.S., Kim, Y.T., Lee, Y.B., Kim, H.C., Shin, S.H., and Cheong, C. 2009. Spectrum of infrasound and low frequency noise in passenger cars. *J Korean Phys Soc.* 55(6):2405–2410.

Kansal, A., Hsu, J., Srivastava, M., and Raghunathan, V. 2006. Harvesting aware power management for sensor networks. In *Automation Conference, 2006 43rd ACM/IEEE.* Moscone Center, San Francisco, CA, 24–28 July 2006, 651–656. doi:10.1145/1146909.1147075.

Khan, F.U. and Izhar. 2013. Acoustic-based electrodynamic energy harvester for wireless sensor nodes application. *Int J Mater Sci Eng.* 1:72–78.

Khan, J.A., Qureshi, H.K., and Iqbal, A. 2015. Energy management in wireless sensor networks: A survey. *Comput Electron Eng.* 41:159–176. doi:10.1016/j.compeleceng.2014.06.009.

Kishi, M., Nemoto, H., Hamao, T. et al. 1999. Micro thermoelectric modules and their application to wristwatches as an energy source. In *18th International Conference on Thermoelectrics.* Baltimore, MD, 29 August–2 September 1999, 301–307.

Kiziroglou, M.E., Samson, D., Becker, T., Wright, S.W., and Yeatman, E.M. 2011. Optimization of heat flow for phase change thermoelectric harvesters. In *PowerMEMS.* Seoul, Korea, 15–18 November 2011.

Kiziroglou, M.E., Wright, S.W., Toh, T.T., Mitcheson, P.D., Becker, Th., and Yeatman, E.M. 2013. Design and fabrication of heat storage thermoelectric harvesting devices. *IEEE Trans Ind Electron.* 61(1): 302–309.

Korean Meteorological Administration Center (KMAC). 2012. Past Wind Records on Jindo Areas. Accessed 29 February 2012. Available at: http://jindo.kma.go.kr/.

Koyvanich, K., Smithmaitrie, P., and Muensit, N. 2015. Perspective microscale piezoelectric harvester for converting flow energy in water way. *Adv Mater Lett.* 6(6):538–543.

Lee, H.J., Sherrit, S., Tosi, L.P., Walkemeyer, P., and Colonius, T. 2015. Piezoelectric energy harvesting in internal fluid flow. *Sensors.* 15(10):26039–26062.

Li, Y. and Shi, R. 2015. An intelligent solar energy-harvesting system for wireless sensor networks. *J Wireless Commun Netw.* 2015:179, Springer.D

Li, X., Wu, Y., Li, G. et al. 2010. Development of wireless soil moisture sensor base on solar energy. *Trans CSAE.* 26(11):13–18.

Liu, Y.J., Li, X.H., Guo, H.J. et al. 2009. Electrochemical performance and capacity fading reason of $LiMn_2O_4$/graphite batteries stored at room temperature. *J Power Sources.* 189(2):721–725.

Liu, G., Mrad, N., Xiao,G., Li, Z., and Ban, D. 2011. RF-based power transmission for wireless sensors nodes. In *International Workshop on Smart Materials, Structures & NDT in Aerospace Conference.* Canada, 14–17 June 2011.

Liu, F., Phipps, A., Horowitz, S. et al. 2008. Acoustic energy harvesting using an electromechanical Helmholtz resonator. *J Acoust Soc Am.* 123(4):1983–1990.

Mieszkowski, M.R. 2004. Excessive vehicle noise impact and remedies. In *Digital Recordings co.* Halifax, Nova Scotia, Canada, 7 May 2004.

Morarka, A.R. and Ghaisas, S.V. 2016. High Sensitivity Fluid Energy Harvester. Available at: https://arxiv.org/abs/1607.00808.

Moser, A., Erd, M., Kostic, M., Cobry, K., Kroener, M., and Woias, P. 2012. Thermoelectric energy harvesting from transient ambient temperature gradients. *J Electron Mater.* 41(6):1653–1661.

Moser, C., Thiele, L., Brunelli, D., and Benini, L. 2007. Adaptive power management in energy harvesting systems. In *Proceedings of the Conference on Design, Automation and Test in Europe.* New York, NY: ACM Press, 16–20 April 2007, 773–778.

Mullen, P., Siviter, J., Montecucco, A., and Knox, A.R. 2015. A thermoelectric energy harvester with a cold start of 0.6°C. In *12th European Conference on Thermoelectrics, Elsevier.* Madrid, Spain, 24–26 September 2014, 823–832.

Naveen, K.V. and Manjunath, S.S. 2011. A reliable ultra-capacitor based solar energy harvesting system for wireless sensor network enabled intelligent buildings. In *IEEE 2nd International Conference on IAMA.* Rajalakshmi Engineering College, Chennai, India, 07–09 September 2011, 20–25. doi:10.1109/IAMA.2011.6048997.

Ostaseviciusa, V., Markeviciusb, V., Jurenasa, V. et al. 2015. Cutting tool vibration energy harvesting for wireless sensors applications. *Sensors Actuat.* 233:310–318.

Park, J.W., Jung, H.J., Jo, H., and Spencer, B.F., Jr. 2012. Feasibility study of micro-wind turbines for powering wireless sensors on a cable-stayed bridge. *Energies.* 5(9):3450–3464.

Priya, S., Chen, C.T., Fye, D., and Zahnd, J. 2005. Piezoelectric windmill: A novel solution to remote sensing. *Jpn J Appl Phys.* 44:104–107.

Raghunathan, V., Kansal, A., Hsu, J., Friedman, J., and Srivastava, M. 2005. Design considerations for solar energy harvesting wireless embedded systems. In *Proceedings of the 4th International Symposium on Information Processing in Sensor Networks* Los Angeles, CA, 24–27 April 2005, 457–462. doi:10.1109/IPSN.2005.1440973.

Ramadass, Y.K. and Chandrakasan, A.P. 2011. A battery-less thermoelectric energy harvesting interface circuit with 35 mV startup voltage. *IEEE J Solid-State Circuits.* 46(1):333–341.

Rancourt, D., Tabesh, A., and Fréchette, L.G. 2007. Evaluation of centimeter-scale micro windmills: Aerodynamics and electromagnetic power generation. In *Proceedings of the Technical Digest PowerMEMS.* Freiburg, Germany, 28–29 November 2007 27–29.

RF Monolithics, Inc. 2010. Sizing Solar Energy Harvesters for Wireless Sensor Networks Application Note M1002. n.d. Available at: http://wireless.murata.com/media/products/apnotes/anm1002.pdf.

Rice, J.A., Mechitov, K., Sim, S.H. et al. 2010. Flexible smart sensor framework for autonomous structural health monitoring. *Smart Struct Syst.* 6:423–438.

Roy, G., Matagne, E., and Jacques, P.J. 2013. A global design approach for large-scale thermoelectric energy harvesting systems. *J Electron Mater.* 42(7):1781–1788.

Saggini, S., Ongaro, F., Galperti, C., and Mattavelli, P. 2010. Supercapacitor- based hybrid storage systems for energy harvesting in wireless sensor networks. In *Proceedings of the 25th Annual IEEE Applied Power Electronics Conference and Exposition (APEC'10)* Palm Springs, CA, 21–25 February 2010, 2281–2287. doi:10.1109/APEC.2010.5433554.

Samson, D., Kluge, M., Becker, T., and Schmid, U. 2011. Wireless sensor node powered by aircraft specific thermoelectric energy harvesting. *Sensors and Actuat Phys.* 172:240–244.

Samson, D., Otterpohl, T., Kluge, M., Schmid, U., and Becker, T. 2010. Aircraft-specific thermoelectric generator module. *J Electron Mater.* 39:2092–2095.

Seah, W.K.G., Zhi, A.E., and Tan, H.-P. 2009. Wireless sensor networks powered by ambient energy harvesting (WSN-HEAP): Survey and challenges. In *Proceedings of the 1st International Conference on Wireless Communication, Vehicular Technology, Information Theory and Aerospace and Electronic Systems Technology, Wireless (VITAE '09).* Aalborg Congress and Culture Centre, Aalborg, Denmark, 17–20 May 2009, 1–5. doi:10.1109/WIRELESSVITAE.2009.5172411.

Settaluri, K.T., Lo, H., and Ram, R.J. 2011. Thin thermoelectric generator system for body energy harvesting. *J Electron Mater.* 41(6):984–988.

Sherrit, S. 2008. The physical acoustics of energy harvesting. In *IEEE International Ultrasonics Symposium Proceedings.*Beijing, China, 2–5 November 2008, 1046–1055.

Shigeki, S., Tai, T., Itoh, H. et al. 2010. Lead zirconate titanate acoustic energy harvester proposed for microelectromechanical system/IC integrated systems. *Jpn J Appl Phys.* 49(4).

Simjee, F. and Chou, P.H. 2006. Everlast: Long-life, supercapacitor-operated wireless sensor node. In *ISLPED '06: Proceedings of the 2006 International Symposium on Low Power Electronics and Design.* New York, NY: ACM Press, 4–6 October 2006, 197–202. doi:10.1145/1165573.1165619.

Stanley-Marbell, P. and Marculescu, D. 2007. An 0.9 × 1.2, low power, energy-harvesting system with custom multi-channel communication interface. In *Design, Automation and Test in Europe Conference and Exhibition.* Nice Acropolis Nice, France 16–20 April 2007, 15–20. doi:10.1109/DATE.2007.364560.

Su, J., Leonov, V., Goedbloed, M. et al. 2010. A batch process micromachined thermoelectric energy harvester: Fabrication and characterization. *J Micromech Microeng.* 20(10):104005.

Sukesha and Kaur, H.P. 2014. Home automation and energy harvesting in wireless sensors network. *Int J Innov Res Comput Commun Eng.* 2(5):4460–4469.

Takahashi, K., Saitoh, M., Asakura, N. et al. 2004. Electrochemical properties of lithium manganese oxides with different surface areas for lithium ion batteries. *J Power Sources.* 136(1):115–121.

Tan, Y.K. and Panda, S.K. 2011. Self-autonomous wireless sensor nodes with wind energy harvesting for remote sensing of wind-driven wildfire spread. *IEEE Trans Instrum Meas.* 60:1367–1377.

Taneja, J., Jeong, J., and Culler, D. 2008. Design, modeling, and capacity planning for micro-solar power sensor networks. In *Proceeding of the 7th IEEE International Conference on IPSN.* St. Louis, MI, 22–24 April 2008, 407–418.

Taylor, R., Liu, F., Horowitz, S. et al. 2004. Technology development for electromechanical acoustic liners. In *Active 04 Paper* 04-093.

Tomioka, S., Kimura, S., Tsujimoto, K. et al. 2011. Lead zirconate titanate acoustic energy harvesters with dual top electrodes. *Jpn J Appl Phys.* 50(9).

Venkatasubramanian, R., Siivola, E., Colpitts, T., and O'Quinn, B. 2001. Thin-film thermoelectric devices with high room-temperature figures of merit. *Nature.* 413:597–602.

Wang, Z., Leonov, V., Fiorini, P., and Van Hoof, C. 2009. Realization of a wearable miniaturized thermoelectric generator for human body applications. *Sensors Actuat A Phys.* 156(1):95–102.

Ward, J.A., Bharatula, N.B., Lukowicz, P., and Troster, G. 2006. Maximum power point tracking for on-body context systems. In *IEEE International Symposium on Wearable Computers (ISWC 2006).* Switzerland, 11–14 October 2006, 135–136. doi:10.1109/ISWC.2006.286363.

Wu, L.Y., Chen, L.W., and Liu, C.M. 2009. Acoustic energy harvesting using resonant cavity of a sonic crystal. *Appl Phys Lett.* 95(1).

Xiang, Y. and Pasricha, S. 2015. Run-time management for multicore embedded systems with energy harvesting. *IEEE Trans Very Large Scale Integr (VLSI) Syst.* 23(12):2876–2889. doi:10.1109/TVLSI.2014.2381658.

Xu, F., Yuan, F., Hu, J., and Qiu, Y. 2010. Design of a miniature wind turbine for powering wireless sensors. In *Proceedings of the SPIE Annual Symposium on Smart Structures and Materials and Nondestructive Evaluation and Health Monitoring.* The Moscone Center San Francisco, CA, 23–28 January 2010, 7–11.

Yokoyama, Y. and Hashimoto, K. 2010. Development of low-noise air conditioning ducts. *JR East co Technical Review*, Japan, 16.

Zhang, C., He, X.F., Li, S.Y., Cheng, Y.Q., and Rao, Y. 2015. A wind energy powered wireless temperature sensor node. *Sensors.* 15(3):5020–5031.

Chapter 7

Data Aggregation in Wireless Sensor Networks

Prashant Shukla, Vidushi Sharma, and Anuradha Pughat

Contents

7.1   Introduction 165
7.2   Elements of Data Aggregation 167
7.3   Energy-Efficient Data Aggregation Techniques 169
    7.3.1   Centralized Approach 170
    7.3.2   In-Network Aggregation 170
        7.3.2.1   In-Network Data Aggregation with Size Reduction 171
        7.3.2.2   In-Network Data Aggregation without Size Reduction 171
    7.3.3   Tree-Based Approach 171
        7.3.3.1   Synchronous Tree-Based Aggregation 172
        7.3.3.2   Asynchronous Tree-Based Aggregation 172
    7.3.4   Cluster-Based Data Aggregation 173
7.4   Security in Data Aggregation 174
    7.4.1   Query Diffusion 174
    7.4.2   Aggregation 174
    7.4.3   Verification 175
7.5   Privacy Preserving Data Aggregation 175
    7.5.1   Perturbation 176
    7.5.2   Shuffling 176
    7.5.3   Privacy Homomorphism 176
7.6   Summary and Challenges in Data Aggregation 177
References 177

7.1 Introduction

Usually, sensor networks are densely deployed to cover vast land spans or geographical areas of interest. The primary target for wireless sensor networks (WSNs) is to gather data and provide information about the environment to the neighboring nodes. The neighboring nodes may generate highly correlated and redundant data with a focus on event detection application. This data is huge and, sometimes, the same events are likely to be gathered and transmitted by other nodes too. Therefore, wherever a gigantic amount of data is to be produced or processed, data aggregation needs to be associated with the system. In WSNs, it is one of the most important tools of data analytics and decision-making. We can say that data analysis is essential at the end of any processed or produced data. In WSNs, data aggregation is required because of two main reasons. The first reason is the large amount of data produced by each node and the large number of nodes in each network, which gives a stack of data to process and analyze. This data is required to be converted into information relevant and valuable to the data consumer. The second reason is the energy optimization in WSNs, as this has been too viable in recent years. Data aggregation reduces the amount of transmission and processing and, thus, the energy use. Understanding the concept of aggregation first, we have to ask a simple question: Why is sending aggregated data required instead of sending raw data? Let us consider that we are sending entire raw data to the base station or sink. The answer to the question is that the very low data rate in WSNs can create a bound even if the connection is very good. On the other hand, the wireless communication consumes a large amount of energy, whereas WSNs have limited energy as battery supply. Therefore, the aggregation is performed instead of sending raw data packets to the sink nodes.

A simple example of a sensor network with three nodes presenting data aggregation is shown in Figure 7.1. Therefore, the two nodes in the network are taken as source nodes (Sr1 and Sr2) and the third as sink node (Si).

Let us assume *d* is the distance between two nodes and it is too low. The nodes Sr1 and Sr2 are deployed to gather similar data and send that to Si. The node Si consumes twice the energy also in receiving and processing the similar data from both the nodes. Another example includes one more node, i.e., C, which works as an aggregator and only receives data from other two nodes and sends it to the sink (see Figure 7.2).

Now, the nodes Sr1 and Sr2 are gathering the data and sending too, and they are transmitting it over half of the distance than in the prior case (Figure 7.1). This explains the importance of data aggregation in WSNs. In general, the batteries supply power to the nodes and this makes energy consumption the biggest concern and an important issue in WSNs. Research shows that communication is considerably more energy consuming than computation. The transmission cost of a single bit information is a thousand times more as compared to energy spent in executing one

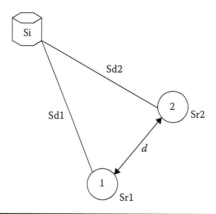

Figure 7.1 Two-node data aggregation scenarios.

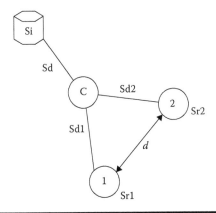

Figure 7.2   Data aggregator between source and sink nodes.

instruction. The main idea is to perform in-network processing to reduce communication cost. Thus, we can say that in spite of processing cost, data aggregation is more valuable than reading, collecting, and communicating raw sensor data.

By definition:

Data aggregation in WSNs is defined as the process of gathering data from multiple sensors on intermediate nodes using SQL queries or mathematical functions in order to eliminate redundant transmission and provide fused information only to the base station.

Therefore, reduced network traffic, reduced energy consumption, and enhanced lifetime are the results of data aggregation on sensor network. From here onward in the rest of the chapter, data gathering with aggregation will be referred to as data aggregation or aggregation, base station as sink, and source nodes will be referred to as nodes.

7.2 Elements of Data Aggregation

Various performance parameters for energy-efficient data aggregation are accuracy of data, reliability, correlation coefficient, detection of false alarm, data redundancy, latency, power consumption, and lifetime of the network (Krishnamachari et al. 2002). Aggregation, like any other mathematical concept, depends upon a few elements or factors, which influence the outcome of it, each having equal importance in the construction of the technique. These elements for aggregation are network architecture, aggregation function, data representation, and aggregation resources. The power consumption depends on data aggregation elements, network time period, latency, and data accuracy. In order to explain them collectively as factors of data aggregation, an example of a metropolitan area is given.

Figure 7.3 depicts an example of an industrial application of WSNs of air quality index (AQI) spread over a metropolitan area. It uses nine nodes and one sink. The network has a data-centric network architecture where AQI is sent to sink, which has an aggregation function AVERAGE (Equation 7.1). Aggregation nodes are $Sn = S4, S5, S7, S8$. The node $S4$ aggregates the data from $S1, S2, S3$, and $S4$.

$$S4 = \frac{S1 + S2 + S3 + S4}{4} \tag{7.1}$$

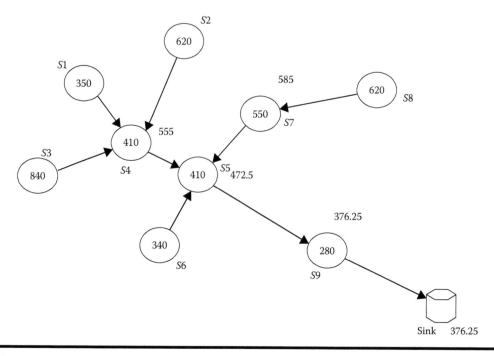

Figure 7.3   Sensor network of air quality index (AQI) of a metropolitan.

Equation 7.2 gives a generalized relation to this type of network.

$$Sn = \frac{Si + ... + Sn}{n} = \frac{\sum_{x=i}^{n} Sx}{n} \tag{7.2}$$

The aggregated value of the sink is 376.25, similar to the value at *S9*. This value is close to many other values, but way too far from some other values. Thus, there should be a format of representation, which can show all aggregated nodes as shown in Figure 7.4.

In the figure, we can see, while the sink is receiving four values, it is getting an idea of the entire area's AQI and further can divide it as sectoral data, which in a scheduling format can also be represented as a quantile digest. For the very same example, we can check out resources, which have higher trade-off value over another. The energy consumed by any network is directly proportional to the messages running through the network. If N_e is the energy in network, N_{nm} is the number of messages running through the network, and N_{acc}, the accuracy in the network, then

$$N_e \propto N_{nm} \propto \frac{1}{N_{acc}}. \tag{7.3}$$

Concluding from that: if the number of messages is low, there will be lesser energy consumed by sensor networks—but this occurs at the cost of accuracy.

There will be a trade-off between data aggregation and accuracy of data arriving at the sink node (Equation 7.4). From this, it is clear that a number of hops affect the performance of any

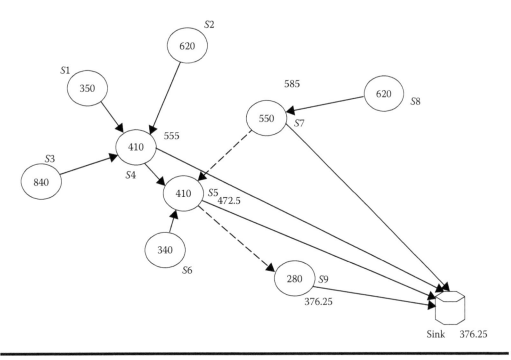

Figure 7.4   Representation of entire AQI at network.

network. The latency of a network (N_{lt}) in which data aggregation is applied is related to the bandwidth of the network (N_b) (see Equation 7.5).

$$N_e \propto \frac{1}{N_{acc}} \tag{7.4}$$

$$N_{lt} \propto \frac{1}{N_b}. \tag{7.5}$$

In a data-centric approach, the data aggregation is more energy efficient than other approaches. The example shows that without aggregation, there would have been hopping from the source to the sink directly, but after aggregation, there are only four hops to the sink directly. The other five hops will travel only half the distance. The data aggregated network is 66.6% more energy efficient than the network without data aggregation. There are various energy-efficient data aggregation techniques available; a few are discussed in the next section.

7.3 Energy-Efficient Data Aggregation Techniques

One possible way of improving the reliability of WSNs is to deploy a few redundant data nodes in the sensing area. These nodes sense similar data and forward that data to the sink nodes. These additional sensor nodes reduce the chances of network failure at the cost of extra bandwidth use and energy consumption in communicating and processing redundant data. There is a trade-off between reliability and energy consumption. Another way is to deploy a few nodes in the routing path, which performs the energy-efficient data aggregation of the sensed data from its neighbors and transmits the reduced data. This reduces redundancy and a huge amount of power consumption. Various data aggregation approaches are centralized data aggregation, in-network data aggregation, tree-based data aggregation, cluster-based data aggregation, etc.

7.3.1 Centralized Approach

The centralized approach is better known as an address-centric approach because in this approach every node transmits packets to a central node by a shortest possible path via multihop protocol. Each one of the wireless nodes that captures data broadcast those packets to the leader, which does not have power as the primary concern. The leader now processes data with aggregation queries, so that redundant data can be pointed out and eliminated. By applying SQL queries, the leader ensures that the database is not only getting unique data but also true data. A few of the data packets that are transmitted may be similar, but a few others may have some additional headers added to them. Now, true data would be a combination of both, and the leader makes sure that the query also takes various test cases into consideration. Therefore, for better-aggregated data, every node falling in-between the central node and the other nodes must transmit the data packets directly to a central node without addressing them to any other node.

Thus, other nodes are going to receive the data packet, but because they are being addressed only to the centralized node, they won't be getting processed. These nodes will just get passed on to the next and to the next till they reach the designated address. There will be a large number of packet transmissions between nodes for getting a single data record aggregated. In the above approach, the best case would be the addition or summation of every node lying on the edges in the network.

7.3.2 In-Network Aggregation

In contrast to what we learned about a centralized approach or address-centric approach, in-network aggregation is a decentralized approach. In this approach, instead of data processing through a central node, each intermediate node performs the same task but on a smaller scale. The in-network aggregation scheme does not provide any centralized processing facility (CFP). Here, each intermediate node functions as an independent node in the network (Fasolo et al. 2007). There is no acknowledgment, i.e., a one-way communication. The very basic objective of reducing resource consumption (in particular, energy) is fulfilled; therefore, network lifetime increases. A generalized aggregation, i.e., tiny aggregation (TAG), approach is developed specifically for TinyOS-based sensor motes. It uses a declarative interface to collect and aggregate the data and distributed aggregation queries to make the network energy efficient (Madden et al. 2002). The sensor data flows upward in the tree (from node leaf to the parent) to reach the user. The in-network TAG approach reduces the number of packets transmitted, reduces latency, and enhances the network lifetime. Each mote sends one message per epoch. Therefore, one big advantage of the TAG approach is to put the processor on energy-saving mode during its idle time between two time slots, i.e., epochs. This requires a synchronized clock that wakes up the node when required. Further, the sleep interval and resolution of the synchronization need to be decided for efficient energy saving. The performance of TAG in flat and single-hop networks is more or less similar to that in the centralized approach, but reduces the number of transmissions when nodes are deployed in a line. For n number of nodes on a line, centralized approach transmits $n^2/2$ records, while TAG transmits n records. Thus, we can say that the communication cost and bandwidth of an in-network approach are lesser than in a centralized approach. The model can be explored for event-driven sensor node applications. Now, there are two ways for in-network aggregation, i.e., in-network aggregation with size reduction and in-network aggregation without size reduction.

7.3.2.1 In-Network Data Aggregation with Size Reduction

Aggregation with size reduction refers to the process of a node gathering data and transmitting it without adding any address to the packets. Further, when a packet is received by a neighboring node, that node combines it with the data it has and compresses the data packet so that the packet length gets reduced and it can be transmitted or forwarded toward the sink.

7.3.2.2 In-Network Data Aggregation without Size Reduction

In-network data aggregation without size reduction refers to a node merging data packets received from multiple neighbors. Instead of calculating the length size or compressing the data, the data packets are transmitted directly.

7.3.3 Tree-Based Approach

The in-network and centralized approaches are the broadcast approaches. There is no predefined structure for these approaches, while a tree-based approach sensor network has a spanning tree-like structure. Here, the branches or leaves of the tree (differ based on the terminology by various researchers) are considered as nodes and the sink is defined as the root of the spanning tree.

The network works in a similar manner as one will get to the root of a spanning tree, starting from the top, i.e., leaves or nodes, and each node will have a parent node where it has to transfer data to. This flow keeps on running through all the nodes to the sink or root. The aggregation is carried out as we move to the roots of the spanning tree. The nodes on the very end of the tree or the top-most leaf, which is not the parent node for any other leaf, gather data and send it to its parent. Now, a parent node has multiple leaves and they are located nearby. Therefore, the data gathered by these nodes will be alike in most cases. All these leaves or nodes route their data to their parent node. If the parent node is very busy in receiving data packets from child nodes, then it can drain out the power before it even starts forwarding them. Therefore, a sampling period is set in a tree-based aggregation, which provides the energy efficiency and extended lifetime. This time period is divided into smaller time intervals. A parent node specifies the time interval during which it will be receiving data packets from its child nodes or leaves. This time scheme is included in the query and sent to the child node from their parent starting from the bottom of the tree, i.e., sink. Now, it becomes clear that if the sink sends its child node a time interval, then those nodes will send the time interval ahead of the sink's time interval. This time interval includes the processing time in data gathering, and this interval window should not be too short because in that case the parent would not be able to receive data packets from all the child nodes. As soon as a child receives this query with the time interval defined in it, the node starts forwarding the packets. This time period or time-sampling duration is divided by the depth of the spanning tree formed to get time intervals. The *greedy aggregation* (Harris et al. 2007) uses the shortest path for data aggregation and the network reliability depends on the connection between the nodes in the the shortest path. The network fails when a link between nodes is broken. This algorithm relates the energy efficiency with the length of the shortest path. A dynamic balanced spanning tree (DBST) algorithm selects the route nodes on the basis of their residual energy, distance, link weight, and node weight (Avokh and Mirjalily 2010). This approach features traffic load balancing through the dynamic capability of routing tree and minimizes the energy consumption at the cost of delay in reformation of trees. The support vector machine-based data redundancy elimination (SVM-based DRE or SDRE) algorithm classifies redundant data and then eliminates the correlated data. The elimination can be threshold dependent and rejects values above threshold level. In an SVM-based data aggregation tree (DAT), each parent node connects to the gateway node (Patil and Kulkarni 2013). This algorithm eliminates redundancy and improves the energy efficiency and performance of the network. Khedo et al. (2010) proposed the redundancy elimination for the accurate data aggregation (READA) algorithm that first groups data and then compresses it for removing duplicate data before sending to the base station. It provides improvement in data accuracy and energy efficiency. Another energy-efficient data aggregation protocol is spatial and temporal multiaggregation for state-based sensor data (STMA). It uses spatial and temporal data aggregation to reduce the number of transmitted bits and thus the traffic load and energy (Kim and Kwon 2004). The tree-based aggregation is further divided into two types of data aggregation, which is based on the node synchronization. One is the synchronous tree-based aggregation and another is the asynchronous tree-based aggregation.

7.3.3.1 Synchronous Tree-Based Aggregation

The tree-based aggregation approach requires a sync pulse of time interval between adjoining levels of the nodes. Over here, sync is defined as the simultaneous time interval provided to the same-level nodes by its parent node. As in a tree-based aggregation, a node is supposed to transmit data packets during the time interval assigned to it by its parent node. During that sampling time interval, the exact time when a node is supposed to transmit is not mentioned. Therefore, the approach can lead to a problem of all or a few nodes transmitting at the same time during that interval. This further leads to the problem of a high amount of energy consumption at the parent node because of receiving multiple similar packets at the same moment and sitting idle for most of the time. To solve this problem, instead of sending all the packets at the very same moment, the sending time of the nodes is randomized within the specified time interval. The nodes are selected randomly to transmit packets when they have gathered the sufficient data. This solves the energy problem and also the parent node gets time to process the received data packets, but packet loss is possible if timeout occurs in order to make this process periodic. This order can be set either periodic or per hop.

7.3.3.2 Asynchronous Tree-Based Aggregation

As the name states, in this tree-based aggregation approach, a spanning tree is formed, but there is no sync between the adjoining levels of the tree in the network. Therefore, in the basic asynchronous tree-based aggregation protocol, each parent node sets a timeout so that there is no data loss. In each level, the parent nodes wait for the maximum timeout to occur or collect data packets from all the adjoining child nodes, whichever occurs first. In this scheme, the timeout value is selected adaptively, i.e., transmission time taken by each level node or combining and finding average of data at all the nodes at each level is set to the timeout value. Therefore, the parent node aggregates this data when it has heard from all the child nodes at a level. When timeout occurs, it combines that with its own data and sends it to the next level.

Between asynchronous and synchronous tree-based aggregation, the asynchronous tree-based aggregation outperforms the synchronous approach as there is near zero data loss. This can achieve the possible highest network lifetime, when level timeout is adaptively calculated. As no time sampling is used, the overhead due to the time interval decreases too.

7.3.4 Cluster-Based Data Aggregation

The various nodes in a communication path perform data aggregation in flat networks, while cluster heads perform data aggregation in hierarchical networks. When compared to flat networks, hierarchical networks result the lowest latency in transmission and are more reliable. In hierarchical networks, the breakdown of the network occurs when almost all cluster heads in the network break down. The cluster-based aggregation approach refers to sensor networks where clustering is being done for data communication either via single hop or multihop. Clustering in the sensor networks is defined in a schema, where the entire network is divided into smaller networks or groups, called clusters. Basically, the clusters are divided according to the locations of the nodes. The neighboring nodes form a cluster and within a cluster the node that is either most power efficient or nearest to the sink is selected as a cluster head. This type of selection for improving energy efficiency is called node heterogeneity. The communication in a cluster-based network is performed very similar to the centralized approach; just over here, the nodes do not broadcast, they unicast the data. They gather the data and transmit it to the cluster head. The cluster head then does the processing and applies queries in order to perform aggregation over the cluster. In recent years, various cluster-based data aggregation protocols have been introduced. This approach is more energy efficient and highly data accurate with the least overhead. The cluster head can also play the role of data aggregator. Many researchers have introduced this to various protocols for data communication in WSNs.

In a sensor network where a cluster-based data aggregation protocol is implemented, sensor nodes have the capability to adjust their power level by transmitting the data packets to a distance according to their requirements. For communication between the sensor nodes of a cluster and its cluster heads, the single-hop communication is used. Some nodes appoint themselves as cluster heads; this selection is usually done based on a random function. Further, these elected cluster heads broadcast this message to the entire network and nodes join their cluster based on their signal level. Now these cluster heads receive the data periodically from their cluster nodes and work as an aggregator and aggregate the data packets and send them to the sink nodes. This is done using code division multiple access (CDMA), where the aggregated data coming from two different clusters do not coincide. An energy-efficient clustering approach reduces the energy consumption during cluster head selection or rotation (Meng et al. 2013; Kumar et al. 2009). In addition, the sink node makes sure that the cluster head keeps on changing in order to increase the lifetime of clusters.

Although the sink node keeps on rotating the role of the cluster head among the nodes of a cluster, this approach prolongs the network's life in a very efficient way by using "id" to form a distributed cluster formation. Here, the cluster head is selected periodically on the basis of how much energy is left in that sensor node. Although this increases network lifetime, the periodic change of the cluster head has a negative impact on the aggregation. There may be data loss in the iteration if aggregated data is transmitted before the cluster head changes. In order to fix this problem, a delay is added, which makes the network to be delay sensitive. This creates a similar scenario such as in asynchronous tree-based aggregation; it waits for all the cluster members to respond and a delay to get completed to make sure that data is aggregated and transmitted. A cluster approach for energy-efficient spatial correlation-based data gathering uses spatial correlation for cluster formation and threshold value for reducing data communication (Tharini et al. 2016). The sensors transmit the data to the base station only when their predicted value crosses the threshold value. There are many more approaches to data aggregation (Zhu et al. 2010; Liu et al. 2011; Dasgupta et al. 2003; Yuea et al. 2012).

Data aggregation is done using an aggregation function, which is a mathematical function that takes raw gathered data from sensors as input and generates a fused formed data as output. Because of this fused formed data or digest data, security comes into the picture. For example, if we want to aggregate the data using MAX, AVERAGE, or SUM of a gathered data, and there is a malicious nodes in the network, these functions mentioned are unsecured. They do not filter legitimate nodes from malicious nodes, and if a malicious node adds only one false data, the entire digest data will be biased. Thus, in order to secure data aggregation in sensor networks a secure data aggregation function is required (Alzaid et al. 2008; Chan and Castelluccia 2011; Othman et al. 2013). The next section discusses the security aspects in data aggregation.

7.4 Security in Data Aggregation

There are two primary reasons why security is important in data aggregation. One is the falsification of end results after data aggregation as a single malicious node is capable of biasing entire aggregation results. Data privacy can be breached if a malicious node performs a man in the middle (MITM) attack, sniffs data, and forwards it to the sink too. At the same time, that will be a breach of confidentiality of data.

Now, we will focus on the first objective of secure data aggregation and discuss the various phases of secure data aggregation in order to prevent falsification of the end data. Then, we will discuss privacy prevention in data aggregation. There are multiple ways to secure the processing of aggregation from producing falsified digest data, such as cryptography, quantiles aggregation, RANBAR, voting, and verification. All of these processes can be categorized into three phases as query diffusion, aggregation, and verification.

7.4.1 Query Diffusion

It is the first phase in which the base station broadcasts SQL query in the network. The query passes through various nodes. It is an energy-consuming process because there is a high possibility that a the node gets the same query twice or thrice depending upon the location. Therefore, the query diffusion phase is combined with localization of the nodes. As mentioned above, the low-level queries are insecure. Therefore, in order to construct a secure data aggregation structure, the query needs to be more complex (Yu et al. 2006).

7.4.2 Aggregation

In the second phase, the nodes (that fulfill a criteria of the discriminated query) form clusters or spanning tree depending upon the query. For example, suppose it is a tree-based aggregation, then the query diffusion takes place from the sink node to the very last leaf nodes. The acknowledgment will form a spanning tree, which will have a parent node and a leaf node as per the query. If the distance between the parent and the leaf node is greater than the distance between the leaf nodes and the sink, then the leaf node would not be the part of the spanning tree. Instead, it will directly send its sample to the sink node here instead of sending to the parent node. However, this becomes more energy consuming, whereas the primary objective of the aggregation is to make the sensor network energy efficient. Therefore, a new approach is followed, which not only uses an aggregator but also a forwarder. After aggregation, the third phase comes onto the scene, where secure data flow starts, i.e., verification.

7.4.3 Verification

In order to be certain about securing the data flow in the network, the base station needs to verify every aggregated data from all the nodes (Mahimkar and Rappaport 2004). This verification can be done in multiple ways. One way is to use a sink that can verify every node, which sends the data to the aggregator. Another way is to verify data aggregated at the aggregators only, at the cost of energy. Therefore, instead of the sink verifying the nodes, the nodes can attest a security token with data packet, which aggregators can verify and attest a security token that the sink can verify. The problem with this approach is a single malicious aggregator can compromise the entire network, as the sink will verify only the security token attested to it and not the data in order to save energy. To solve this problem, another approach that can be added to the existing one is that the sink should not accept the aggregated data from an aggregator unless the leaf nodes verify the aggregator. Still, if there is sniffing performed on the network, then not all these techniques would be useful, as data privacy is still breached. This is discussed in the next section.

7.5 Privacy Preserving Data Aggregation

WSNs are not conceptualizing intercommunication for smart devices in today's and future infrastructure due to which sensor networks have various applications like health monitoring, military surveillance, data logging, industrial monitoring, etc. Each of these applications monitors or logs huge amounts of data via sensor nodes. This makes data aggregation a requirement for these applications. However, most applications have uttermost need of data privacy during aggregation. *Why do we need privacy preservation?* There are various reasons due to which data privacy with aggregation becomes a necessity. Some of them are listed below:

- Misuse of data is the first and most critical reason behind the concept of privacy preservation, for example, health monitoring applications, where a hospital periodically monitors the patients' various health standards like blood pressure, sugar levels, etc. and stores at database (sink). This becomes exclusively private information, which a hospital has to keep private except from doctor or patient.
- Location of army vehicles in a battlefield should be kept private.
- Revealing incompetence is another reason behind making these datasets private, for example, a company that provides household monitoring systems XYZ like thermostat, water monitoring, etc. The company is not preserving the data of the user and a competitive firm uses the personal data of users, as data was not having end-to-end encryption. This can bring down XYZ's business.
- It is critical because the companies, government agencies, or any other public, private, or non-government organizations are not allowed to violate the privacy of any country. Legislation bounds them to preserve the privacy of the users.

Privacy preservation has only two concerns, i.e., internal and external. Solution of maintaining internal privacy lies in securing the network and making all the nodes as trusted nodes. To maintain external privacy, privacy-preserving data aggregation (PPDA) is implied (He et al. 2007; Erkin 2013; Xu et al. 2015). PPDA is basically categorized into two types of protocols: homogeneous protocols and heterogeneous protocols. This categorization is based on the types of nodes in the network. If all the nodes have the very same resource, then homogeneous protocols are applied; if there is more than one type of nodes in the network such as aggregator nodes and leaf nodes, then heterogeneous protocols are applied. Further, the classification is based on the type of network, cluster, tree based, centralized, or in-network. At the end, they have been divided into two types, i.e., end-to-end and hop-by-hop. In end-to-end encryption, the entire communication is encrypted. Apart from the sink and node, no one can decrypt the packets. This makes the aggregation difficult to perform, but decreases communication overhead and guarantees privacy preservation. On the other hand, in a hop-by-hop scheme, the sensor sends encrypted data packets to aggregators and over their aggregator decrypts; it aggregates data, re-encrypts the aggregated data, and sends it to the sink. This technique does not guarantee privacy preservation because the data can be read at aggregators as they have the key to decrypt packet; also, this approach increases communication overhead, but performs the aggregation in a much better manner.

Further, these protocols are basically of three types used over various kinds of unicast- and broadcast-based networks. These are perturbation, shuffling, and privacy homomorphism. Each technique is described in detail below.

7.5.1 Perturbation

Perturbation means disorder or disturbance. In privacy preservation, it means that the sensor network always has an order or hierarchy, which is the fundamental equation of the network, in a scenario where only the sink knows the network formation. Knowing the data flow, we are not able to tell which is the route to an aggregator or to the sink, and the network is called a perturbed sensor network. If every node divides its data into a polynomial of order $k - 1$ using any scheme and k number of nodes sends it to all other nodes using shared key, then after a sensor sums all the received polynomial and sends it up to the aggregation level. The aggregator inverses the matrix and resolves it into separate packets and aggregates them without knowing from where any of the packets came and forward level up to the sink. This is a good privacy-preserving technique, but it increases the calculation overhead.

7.5.2 Shuffling

As the word describes, the data is sliced into the number of members of the nodes falling under the aggregator; then that node keeps one slice and distributes the other slices after encrypting with a private key to the rest of the nodes. By this shuffling, the exact origin of the data cannot be recognized, and further because of these slices getting combined at aggregator with separate keys, even if a aggregator is malicious, then too it will receive encrypted packets in parts only. Shuffling technique increases the privacy preservation, but because of multihop, energy efficiency is affected.

7.5.3 Privacy Homomorphism

This is considered to be the most energy-efficient privacy preservation technique because of arithmetic operations done on encrypted data, whereas decryption is not required, which decreases the computation overhead too (Diana and Vijayakumar 2014; Domingo I Ferrer 1996). In privacy homomorphism, every leaf node shares a separate key with the sink, but at the same time function adds a message with a key. This message when added up forms a key at the aggregator, which is used to aggregate the data by summing up; further the aggregator forward the message to the sink too.

This is a good technique for privacy preservation and energy efficiency, but not scalable as the sink has to keep the different key of the nodes. As the size of the network increases, the number of keys in a sink's database also increases.

7.6 Summary and Challenges in Data Aggregation

Data aggregation is the most important technique for enhancing the lifetime of sensor networks. Various protocols are proposed in the literature, which shows the increase in the lifetime of the WSN. In this chapter, the various types of data aggregations in WSNs are discussed. The data aggregation protocols are not classified on the basis of their transmission type; these protocols are classified on the basis of their network architecture too. Apart from that, various elements that play crucial roles in data aggregation techniques have been discussed with an example in order to explain it.

- We are aware that lifetime is one among the primary concerns of sensor networks and to enhance it a systematic study of the relation between energy efficiency and system lifetime can be termed as road to future research.
- Another area worth exploring is by analyzing the limits of the lifetimes of sensor networks. Existing work has provided limits of lifetimes for sensor networks with specific network architectures, source behaviors, and application usage.
- It will be a great challenge to generalize sensor networks for data aggregation in telecommunications by working on the mobility factor of sensor networks.
- Security is another eminent issue in data aggregation applications that has been unexplored for a long time period. Integrating security as an essential component of data aggregation protocols can be one of the interesting problems for future research.
- Data aggregation in dynamic environments serves various challenges and this can serve as worthy future research work.
- Another interesting domain of research is the application of source coding theory for data-gathering networks. Power savings in data aggregation become crucial depending upon the fact that WSNs have resource constraints.

In this chapter, after going through various protocols, it becomes clear that the power efficiency of a data-centric approach for networking in WSNs in comparison to an address-centric approach is excessively higher. The optimization in data aggregation will only decrease the minimum number of edges in the tree, which should be compared when sources send information to the sink for the shortest path. Now, we conclude that the distance between the sink and the sources should be large compared with the distance between the sources; then optimal data-centric protocols give *k*-folds savings over address-centric solutions.

References

Alzaid, H., Foo, E., and Gonzalez Nieto, J. 2008. Secure data aggregation in wireless sensor network: A survey. In *Proceedings of the Sixth Australasian Conference on Information Security-Volume 81*. 4–7 December, 2006, Taipei, 81 (January): 93–105. doi:10.1109/PDCAT.2006.96.

Avokh, A. and Mirjalily, G. 2010. Dynamic balanced spanning tree (DBST) for data aggregation in wireless sensor networks. *Network*. 1:391–396.

Chan, A.C.-F. and Castelluccia, C. 2011. A security framework for privacy-preserving data aggregation in wireless sensor networks. *ACM Trans Sens Netw*. 7(4):1–45. doi:10.1145/1921621.1921623.

Dasgupta, K., Kalpakis, K., and Namjoshi, P. 2003. An efficient clustering-based heuristic for data gathering and aggregation in sensor networks. In *IEEE Wireless Communications and Networking Conference, WCNC*. 16–20 March, 2003, New Orleans, LA, USA, 3:1948–1953. doi:10.1109/WCNC.2003.1200685.

Diana, B.S. and Vijayakumar, S. 2014. A concealed data aggregation using privacy homomorphism in wireless sensor networks. *Ijirset.com*. 3(3):2404–2410.

Domingo-Ferrer, J. 1996. A new privacy homomorphism and applications. *Inform Process Lett.* 60(5):277–282. doi:10.1016/S0020-0190(96)00170-6.

Erkin, Z. 2013. Privacy-preserving data aggregation in smart metering systems: An overview. *IEEE Signal Process Mag.* 30(2):75–86. doi:10.1109/MSP.2012.2228343.

Fasolo, E., Rossi, M., Widmer, J., and Zorzi, M. 2007. In-network aggregation techniques for wireless sensor networks: A survey. *IEEE Wirel Commun.* 14(2):70–87. doi:10.1109/MWC.2007.358967.

Harris, A.F., Kravets, R., and Gupta, I. 2007. Building trees based on aggregation efficiency in sensor networks. *Ad Hoc Netw.* 5(8):1317–1328. doi:10.1016/j.adhoc.2007.02.021.

He, W., Liu, X., Nguyen, H., Nahrstedt, K., and Abdelzaher, T. 2007. PDA: Privacy-preserving data aggregation in wireless sensor networks. *IEEE INFOCOM 2007—26th IEEE International Conference on Computer Communications.* 6–12 May, 2007, Anchorage, AK, 2045–2053. doi:10.1109/INFCOM.2007.237.

Khedo, K., Doomun, R., and Aucharuz, S. 2010. READA: Redundancy elimination for accurate data aggregation in wireless sensor networks. *Wirel Sens Netw.* 2:300–308. doi:10.4236/wsn.2010.24041.

Kim, H.S. and Kwon, W.H. 2004. Spatial and temporal multi-aggregation for state-based sensor data in wireless sensor networks. *Telecommun Sys.* 26:161–179. doi:10.1023/B:TELS.0000029037.36000.6e.

Krishnamachari, L., Estrin, D., and Wicker, S. 2002. The impact of data aggregation in wireless sensor networks. In *Proceedings 22nd International Conference on Distributed Computing Systems Workshops.* 2–5 July, 2002, Vienna, Austria, 575–578. doi:10.1109/ICDCSW.2002.1030829.

Kumar, D., Aseri, T.C., and Patel, R.B. 2009. EEHC: Energy efficient heterogeneous clustered scheme for wireless sensor networks. *Comput Commun.* 32(4):662–667. doi:10.1016/j.comcom.2008.11.025.

Liu, X., Zhao, H., and Li, X. 2011. EPC: Energy-aware probability-based clustering algorithm for correlated data gathering in wireless sensor networks. In *Proceedings-International Conference on Advanced Information Networking and Applications, AINA.* Washington, DC, 22–25 March, 2011, 419–426. doi:10.1109/AINA.2011.45.

Madden, S., Franklin, M.J., Hellerstein, J.M., and Hong, W. 2002. TAG: A Tiny AGgregation service for ad-hoc sensor networks. In *Proceedings of the 5th Symposium on Operating Systems Design and Implementation,* Boston, Massachusetts—December 09–11, 2002, 36(SI):131–146. doi:10.1145/844128.844142.

Mahimkar, A. and Rappaport, T.S. 2004. SecureDAV: A secure data aggregation and verification protocol for sensor networks. *IEEE Global Telecommunications Conference, 2004. GLOBECOM '04.* 29 November–3 December, 2004, Dallas, Texas, USA, 4:2175–2179. doi:10.1109/GLOCOM.2004.1378395.

Meng, J.-T., Yuan, J.-R., Feng, S.-Z., and Wei, Y.-J. 2013. An energy efficient clustering scheme for data aggregation in wireless sensor networks. *J Comput Sci Technol.* 28(3):564–573. doi:10.1007/s11390-013-1356-y.

Othman, S.B., Trad, A., Youssef, H., and Alzaid, H. 2013. Secure data aggregation in wireless sensor networks. In *IEEE International Conference on Trust, Security and Privacy in Computer and Communications.* Melbourne, Australia, 16–18 July, 2013, 12:188–195. doi:10.1109/MedHocNet.2013.6767410.

Patil, P. and Kulkarni, U. 2013. SVM based data redundancy elimination for data aggregation in wireless sensor networks. In *2013 International Conference on Advances in Computing, Communications and Informatics (ICACCI).* 22–25 August, 2013, Mysore, Karnataka, India, 1309–1316. IEEE. doi:10.1109/ICACCI.2013.6637367.

Tharini, C., Tharini, C., and Vanaja Ranjan, P. 2016. An Energy Efficient Spatial Correlation based Data Gathering Algorithm for Wireless Sensor Networks. Accessed 17 December. Available at: http://citeseerx.ist.psu.edu/viewdoc/summary?doi=10.1.1.206.3996.

Xu, J., Yang, G., Chen, Z., and Wang, Q. 2015. A survey on the privacy-preserving data aggregation in wireless sensor networks. *China Commun.* 12(5):162–180. doi:10.1109/CC.2015.7112038.

Yu, W., Nam Le, T., Lee, J., and Xuan, D. 2006. Effective query aggregation for data services in sensor networks. *Comput Commun.* 29(18):3733–3744. doi:10.1016/j.comcom.2006.06.020.

Yuea, J., Zhang, W., Xiao, W., Tang, D., and Tang, J. 2012. Energy efficient and balanced cluster-based data aggregation algorithm for wireless sensor networks. *Procedia Eng.* 29:2009–2015. doi:10.1016/j.proeng.2012.01.253.

Zhu, Y.H., Wu, W.-D., Pan, J., and Tang, Y.P. 2010. An energy-efficient data gathering algorithm to prolong lifetime of wireless sensor networks. *Comput Commun.* 33(5):639–647. doi:10.1016/j.comcom.2009.11.008.

Chapter 8

Sensor Network Security

Aarti Gautam Dinker and Vidushi Sharma

Contents

8.1   Introduction 180
8.2   Security Aspects in WSNs..181
    8.2.1   Security Goals 181
    8.2.2   Performance Metrics..182
    8.2.3   Security Limitations in WSNs 182
8.3   Vulnerable Components...183
    8.3.1   Base Station Security 183
    8.3.2   Sensor Node Security...184
8.4   Attacks in WSNs...185
    8.4.1   Adversary's Capability-Based Attacks 185
    8.4.2   Information in Transit-Based Attacks 185
    8.4.3   Host-Based Attacks..186
    8.4.4   Network-Based Attacks..186
        8.4.4.1   Physical Layer Attacks 186
        8.4.4.2   Data Link Layer Attacks 187
        8.4.4.3   Network Layer Attacks 188
        8.4.4.4   Transport Layer Attacks 188
        8.4.4.5   Application Layer Attacks..189
8.5   Security Mechanisms 189
    8.5.1   Attacks-Based Security Schemes 189
8.6   Cryptography...191
    8.6.1   Symmetric Key Cryptography 192
    8.6.2   Asymmetric or Public Key Cryptography 192
8.7   Key Management 194
    8.7.1   Symmetric Key Management..194
        8.7.1.1   Entity-Based Schemes..194
        8.7.1.2   Pairwise Key Predistribution Scheme.................................195
        8.7.1.3   Pure Probabilistic Key Predistribution Schemes................196

8.7.1.4 Polynomial-Based Key Predistribution Schemes 196
8.7.1.5 Matrix-Based Key Predistribution Schemes 196
8.7.1.6 Tree-Based Key Predistribution Schemes 197
8.7.1.7 Combinatorial Design-Based Key Predistribution Schemes 198
8.7.1.8 Exclusion Basis System-Based Key Predistribution Schemes 198
8.7.2 Asymmetric Key Management Schemes 198
8.7.2.1 RSA-Based Asymmetric Encryption System 199
8.7.2.2 ECC-Based Asymmetric Encryption System 199
8.7.2.3 ID-Based Key Agreement Schemes 199
8.7.2.4 Hybrid Schemes 200
8.8 Authentication and Integrity in WSNs... 200
8.8.1 One-Hop Authentication 201
8.8.2 Multihop Authentication 201
8.8.3 Broadcast Authentication 201
8.9 Secure Routing.. 202
8.10 Secure Location... 202
8.10.1 Secure Location Scheme with Beacons 202
8.10.2 Secure Location Scheme without Beacons 203
8.11 Secure Data Aggregation.. 203
8.11.1 Plaintext-Based Scheme 203
8.11.1.1 Scheme Defending Against One Compromised Node.......................... 204
8.11.1.2 Bidirectional Authentication Schemes 204
8.11.1.3 Neighbor's Certificate Schemes.. 204
8.11.1.4 Statistical Method 204
8.11.2 Cipher-Based Scheme 204
8.12 Open Issues and Challenges.. 205
References 206

8.1 Introduction

Currently in our vicinity, we can notice the growth in the number of such devices which are being used for monitoring our environment. The task of interconnecting such embedded devices for achieving various purposes is accomplished with the help of the Internet of Things (IoT). The IoT relies on wireless sensor networks (WSNs) for ensuring connectivity between nodes on the network architecture level. Some applications often require cryptographic security mechanisms to secure sensitive information. Securing WSNs is inevitable due to their inherent nature of constraint resources, unreliable communication, isolated environment, and strenuous deployment. The unattended and exhausting deployment scenario poses the risks of any unauthorized access or execution of security attack (Dinker and Sharma 2016) in the network. The unauthorized admittance of an adversary into the network may result in active or passive attack. In an active attack, an adversary may directly affect the network by replaying, modifying, delaying, redirecting the information, or injecting the fabricated data into the network or simply destroying the data. On the other hand, in a passive attack, the attacker monitors or listens to the wireless medium to extract or steal the important information. The malicious activities of an adversary influence the security of the data on the network in terms of its secrecy, accuracy, and integrity. These problems are due to the intruder who has exploited the vulnerabilities and threats of the WSNs and performed an attack to gain access to the information. To prevent such security attacks, there are security services or goals to be achieved, namely confidentiality, integrity, availability, privacy, etc. (Chen et al. 2009). To achieve security in WSNs is very complicated due to the constrained environment; thus studies and research are being carried out to address these security issues. This chapter presents the security goals along with the limitations of the WSNs. Various types of threats, vulnerabilities, and attacks in WSNs are discussed and an outline of the defense mechanisms on the basis of networking protocol layers is given. The motive is to present existing security techniques and mechanisms deployed in WSNs and give an insight into the remaining undefended issues and challenges in the area of WSN security.

8.2 Security Aspects in WSNs

WSNs are unique in their characteristics; hence, their security concerns and requirements are different. To understand them, we need to first delve into the basic concepts of security so that the security goals of WSNs can be achieved.

8.2.1 Security Goals

In WSNs, the communication design is multihop networking where the transmission model can be many-to-one, one-to-many, and other local unicast or broadcast among neighboring nodes. All these correspondences should be guarded and tamper proof. So, WSNs have the following security goals and most of the techniques emphasize accomplishing these goals (Tanenbaum 2013):

Confidentiality: Confidentiality or secrecy is the assurance of keeping information inaccessible to an unauthorized entity (Gurudatt et al. 2013). Confidential data is impenetrable and does not let the intruder reveal its content and its meaning to be used in malicious action.

Integrity: Integrity is to preserve data on the network from illegitimate modification or fabrication by the adversary. Integrity measures confirm that the data is free from any kind of alteration and changes by an attacker.

Availability: Availability is to ensure that wanted or required services may be permitted and obtainable to the allowable entity. Availability ensures the survivability of network services to legal users when demanded, irrespective of any obstacle in the network.

Access control: This security requirement is necessary to prevent unauthorized and illegal admittance and take possession of the information and network resources.

Data origin and entity authentication: This goal achieves security by authenticating the beginning point of a correspondence and the identity of communicating user/node/base station (BS). Authentication facilitates a node to make certain the distinctiveness of the peer node with which it is communicating.

Nonrepudiation: Nonrepudiation is the measure by which the corresponding entity/node cannot contradict the transmission of a message it has previously dispatched.

Authorization: Authorization establishes that only an authorized and legitimate entity/node can get access to network services or resources.

Privacy: It ensures that only the authorized entity will have the access to data.

Freshness: It may be about data freshness and key freshness. Since all WSNs deals with time varying measurements, there is a requirement of the data to be fresh, recent, correct, and unrevealed.

Moreover, in WSNs, as per requirement, new sensors are added whenever previous nodes wear out, so the secrecy of data going either forward or backward is also crucial for security. In *forward secrecy*, a node must not be endorsed to receive or send or know the messages which will be transmitted in the future in the network after it disintegrates from the network. In *backward secrecy*, a newly associated sensor node must not have access to any messages earlier sent on the network. In addition to the above-mentioned security goals or requirements, WSNs have performance-specific requirements based on their areas of applications as follows:

■ **Self-organizing:** This property of the WSNs enables them to reorganize the network as per conditions. Whenever there is a situation of any node draining or node failure, the network has the capability to reshape itself and continue the communication.
■ **Scalability:** This capability of WSNs enables them to sustain a large number of nodes. It helps in continuing the operation of the network, even when new nodes are added to it.
■ **Time synchronization:** To avoid collision and traffic manipulation in the network there is a requirement for proper time synchronization.
■ **Efficiency:** It is the best possible performance by network nodes as per on hand storage, processing, power, and communication resources available. This ability is expected in security schemes so that the lifetime of networks can be maximized.
■ **Survivability:** This property provides the network with a minimum level of facility in times of outages and failures or attacks.

8.2.2 Performance Metrics

Besides security goals and implementing performance-specific requirements, there are some metrics which help in evaluating an appropriate and efficient security scheme. In addition to the above-discussed security and performance requirements, an adequate and competent security scheme should also meet the following metrics to be successfully deployed in WSNs (Shi and Perrig 2004):

■ **Resilience:** Resilience is the ability of the network to provide and maintain an acceptable level of security service in case some nodes are compromised.
■ **Resistance:** Resistance is the ability to prevent the adversary from gaining full control of the network by node replication attack (Deb et al. 2003) when some of the nodes are attacked.
■ **Flexibility:** In consideration of the WSNs, application environment and unstable mission objectives, flexibility and self-organization (such as sensor networks fusing, nodes leaving and joining, etc.) become vital aspects while securing the network.
■ **Robustness:** A security scheme is said to be strong and substantial if it has the capability to continue to administer despite irregularities, such as attacks, node failure, etc.
■ **Assurance:** It is the capability of the nodes to propagate appropriate information at different assurance levels to the end-user (Li and Gong 2013). A security scheme must be designed in such a way to permit the network to deliver a different level of information with considerations to different desired reliability, latency, etc., with a different cost.

8.2.3 Security Limitations in WSNs

WSNs are wireless networks that have several constraints as compared to other similar networks. Some of these constraints include (Shi and Perrig 2004) the following:

■ **Limited resources:** The scarcity of resources makes implementation of security techniques difficult as they need a certain amount of resources for operations, e.g., data memory and processing power.

■ **Unreliable communication:** The security of the network greatly depends upon the defined protocols and communication medium which is wireless in nature.

■ **Unattended operation:** The sensor nodes may be left unattended for long periods as per the type of application of the particular WSN which make them more vulnerable to many attacks.

The overall objective of defining these limitations and performance criteria is to achieve the following:

■ To minimize the resource consumption and maximize the level of security performance

■ To identify security attacks on the WSN channel, including passive as well as active interference

■ To evolve less complex security schemes best suited for wireless communication, sensor network as the traditional wired and wireless ad hoc schemes do not work properly

■ To develop security schemes which are able to handle increased complexity as a result of large-scale deployment and node mobility

■ To propose a security scheme to manage dynamic topology of the network in view of node addition and node wearing out

8.3 Vulnerable Components

Security issues mainly exist as a result of the constrained environment of the WSNs which make them vulnerable to threats and provide a basic platform for execution of security attacks. The probability of occurrence of any attack is greater within WSNs than that of any other type of networks considering their placement environments and resource inhibitions. In WSNs, BSs are usually assumed to be trustworthy and authoritarian but they are also subjected to various attacks. Sensor nodes are another vulnerable point of attack, which requires investigation.

8.3.1 Base Station Security

The BS is the gateway of a WSN, which connects WSN to the outside world. A BS enables a WSN to communicate all the processed information to the outer world via wireless medium. In general, the BS has more computational and communication capabilities which make it more resilient to malicious activities like security breaches and attacks. The traditional security mechanisms for WSNs consider that the BS is secure and robust as compared to other nodes in the network. However, it may not be so. If the adversary has more powerful and capable devices to breach security, the BS may become a failure point. Hence, securing the BS is essential and requires more vigilance. Various researchers have come up with several schemes (Deng et al. 2003) to protect the BS from security attacks. One of the proposed methods promoted the deployment of multiple BSs to administer resistance against individual BS failure. There is another method which attempts to obscure the identity of the BS. The identity of the BS is hidden by encrypting the packets and the address filed in the packet headers using a pairwise shared key, between two neighboring sensor nodes. The address field does not carry node IDs in it, but a hash function which is utilized for construction of several anonyms for all the nodes. These anonyms of the nodes are used as either their source addresses or destination addresses. The generation and distribution of the pairwise shared keys are done by the BS during the network topology formation phase. Another method proposes relocation of the BS so that location tracking of the BS is difficult for the attacker. An attacker needs to know the place where the BS has been installed in order to launch an attack. The attacker can attain the objective of finding the location of the BS by analyzing traffic of the network. Henceforth, it is very essential to obscure the traffic flow pattern and routes. To prevent this three methods are proposed. In the first scheme, a multihop path is selected by the originating node for each data flow which renders an attacker unfamiliar with the path from which traffic may flow. The second scheme suggests the random creation of fake paths to confuse attackers. In the last scheme, multiple random areas are designed or created for communication activity to conceal the real location of the BS in the network. Some researchers suggested a location of the BS far from the area of activity, but it leads to increased overheads. To reduce the overhead of long path data transmission, the BS assigns mobile sinks which take up the task of collection and processing of data. Sinks in this case also handle other activities such as reprogramming of local sensor, compromised sensors identification and revocation, and other network maintenance. However, the mobility of the sinks is itself a security challenge because a mobile sink with too many privileges will attract the adversary's attention for performing attack and compromise. An adversary can effortlessly intrude in and access other nodes or even take over the whole sensor network through a compromised mobile sink. Thus, security mechanisms that can tolerate mobile sink compromises are essential. Researchers in Deng et al. (2005) proposed a mechanism to confine the privilege of a mobile sink without impeding its capability of carrying out authorized procedures for an assigned task. Each mobile sink carries an authentication service and may be called as authenticator for each task it may administer. Each sensor node is able to verify the authenticator from the mobile sink before performing the assigned task. When a mobile sink is behaving maliciously, the BS can revoke the ID of the compromised mobile sink in the network. So, it is imperative that the BS as well as sink needs to be secured against attacks. At the same time, the nodes of the network should also be protected as the data collection begins, propagates, and is forwarded from them only.

8.3.2 ■ Sensor Node Security

A sensor node is the smallest unit of WSNs which has low cost, low power, and low storage space. These nodes notice the occurrence of any real-world event of interestand process and communicate that information to the next node or level to be available to the end user for predetermined purposes such as healthcare, defense, monitoring, etc.

The resource limitations of these nodes may result in security breaches or attacks such as node capture, node replication, etc. The attacks can be performed from outside or inside of the network and categorized as external attacks and internal attacks. The external or outsider attack is performed when an unauthorized node which is not a legitimate member of the network attacks it. (Shi and Perrig 2004). External have two categories: passive and active. Passive attacks deal with unauthorized monitoring or "listening" to the information of packets in the channel. Active attacks, which are performed externally, interrupt working on the network by intercepting the communication channel, fabrication, or replay of data packets, denial-of-service (DoS) attack, jamming, etc. Internal or insider attack is performed when the attacker node is from the legitimate nodes. An adversary can perform an internal attack by compromising the sensor node. The compromised nodes are more harmful than the disabled nodes because they act maliciously and are hazardous to the network as they seek to disturb or incapacitate the network (Pandey and Tripathi 2010). It is very difficult and complex to find and recover from internal attacks, which leads to increased security problems and challenges (Krauss et al. 2007). When a node is compromised, it can act in many different ways, which are given below:

- It can give away important information such as encrypted data, etc. to the attacker.
- False information can be injected in the network through a malicious node.
- Such nodes are capable of giving wrong information about normal nodes as compromised nodes to the BS.
- The compromised node can also be used to launch routing attacks like selective forwarding, black hole, etc., to alter the routing path which may lead to collision, dropping of packets, etc.
- It may demonstrate uninformed behavior and may conspire with other nodes.

There are various ways in which a malicious or compromised node can behave and perform security attacks as mentioned above. So, it is necessary to prevent the node capture or compromise and devise node-resilient security schemes. The security issues and attacks lead to a demand for creation, implementation, and practice of security mechanisms and schemes. If no attack occurs then there is no need for security.

8.4 Attacks in WSNs

Attacks are categorized on the capacity of the attacker, information in transit, and host- and network-based (Sastry et al. 2013; Dinker and Sharma 2016).

8.4.1 Adversary's Capability-Based Attacks

In WSNs, the attacker may illegally intrude and affect the traffic directly or indirectly for some malicious purposes. An adversary, who may try to eavesdrop the wireless medium without directly affecting the information transmitted is called a "passive attacker." And an adversary called an "active attacker" may try to delay, replay, or inject fabricated messages in the original data stream. The attack can be performed from outside or inside the network as per the origination of attack. The attacks that are sourced from sensors, which are not part of the WSN, are outsider attacks and insider attacks are performed by genuine sensors, which got compromised. The attacker may utilize various devices which have similar or higher performance capabilities as that of nodes of the target WSN to launch an attack. When the attacker attacks the network with a small number of sensors with equivalent capacities as the WSN sensors, it is known as a mote-class attack. When devices with superior transmission range, processing power, etc., are used by an attacker, then laptop-class security attacks are performed.

8.4.2 Information in Transit-Based Attacks

In WSNs, sensors examine the occurrence of an event and subsequent changes in parameters and other values and forward them toward the sink. The information sent to the sink is the processed report of the network activity, which should reach the sink correctly and completely. This information if stolen by an attacker can be misused to gain unwanted advantage and compromise the nodes in the network. The aim of the adversary is to disseminate false information and deceive network users. The attacker can attack the information traveling on the network and can perform replay attack, DoS (Wood and Stankovic 2002), etc.

8.4.3 Host-Based Attacks

The host-based attacks are the attacks in which the system or the entity using that system may be corrupted and behave in an unexpected manner. This attack can be divided into user compromise, hardware compromise, and software compromise. In the case of user compromise, the entities that are accessing the network are misled and made to reveal the credentials, key material, etc. Hardware compromise is done by the attacker when he either tampers with the hardware machinery of the node or captures the node itself to destroy the node and information stored in it. In software compromise, the adversary tries to intrude in the network to access the 0 node's software running inside it to launch a malicious attack. By compromising the node, the attacker can capture the node and perform attacks like selective forwarding by node, Sybil attack (Newsome et al. 2004), etc.

8.4.4 Network-Based Attacks

The network-based attacks are the attacks which are being performed on the communication layers of the network protocol. Such attacks may be performed from inside or outside the network in which sometimes the attacker does not want to cause direct loss like modification, destruction, or fabrication of information but wants to access the network for his own advantage. Then, he may deviate the protocol from its planned behavior. Based on the network/protocol stack, the layerwise problems are as follows.

8.4.4.1 Physical Layer Attacks

The physical layer of the communication protocol is responsible for connecting two systems for data transfer or communication. This layer supports the source and destination to communicate through the wires or other networking medium. If this layer is attacked, it may result into total unavailability of services to the user. These attacks can be of following types:

1. **Jamming:** In this attack, the attacker tries to disrupt the network services by sending a signal of high energy toward the target which results in unavailability of the required signals. Further, these attacks can be of four types, the first of which is called constant jamming; in this, the packets get corrupted as soon as they are transmitted. Second is deceptive jamming in which network is misguided with a constant data stream similar to the legitimate data stream being sent into the network. The third type is random jamming; in this, the energy loss is prevented by randomly interchanging jamming and sleep or inactive state. The fourth is reactive jamming, and, as the name suggests, it reacts by broadcasting a jam signal whenever any data transmission happens.
2. **Tampering or destruction:** In this, the target node can be physically captured or tampered with, by the attacker. By doing this, the attacker can gain access to the content of the node which can be encryption/decryption keys, algorithms, or similar information.

3. **Radio interference:** This is basically the hindrance in communication caused by attacker using radio signals. This action of the adversary can be regular or irregular, creating problems in sending and receiving of the data. The classification of the security attacks is given in Figure 8.1.

8.4.4.2 Data Link Layer Attacks

This layer has the function of transmitting data on a physical link and provides networking media. It is generally associated with network access, topology, packet delivery, flow control, etc. So, on this layer the attacker can perform attacks to interrupt these functions of the data link layer and those attacks may include the following:

1. **Collision:** This type of situation occurs when two nodes in the network try to transfer data at the same time and at identical frequencies. Such simultaneous data transfer may result in the collision of the packets, which may further change the data content of the packets. This change causes an error in the data called checksum error that occurs at the destination. Such kind of errors will be noticed as rejected packets due to checksum mismatch at the receiving end, which is a loss of resources.
2. **Continuous channel access or exhaustion:** Such kind of situation may arise due to the presence of a compromised or malicious node in the network. This node disturbs the media access control protocol (MACP), by constantly engaging the channel. By doing this, the attacker node keeps the channel busy and so the other nodes starve for using the channel.
3. **Unfairness:** The unfairness in the network results due to repeated exhaustion of the nodes, occurrence of collisions at MACP layer, or an unauthorized/over usage of privilege methods

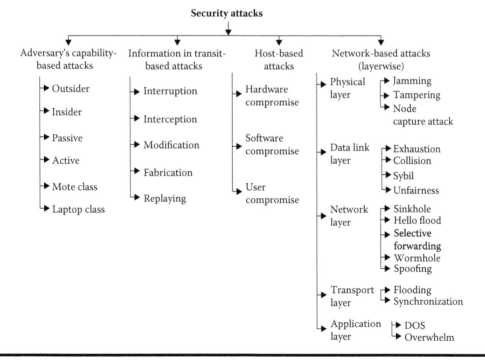

Figure 8.1   Security attacks in WSNs.

of cooperative MACP layer. This condition in the network can be called as an inequitable DOS attack which results in degraded performance.

4. **Sybil attack:** In this attack, the compromised or malicious node behaves as the legitimate node and poses as different nodes at the same time. The attacker node creates fake communication with other nodes as it can have multiple identities at the same time in the same network.

8.4.4.3 Network Layer Attacks

This communication layer is responsible for addressing, routing, and end-to-end delivery of the packets. So, the attacker can create problems in routing or packet delivery, etc., and affect the network operation and security through following the attacks:

1. **Sinkhole:** This attack is experienced when almost the whole traffic is lured toward the malicious sensor node, which creates a scenario of the symbolic sinkhole with attacker node being in charge at the core. This attack results in a loss of resources.
2. **Hello flood:** This attack uses the Hello packets sent by nodes, which are normally required in various protocols to show their existence in the neighborhood. When a node accepting these packets means it is in radio range of the sender node and start to communicate with this node, whereas actually this node may be very far from it. This situation creates a lot of confusion within the network and disturbs the flow of traffic and topology.
3. **Selective forwarding/black hole attack/neglect and greed:** This attack is experienced when the attacking node receives the packets, but refuses to forward certain packets or messages and drop them in the middle. When the attacker node dumps each and every packet through it, then it is called black hole attack. On the other hand, if the node decides to forward some packets and not to forward some others, then selective forwarding occurs. This attack creates confusion and loss of resources in the network.
4. **Node capture:** This attack is associated with the node's physical tampering and confining by the attacker. By taking over the node the attacker can use it to launch malicious activities and disrupt the operation of whole network.
5. **Wormhole attack:** In this attack the adversary tries to compromise the node and routes the data packets over a less secured path and collect the traffic in that portion of the network. Then the attacker replays that collected traffic in some other network division.
6. **Spoofed, altered, or replayed routing information:** In this attack, the intruder tries to spoof, that is, create false copy, alters, or replays the routing information of the network to create ambiguity and faults in the network traffic. These confusions can make loops in route, shorten or lengthen the routes, divide the network into different divisions, etc. (Gurudatt et al. 2013).

8.4.4.4 Transport Layer Attacks

This layer of the communication protocol stack helps in transmission of the packet to the destination and reassembling them. To disturb the data delivery and create vagueness, the attacker can perform the following attacks:

1. **Desynchronization attack:** In this attack, the adversary's aim is to exhaust nodes of the network of their energy and disturb the synchronization at sending and receiving end of messages. Here, the invader tries to corrupt the messages, which create mismatch at the

reception of packet. Due to this, there is a request to resend the neglected message packets, which if again captured by the attacker will result into no communication at all.

2. **Flooding:** In this, the compromised node continuously sends request to make new connections, till the instance the available resources for each connection are consumed. It creates severe resource restrictions for genuine nodes (Kavitha and Sridharan 2010).

8.4.4.5 Application Layer Attacks

This layer acts as the interface for different user applications and enables access to the Internet. This layer introduces synchronization, data integrity, and error control. To interrupt the interoperability and functionality of this layer, an attacker can perform the following attacks:

1. **Path-based DOS attack:** In this attack, the adversary intercepts the network and puts in spurious or old packets at the leaf nodes. This action of the adversary does not let the legitimate traffic flow in the network because the lack of resources prevents other nodes also from sending reports to the sink.
2. **Overwhelm attack:** Here, the adversary tries to overtake network nodes with sensor stimuli and due to this, the nodes generate a lot of data and forward these toward the BS. These attacks deplete network bandwidth and energy.

The effect of these attacks performed at the communication protocol layers may lead to node outage, service interruption, interception, data modification, data unfairness, eavesdropping, etc. The network may be affected partially or fully. To overcome these attacks and other issues, there are many security mechanisms proposed and are discussed in the rest of this chapter.

8.5 Security Mechanisms

Security has become a key issue for WSNs as the nodes are deployed in a hostile environment with limited energy, smaller memory, and processing power, which have resulted in the security attacks on WSNs. The motive of any security mechanisms is to detect, prevent, and recover from the security breaches as and when they occur. To secure the WSNs effectively, the security mechanism must fulfill certain criteria like resiliency, fault-tolerance, energy efficiency, scalability, flexibility, self-healing, etc. A security scheme should propose choices in view of desired qualities like reliability, latency, etc. (Karlof and Wagner 2003). There are several security mechanisms proposed by various researchers to encounter attacks and other issues related to the security in sensor networks. Broadly, the security schemes may be based on cryptography, frameworks, secure routing, intrusion detection system, secure groups, authentication, key management, etc. (Zhou et al. 2008). The security scheme should have the ability to combat the attacks and reduce the impact of the attacks on the network so that it can continue the network operations. In this section, we have discussed the security mechanisms based on the attack and their effects on the network.

8.5.1 Attacks-Based Security Schemes

An attacker can launch an attack in a WSN as per his capability and resources. On the basis of his ability and resource capability, he can launch attack from either inside or outside the network with devices of similar or more functionality. The WSN can be protected from such kind of attacks by deploying strong and robust security mechanisms. Such mechanisms are designed in consideration of the constrained capabilities of WSNs. The attacker may have an intention to affect the network by simply eavesdropping, compromising the nodes or taking over the whole network itself. When the attacker directly affects the network he can interrupt the network operation, intercept the flowing traffic, inject, or fabricate the information in traffic. These attacks can be prevented by implementing a security scheme which may address intrusion detection and prevention, authentication, tamper resistance, etc. (Huang et al. 2011).

In WSNs, an attack on information in transit results whenever any event occurs and the sensors report it to the sink. The information being sent may be compromised to supply false information to BSs or sinks. This may lead to information interruption, interception, modification, fabrication, and replaying. Such situations threaten the message availability, confidentiality, and integrity. The key motive of such attacks is to obscure or deceive the communicating parties. This results in DOS attacks, flooding attack, etc. These attacks may be prevented by use of spread spectrum for radio communication and limiting the number of connections by a node (Bekara et al. 2008).

The attacks by the intruder such as user compromise, hardware compromise, and software compromise become the target for extracting vital information in the WSN such as passwords, encryption and decryption keys, operating system, and other communications facilitating information; this leads to spoofing, node capture, and compromise, which further lead to more compromised pairwise keys and therefore affect the security of the network.

Security attacks on the network layers target the information exchange happening over the different protocols of the layer and during this communication, the physical layer mostly suffers from jamming problem which involves DOS attack. In order to overcome this problem, the techniques like spread-spectrum for radio transmission can be used in WSN. The issue related to the MAC layer jamming can be managed with admission control mechanisms in which the network layer first records the network area which is jammed and then routing is done to handle that part of the jamming happening in that affected area. Their techniques can unite the arithmetically analyzed received signal strength indicator (RSSI) values, carrier sense time (CST), which is the regular time needed to sense an unoccupied channel, and the packet delivery ratio (PDR) techniques to recognize the type of jamming in the network reliably. The cryptographic technique of symmetric key with delayed disclosure of the keys can be exercised to handle radio interference. The issue of tampering or destruction can also be managed with tamper-proofing the node's physical pack up. The technique of self-destruction, that is, tamper-proofing packages, can also be used to vaporize the memory content of the node as soon as they are physically contacted which further thwarts any outflow of sensitive data. Fault-tolerant protocols must be resilient to tackle such attacks as they are created for WSN (Walters et al. 2006).

When the data link layer is attacked and a malicious node repeatedly requests or sends packets over the channel to prevent the legitimate nodes in the network from accessing the channel, then they are handled in a different manner. This attack can be prevented by exploiting countermeasures like rate limiting MAC access control and stopping the unnecessary requests in the network which prevent the exhaustion due to continuous transmissions. Another technique of time-division multiplexing can be used in which a node is allotted a time slot for transmitting. Another issue is a collision, which can be prevented by using various error-correcting codes. Continuous exhaustion or collision-based attacks on the MAC layer may result in unfairness, which may be called as partial DOS attack, the self-protective measure against this attack is the application of small frames. This will help every single node to stop the channel for a little duration only. Sybil attack, in which a particular compromised node operates as diverged Sybil nodes can be prevented by proper identity verification and authentication (Yin and Madria 2006).

The network layer also suffers from attacks like sinkhole, Hello flood, node capture, selective forwarding, etc., for which there are different security mechanisms, which can be applied to prevent such attacks. For example, geo-routing protocols provide resistance to sinkhole attacks, the reason being its topology. Geo-routing protocols (Hung et al. 2007) have a special topology which is assembled with just localized information, and traffic is forwarded through the BS's actual site that hampers the creation of a sinkhole. In a Hello flood attack, advantage of Hello packets is taken which is a requirement in many protocols to be broadcasted to the neighboring nodes. Authentication and bidirectionality verification of a link are the solutions (Singh et al. 2010). Selective forwarding in WSNs is experienced when a malicious or attacking node refuses to transmit certain packets and lose them. To overcome such an attack, a combination of multipath routing with arbitrary selection of paths to the target node can be used in our network. Another solution for this attack is the braided paths, which allow the packets sent as such on the route, because it involves no common link or two consecutive common nodes or the usage of hidden acknowledgments. A wormhole attack is generally performed in coordination with two attacker nodes having the knowledge of their distance from each other. To protect the node against this attack, either the traffic is always directed along the geographically shortest route to the BS or rigid time synchronization can be deployed amid the nodes, which is difficult practically. Spoofed, altered, or replayed routing attack in the WSN is mainly performed in opposition to a routing protocol for extracting the steering information. Spoofing and alteration can be removed by affixing a message authentication code (MAC) with the packets and by using different cryptographic and authentication mechanisms, which may be used to secure spoofing attacks in WSN (Healy et al. 2009).

A transport layer may be flooded with repeated requests for new connections till all the resources necessary for the functioning of the network are drained. In order to handle this problem, it is to be made sure that the connecting clients express their commitment toward the connection by either answering a puzzle or by limiting the count of associations from a node. Another issue of desynchronization is when an intruder repeatedly forges messages to request the transmission of missed frames. A possible solution is to check the authenticity of each and every packet including control fields sent or received between nodes. Header or full packet authentication (Tajeddine et al. 2014) can overcome this type of attack.

An adversary may attack the application layer by attempting to overpower the nodes of the network with sensor stimulus. The mitigation of such an attack is possible by carefully tuning sensors in such a way that only the specific predefined stimulus triggers the nodes. In order to reduce the effect of this attack to a particular level, the mechanisms of rate-limiting and capable data aggregation can be used. The DOS attack can deprive the network of valid traffic to pass through it. The combination of techniques such as packet authentication and anti-replay may be able to mitigate these attacks (Das 2009).

8.6 Cryptography

Cryptography is the study and art of encrypting the simple data or plaintext and decrypting coded data or cipher text for security from adversary or attack. For the protection of the sensor nodes from different security attacks such as Sybil attack and blackhole attack and maintain data confidentiality and integrity, there is a need for robust encryption techniques to be deployed in the network. The encryption techniques may involve the deployment of both symmetric key ciphers and asymmetric key ciphers. The first one is symmetric key cryptography whose main theoretical frame was recognized by Claude Shannon and published in his classic research paper "Communication Theory of Secrecy Systems" (Shannon 1949). In a symmetric key system, the sender and receiver share and use a common key that is saved and kept secret from others. The sender encrypts a plaintext M with the key K by an encryption algorithm E to get a cipher text $C = E(M, K)$. At the receiving end, the cipher text C and the key K are given as input by the receiver into a decryption algorithm D to get the original readable plaintext $M = D(C, K)$. The other widely used type is the asymmetric key cryptography, which was earlier studied in Diffie and Hellman (1976). In an asymmetric key system, each communicating entity has a pair of keys $\{K_s, K_p\}$ to be used for encryption and decryption. The entity must keep his/her private key K_s secret and safe, while he/she may reveal and publish his/her *public key* K_p. For secure communication, the sender sends a plaintext M to the receiver, the sender uses the receiver's public key, K_p, to encrypt M to get a ciphertext $C = E(M, K_p)$. When the receiver receives the cipher text, only the legitimate receiver can decrypt the ciphertext using his/her private key and yield $M = D(C, Ks)$ because only the receiver knows his/her own private key, K_s. Since a public key is being utilized here, asymmetric key systems are also called public key systems.

8.6.1 Symmetric Key Cryptography

Symmetric key cryptography uses the same key for encryption and decryption as shown in Figure 8.2. The symmetric key systems such as Advanced Encryption Standard (AES) (Daemen and Rijmen 2013), Data Encryption Standard (DES) (FIPS PUB 1993), or Rivest Cipher 5 (RC5) (Rivest 1996), etc., is most logical and efficient to be deployed in such limited resources of the sensor network. This secret key system requires scrambling or substitution operations, hashing, rotation, or shifting, etc., which can be efficiently designed and implemented in hardware or software.

However, the use of DES is rather restricted due to the fact that it had been broken and then there was a requirement of a better cryptographic system for protection of data. In view of the limitations of the DES, other symmetric cryptography systems have been proposed, including Triple DES (3DES) and AES, which are comparatively secure and hard to break. AES is one of the most symmetric cryptographic techniques till date and it is expected that AES will be the leading symmetric-key system for many commercial applications for the next few decades. Most of the security protocols in the literature for WSNs are based on symmetric key technology. But, this may result in the need for more security even at the cost of additional energy and memory where the researchers need to use asymmetric key ciphers in WSN.

8.6.2 Asymmetric or Public Key Cryptography

Asymmetric or public key cryptography uses different keys to encrypt and decrypt namely public and private keys as in Figure 8.3a and b. Asymmetric key systems such as the Rivest–Shamir–Adleman

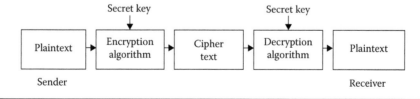

Figure 8.2 Symmetric key cryptography.

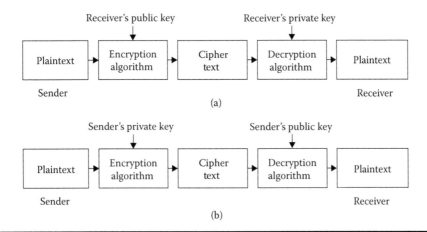

Figure 8.3 Asymmetric key cryptography: (a) Encryption with public key; (b) Encryption with private key.

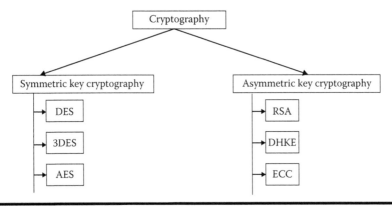

Figure 8.4   The main cryptographic scheme.

(RSA) (Rivest et al. 1978) algorithm, Diffie–Hellman key exchange (DHKE) (Diffie and Hellman 1976), digital signature standard, etc., are very secure and robust when compared with the symmetric key system. The important cryptographic techniques are given in Figure 8.4.

Although they are secure, their implementation in WSN environment results in the consumption of more memory and processing power than symmetric key cryptography (Amin et al. 2008). An asymmetric cryptography technique called elliptic curve cryptography introduced by Koblitz (1987) is said to be comparatively more efficient than other asymmetric key algorithms. Elliptic curve cryptography (ECC), due to its comparatively lesser key size, is being used for authentication purpose in different WSN deployments (Wander et al. 2005). It is worthy to note that the security provided by a 160-bit ECC key is almost the same level of security provided by a 1024-bit RSA key, but the key lengths of ECC make it more preferable and efficient. ECC has attracted many researchers because of its limited key size (160 bits) and greater security (Wei-hong et al. 2009; Shah et al. 2010; Huang et al. 2011, and so on).

Asymmetric key cryptography provides better security, but it requires more computation resources than symmetric key cryptography does. On the other hand, symmetric key cryptography is more energy efficient, but difficult for key deployment and management in WSNs.

Cryptographic methods used in WSNs should meet the constraints of sensor nodes and be evaluated before choosing. These cryptographic systems are used in designing security schemes for data and networks.

8.7 Key Management

Key management is very important and complex, especially in symmetric cryptography structure. Cryptographic methods involve the use of keys (symmetric or asymmetric) and these keys need to be handled carefully. Therefore, there are various researchers who have developed and designed algorithms and frameworks for secure distribution and management of the very critical key material. The key distribution can be done in three ways: randomly, predetermined (predistributed or stored in the node), and hybrid. Typically, key management is the process by which cryptographic keys are generated, stored, protected, transferred, loaded, used, and destroyed. The main objective of key management is to establish and maintain secure channels among the communicating parties (Shanyue and Liqing 2013). Classically, key management schemes use keys for the secure and efficient (re)distribution, and at times, generation of the secure channel communication keys to the communicating parties. Communication of keys may be through pairwise keys, which are used to secure a communication channel between two nodes that are in direct or indirect communications (Eschenauer and Gligor 2002; Liu and Ning 2005), or they may be grouped keys shared by multiple nodes. Network keys (both administrative and communication keys) may need to be changed (rekeyed) to maintain secrecy and resilience to attacks, failures, or network topology changes. Whatever may be the key distribution method, there is a requirement of secure key distribution and management. Sensor network dynamic structure, easy node compromise, and self-organization property increase the difficulty of key management and bring broad research issues in this area. Due to the importance and difficulty of key management in WSNs, there are a large number of approaches focused on this area. Based on the main technique that these proposals use or the special structure of WSNs, we classify the existing schemes as symmetric key management schemes, asymmetric key management schemes, and hybrid key management in WSNs.

8.7.1 Symmetric Key Management

The security mechanisms designed for WSNs due to their constrained capabilities are mostly based on symmetric key management, as these consume less computation power and time. These schemes are devised using different considerations and assumptions for key establishment, key distribution, key revocation, etc. Accordingly, these keying schemes can be classified into various categories as in Figure 8.5. These schemes are discussed as follows.

8.7.1.1 Entity-Based Schemes

Entity-based schemes, which may be called as arbitrated schemes are based on trusted entity for key distributions and key establishment.

- **Master key-based predistribution scheme:** In this scheme, a master key is to be predistributed and stored in each sensor node of the network. A random number and the decided master key communicated within nodes help to establish pairwise keys between each sensor. The

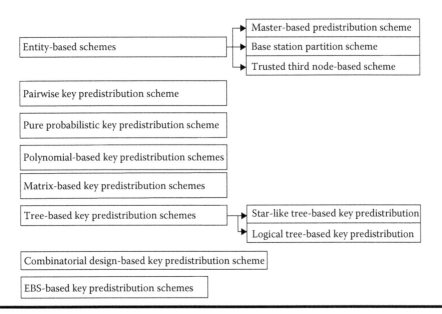

Figure 8.5 Symmetric key management schemes.

advantage of this scheme is that it provides scalability and there is less memory consumption in nodes. However, the disadvantage is that if the master key is compromised, all the pairwise keys are exposed which means that this scheme is not resilient and node authentication is also not considered. An improved scheme presented that the master key should be erased after the establishment of the pairwise keys. Although resilience was improved, scalability and authentication were not there.

- **BS participation scheme:** In this type of scheme, the BS plays a vital role in distribution of the keys to sensor nodes. SPINS is a BS participation scheme presented by Perrig et al. (2002) which enable each sensor to store a shared key with the BS. Whenever two sensors are required to communicate, a pairwise key can be sent by the BS, which is encrypted with the shared key. This scheme provides resiliency, but not scalability.

- **A trusted third node-based scheme:** A peer intermediary's participation for key establishment (PIKE) scheme was given by Chan and Perrig (2005). This scheme relies on a common trusted third node for the key establishment between two nodes. This scheme provides resiliency and scalability as there is no need to store any master keys in the node, but if the trusted third node is compromised, it may lead to other nodes also to be compromised.

8.7.1.2 Pairwise Key Predistribution Scheme

For the secure and distinct pairwise key establishment and distribution between each pair of sensor nodes a scheme was proposed by Chan and Perrig (2003). As per the scheme, a predistributed key is stored in each node before deploying the node in the WSN. This scheme offered good resiliency and authentication, that is, even if one node is captured, the keys of other nodes are safe. But, this scheme is not good for large WSNs as scalability is the issue here and because each sensor has to store a number of keys equal to the number of sensors in the network.

8.7.1.3 Pure Probabilistic Key Predistribution Schemes

The above-discussed schemes have few limitations for large or dynamic WSNs, the reason being that the storage of too many keys, joining and leaving, or rekeying of sensor nodes are complex and expensive in terms of resources. These disadvantages were addressed in the probabilistic key predistribution scheme (Eschenauer and Gligor 2002). In this scheme, communication keys are established in three phases: key predistribution, shared-key discovery, and path-key establishment. For key predistribution, a large pool of P keys is generated and then k distinct keys out of P are drowned and loaded into each sensor node memory. The security of a communication depends upon the key connectivity for discovering the shared key and establishing the path key. Connectivity of the WSN can be analyzed and studied from the random graph theory (Spencer 2000). A graph of n nodes, in which the probability that a connection can be established between two nodes is p, can be called as a random graph. The link can be the shared key in the WSN. This scheme provides good authentication and scalability, but resiliency to sensor node capture is still a concern.

Another scheme called the q-composite keys scheme (Chan and Perrig 2003) used q common keys to form the shared key with a hash function instead of using a single key. With the use of at least q number of keys, this scheme made the network's resistance toward node capture attack stronger and robust. But, the network became weak when there were a large number of nodes under attack. Multipath key reinforcement scheme (Chan and Perrig 2003) strengthened the security by establishing the link key through multiple paths. The limitation of this scheme was the increased communication overhead.

A random key predistribution scheme grouped the nodes using the information of the node implementation by Du et al. (2004). It is noticed that the neighbor nodes are from the same group in general, so the choice of keys can be made from dissimilar pools of keys. The small size of the key pool, results in better key connectivity. This leads to increase in both resiliency and efficiency.

8.7.1.4 Polynomial-Based Key Predistribution Schemes

The basic structure of the polynomial pool-based pairwise key predistribution scheme in WSN was developed by Liu and Ning (2003). Two more key predistribution schemes, one based on random subset assignment and another based on grid were suggested by the same authors. In this scheme, the keys were distributed through polynomials (Blundo et al. 2001). A pairwise key predistribution using polynomial pools utilized a multiple random bivariate polynomials. When the polynomial pool had only one polynomial, the general framework degenerated into the polynomial-based key predistribution. The pairwise key establishment was carried out in three steps: setup, direct key establishment, and path-key establishment. This scheme offered better security and the scalability in WSNs.

Closest polynomial scheme (Liu and Ning 2005) utilized the knowledge of node deployment for distributing keys (predistribution) in a sensor network. This scheme, by utilizing the location information, can prevent node compromise and improve the performance.

8.7.1.5 Matrix-Based Key Predistribution Schemes

A matrix-based group key predistribution scheme (Blom 1985) in which a symmetric matrix Kn^*n stores all pairwise keys of n nodes group, where the key of node i represented by element k_{ij} for securing the link with node j. Each node i stores the ith row of the secret matrix and the ith column of the public matrix G. After network deployment, each pair of nodes i and j can individually figure a pairwise key $k_{ij} = k_{ji}$ by exchanging their columns in plaintext as the key is the product of their own row and the other's column. Their rows are always kept secret. This scheme is said to be *l*-security means if more than *l* rows are compromised, the entire secret matrix can be extracted or broken by an attacker.

Another group-based key management scheme based on Blom's scheme was presented by Yu and Guan (2005) which used sensor deployment knowledge. This scheme deployed the sensors into hexagonal grids as hexagonal arrangement offers the best approximation area. So, sensor nodes which are deployed into a grid were then divided into groups. The careful usage of deployment knowledge and hexagonal grids led to less memory requirement as the number of neighbors for every node reduced considerably. The neighbor to a node can be from its own group and adjacent groups. Here, secret information was distributed instead of secret keys to form pairwise keys for nodes. A unique secret matrix was provided to every group. Then the pairwise key was generated using some other secret matrix for neighboring nodes in adjacent sensor groups.

The Blom's matrix-based key scheme and the random key predistribution scheme were again utilized in another pairwise key predistribution scheme (Du et al. 2005) for WSNs. In this scheme, for less memory consumption for storing key material, a connected graph was used. This scheme assigned keys only to connected graph whereas Blom's scheme needed all the nodes to be linked. Scalability and resiliency issues were addressed in this key distribution mechanism.

8.7.1.6 Tree-Based Key Predistribution Schemes

This key management mechanism predistributed the keys to the nodes arranged in a tree-like structure in the network. This can be further of two types: the star-like tree and binary logical tree.

Star-like tree-based key predistribution schemes: In order to develop resilience against node capture, deterministic key predistribution schemes (Lee and Stinson 2005) were presented. These schemes were based on strongly regular graphs and random graphs correspondingly. A strongly regular graph can have parameters (n, r, λ, μ) on *n* vertices, without loops or multiple edges. In this *r* represent the degree *r* (with $0 < r < n-1$) and is such that any two adjacent vertices have λ common neighbors or μ common neighbors for nonadjacent neighbors. The first scheme is an ID-based one-way function scheme in which the use of a one-way function *h* has reduced the number of stored keys in a node. A unique ID is assigned to each node for calculating secret key and a hash key may also be allocated to it if it has a star-like subgraph. This scheme is advantageous over the probabilistic random scheme as the number of keys computed is reduced by half and it has good resiliency. The shortcoming of this scheme is that it is not suitable for large WSNs.

Logical tree-based key predistribution schemes: A group key management algorithm using the hierarchical binary tree (HBT) for key predistribution in WSN was proposed by Wallner et al. (1999). In this scheme, there were keys associated with a tree structure which is retained only by one group controller, where every node corresponded to a key encryption key (KEK). Every node as a component of the group communicates to a leaf of the tree and keeps a node's KEK from its leaf to tree roots. Then the root of the tree keeps the group key.

A tree is called balanced if each leaf stores log 2*n* keys; here *n* is the number of members/leaves and log 2*n* is the height of the tree. When the group is joined by a new node, a fresh leaf is added to the group tree. Earlier, this logical tree-based key technique was used for multicast key distribution and then it was extended to WSNs (Di Pietro et al. 2003; Lazos and Poovendran 2003, 2002). This technique is categorized as the centralized group key management protocol, which is often not considered ideal for WSNs.

Later, a new secure multicast scheme was presented by combining the directed diffusion (Intanagonwiwat et al. 2000) with the logical key hierarchy for WSNs called LKHW (Di Pietro et al. 2003). This scheme took advantages of both of them. LKHW scheme required establishing a secure group with group initialization before directed diffusion. Here, the security of the data transferred was very important.

8.7.1.7 Combinatorial Design-Based Key Predistribution Schemes

Key predistribution schemes based on the combinatorial design theory (Colbourn and Dinitz 1996) were proposed (Camtepe and Yener 2007). The combinatorial design theory was based on the existence and construction of systems of finite sets whose intersections have specified numerical properties. Several schemes to generate keychains were proposed by the authors for WSN based on the symmetric designs and generalized quadrangles. The main limitation of the combinatorial approach was the difficulty in construction, but the hybrid schemes may provide better scalability and security.

8.7.1.8 Exclusion Basis System-Based Key Predistribution Schemes

The exclusion basis system (EBS) was proposed (Eltoweissy et al. 2006) aiming at improved key management efficiency and reduced overhead in group communications. EBS was based on combinatorial optimization methodology for key management of group communication networks. The EBS approach was considered scalable for a large network and has great flexibility in network management. The limitation of this scheme is that it had collusion problems which were addressed in Huang and Du (2005) another EBS-based scheme called SHELL. It decouples the key management functions and allocates them to multiple nodes to provide attack/failure resilient solutions and avoid uneven resource utilization by nodes. SHELL is scalable, hierarchical, efficient, location-aware, and light-weight for group key management.

Another dynamic key management scheme for WSN called localized combinatorial keying (LOCK) (Eltoweissy et al. 2004) was proposed which was an improved version of SHELL. This scheme performed localized rekeying to minimize the communication overhead by using two layers EBS. LOCK used key polynomials which enhanced resiliency to collusion.

8.7.2 Asymmetric Key Management Schemes

A public key cryptosystem is widely accepted and used for providing security of data and networks in the realm of the Internet. RSA (Rivest et al. 1978) and ECC (Schneier 1996) are two major asymmetric cryptographic key techniques. It is worthy to note that the security provided by a 160-bit ECC key is almost the same level of security provided by 1024-bit RSA key. The relative performance of ECC over RSA becomes higher as the word size of the processor is decreased as in the case of an 8-bit CPU. Both techniques provide robust data security and integrity. But, some researchers believe that these techniques are not very suitable for WSNs because of their key sizes and computational complexities. Even then, many researchers have explored asymmetric techniques for designing key management schemes for WSNs (Gura et al. 2004; Watro et al. 2004; Karlof et al. 2004; Gaubatz et al. 2004). They were also successful in implementing these schemes in the constrained environment of the WSNs. It is said that when a robust security system is required, we have to afford some resource consumption also like in defense communication and surveillance systems.

8.7.2.1 RSA-Based Asymmetric Encryption System

The most common public key algorithm is the RSA cryptosystem, named after its inventors (Rivest, Shamir, Adleman). RSA is a block cipher which uses two exponents e and d, where e is public and d is private. Suppose P is plaintext and C is the cipher text. The sender uses $C = P^e \pmod{n}$ to create a cipher text C from plaintext P. The receiver uses $P = C^d \pmod{n}$ to retrieve the plaintext where n is the product of two distinct prime numbers. This technique provides considerable confidentiality and integrity which lead the application of this technique for securing communication in WSNs. TinyPK (Watro et al. 2004) protocols were implemented for securing sensor networks with proper authentication and key agreement between sensor nodes. The implementation of these protocols was done on UC Berkeley MICA2 motes and TinyOS as a development environment.

DHKE (Diffie and Hellman 1976) was also implemented on the MICA2 platform. The aim of this algorithm was to securely exchange a shared secret key between two parties to be used for encryption of the messages. The authors used DHKE for generating a suitable secret key which can be used as a new or replacement for TinySec key. Two disjoint WSNs can communicate using such a secret key and allow the addition of new nodes into the existing network without any look up or loading of the current key in use. This algorithm promotes scalability and resiliency in the network.

8.7.2.2 ECC-Based Asymmetric Encryption System

ECC (Schneier 1996) is very efficient and is frequently used in the current network and data security mechanisms. Since the set of elliptic curves used are infinite in number, the eavesdropper or attacker finds it difficult to decrypt the data used in communication or other related means. In addition to that, elliptical curve cryptosystems are efficient, as the size of keys used is small. Also, they require less computational power requirements and perform better than RSA for WSNs. Based on ECC a public-key infrastructure (PKI) for key distribution in TinyOS was proposed (Malan 2004). Their scheme demonstrated that the time taken for the generation of a public key and shared secret key distributed among nodes is almost same. The authors also gave a working implementation of DHKE based on the elliptic curve discrete logarithm problem (ECDLP).

Another ECC-based scheme and authentication was proposed by Ren et al. (2007). This scheme was able to successfully authenticate a user out of a set of n sensors if he authenticated any subset of sensors. These researchers proposed many ECC-based n-authentication schemes and other cryptographic techniques including the Bloom filter, Merkle hash tree, and the partial message recovery signature scheme. There are two basic schemes, first one is the certificate-based authentication (CAS) and the second is the direct storage-based authentication. In CAS, each user (not a sensor node) of the WSN assigned a public and private key pair (PK and SK). The user has to sign every message he broadcasts with his SK using a digital signature algorithm such as ECDSA (Hankerson et al. 2004). The sink, which is also known as the network planner in WSNs, has its PK and SK. The sink serves as a certifying authority (CA); it issues each user a public key certificate to prove the user's rights over his public key. The second, the direct storage-based authentication scheme eliminated the existence of public key certificate which reduced the message overhead and computational cost. This scheme enabled all sensor nodes to store their current ID information and corresponding public keys.

8.7.2.3 ID-Based Key Agreement Schemes

ID-based key agreement schemes were proposed and implemented taking ECC as a foundation. The identity-based encryption (IBE) idea was proposed by Shamir (1984). The motivation of

IBE is to get rid of the complexities of the certificate-based public key cryptographic system in which one user has to validate another's public key certificate. An IBE-based scheme (Boneh and Franklin 2001) promoted the usage of an arbitrary data (ID) which can be his/her email id, etc., for computation of the user's public key instead of taking it from a CA-issued certificate. An IBE scheme may typically have four randomized algorithmic steps, namely setup, extract, encrypt, and decrypt. Another ID-based key agreement scheme was proposed (Yang et al. 2006) which computed public and private keys through a function and used a secret master key for secure communication among nodes. An ID-based key management scheme using polynomials (Zhang and Varadharajan 2008) was given in which the BS sets up the necessary system for calculating the public and private keys. Performance of IBE-based schemes was seen as better than the RSA- or DES-based schemes.

It has been proved by various researchers that the public key cryptography-based schemes perform better in terms of security, scalability, and connectivity, the limitation being the computation overhead.

8.7.2.4 Hybrid Schemes

In the light of advantages and disadvantages of symmetric and asymmetric key cryptography, several key management schemes presented by various researchers. There are also researchers (Huang et al. 2003; Zhang and Varadharajan 2008; Kodali and Sarma 2013; Geetha and Kannan 2015) who proposed the hybrid key distribution and generation schemes for WSNs. These hybrid schemes utilize the merits of secret and public key systems which have been mentioned in sections discussed above. In such schemes, it has been noticed that the BS, sink, and cluster heads play an important role, being more resourceful as compared to normal sensor nodes. They may be assigned the duty of performing some cryptographic computations and broadcast the same to other sensors, as per requirement. The hybrid key establishment schemes reduce the high computational cost of the sensors by placing them on the BS side. Such kinds of schemes are very efficient and suitable for large-scale WSNs.

Furthermore, a lot of research is being done for designing next generation sensor nodes, which are expected to have a nonstop power supply and an ultralow-power circuit. So, with such new technologies in sensor systems, public key schemes being more secure will have an edge and will be used extensively.

8.8 Authentication and Integrity in WSNs

When two parties are in a communication protocol, each party remains attentive for the legitimacy of the other party with whom one is communicating. This can be achieved by authentication of the either parties through some predefined means. The authentication and integrity can be accomplished by means of suitable protocol or scheme, etc., deployed with the WSNs. In general, MAC can be attached to the packet to be transmitted to maintain its authenticity and secrecy. MAC may be procured from a symmetric key system by using a shared key between sender and receiver. The sender appends the shared key K with message M and a MAC, $C = H(M\|K)$ is calculated by using hash function H (Bellare et al. 1996). On reception of the packet $\{M, C\}$, a MAC C with the message M and the shared key K and verify if $C' = C$. When $C' = C$, message is authentic and unmodified, if not message has been corrupted. And in the asymmetric key system, the sender can calculate MAC C for a message M using private key. In WSNs, the wireless channel makes them vulnerable to security attacks which may result in data tampering. On the basis of the communicating parties, communication can be of any type such as one-hop unicast, multihop unicast, and broadcast.

8.8.1 One-Hop Authentication

In one-hop unicast authentication, each packet of the message is verified between neighboring nodes at the link layer by the link-layer key. It is not efficient for higher layers to authenticate one-hop unicast. The authentication of receiving a link layer packet is difficult due to fragmentation of data payload until a certain number of packets are received. In addition to it, the fragmentation can attract DOS attack, false data injection in original data to exhaust resources. A shared link-layer key can be used for achieving one-hop authentication. One such architecture called TinySec (Karlof et al. 2004) was implemented for link-layer security, that is, encryption and authentication.

8.8.2 Multihop Authentication

In the case of multihop, the link-layer authentication is considered not secure as the intermediate nodes are not trustworthy and may modify the data payload. So, higher layer, that is, transport layer authentication is required as it maintains end-to-end connection. A multihop key can be computed between two end nodes via multihop path while using a symmetric key system. This key negotiation may be unsuccessful if any of the intermediate nodes on the path is compromised. This issue was addressed by multipath enhancement (Lou and Kwon 2006) combined with secret sharing (Shamir 1979). Multihop key establishments become more secure in asymmetric key infrastructures where the two end nodes will be able to encrypt and decrypt the messages. An authentication framework for hierarchical ad hoc sensor networks (Bohge and Trappe 2003) consisted of mobile sensor nodes, forwarding nodes, and access points. In this, preloaded RSA-based initial certificates were utilized for authentication of sensor nodes and access points. Applications can authenticate new certificates using the timed, efficient, streaming, loss-tolerant authentication (TESLA) protocol, also (Perrig et al. 2008). The shared key-based authentication is more efficient than the certificate-based one.

8.8.3 Broadcast Authentication

Broadcast is desired when some common message needs to be sent to a group of nodes in WSNs. Each broadcast packet should be authenticated so that attackers cannot inject false information. Symmetric key-based techniques are assumed efficient, as they require a single secret key. A one-way hash chain (OHC) (Deng et al. 2003) is one of the basic techniques. The OHC-based approach had limitations such as a threat to the time synchronization procedure (Manzo et al. 2005) and the complex OHC management. Therefore, a suite of security protocols for sensor networks, SPINS (Perrig et al. 2002) have been optimized and implemented which was based on sensor network encryption protocol (SNEP) and micro-version of timed efficient stream loss-tolerant authentication (μTESLA). SNEP includes data confidentiality, two-party data authentication, and evidence of data freshness. μTESLA provides authenticated broadcast for severely resource-constrained environments (Du et al. 2004). It is difficult to improve this technique, but if a more powerful platform can be used, the performance of the protocol may increase. There also arose a need for authenticating local broadcast; then a batch-based broadcast authentication scheme, called BABRA was proposed (Zhou and Fang

2006). In BABRA, packets are broadcasted in batches, and each batch is a burst sequence of packets. A key associated with each batch help in computation of a MAC to be appended with the packets of a batch. After a delay from the end of the batch, this key is disclosed which prevent tampering of the batch. The delayed key disclosure is good for authentication, but it needs a certain number of packets to be buffered by each node. This may result in a flooding DOS attack. This problem can be solved by asymmetric key-based authentication. An asymmetric key-based authentication scheme (Ren et al. 2006) was presented to authenticate packets immediately after reception. In addition to this, two more efficient public key certificate-based schemes are proposed by these authors. One scheme needs to verify whether a user is authorized to access the network using the Merkle tree (Merkle 1980). The other is to use an identity-based signature scheme to authenticate external users. In another scheme (Chang et al. 2006), one-time signatures to authenticate broadcast were proposed.

8.9 Secure Routing

When a packet travels from source to destination, it is being routed through the network and a path is formed till the destination. Routing protocols are the most important factors in deciding a path from the source to the destination and finally delivery of data. The path followed should be secure for preventing any data loss. If any attack occurs and affect routing protocols, high-layer applications also get affected and the whole network may be compromised. So, it is necessary to provide security at the routing level also for proper network functionality. WSNs may face several problems related to energy consumption, data redundancy, data aggregation, security, etc. (Shi and Perrig). To overcome security issues in routing many security countermeasures are discoursed in Karlof and Wagner (2003). The usage of the global key for link-layer encryption and authentication can protect WSNs against external attackers. A trustable BS can detect spoofed node identities if every node shares a unique key with it, which is studied in SPINS. Multipath routing can be used to overcome the selective forwarding attack to increase the data delivery. Such techniques are good for preventing external attacks but internal attacks also needed to be countered. Intrusion-tolerant routing protocol for WSNs called INSENS (Deng et al. 2006), in this, BS collects the authenticated routing information to compute the routing table for every node. INSENS does not support in-network processing, it only checks the BS to node communications. Similarly, other researchers also devised schemes for securing routes of the network (Intanagonwiwat et al. 2000; Pietro et al. 2003; Tubaishat et al. 2004). The secure routing protects the information in transit and prevents attacks on data.

8.10 Secure Location

The information related to the location of the node in some applications of WSNs is very crucial, as it is used in many security schemes. The networks, which are deployed in monitoring and surveillance type of applications, require correct location details or distance between two nodes. Accuracy in location information is required, especially to find the area under attack, etc. The schemes for a secure location may or may not use beacons.

8.10.1 Secure Location Scheme with Beacons

The schemes, which utilize beacons, have sensors with a positioning system like the global positioning system (GPS) to find the whole location of the network. The positioning and ranging mechanisms are designed to find location of the node on the basis of measurements of the received signal strength and the times of flight of radio or ultrasound signals. Even when these methods are measurement-based, the attackers may perform any malicious activity. So, a position verification mechanism called verifiable multialteration (VM) (Capkun and Hubaux 2005), based on distance bounding techniques (Brands and Chaum 1994), was proposed to prevent compromised nodes from altering and reducing the distance measured between neighboring nodes. Another mechanism (Lazos and Poovendran 2004) based on a range overlapping method was proposed as it is cost-effective and more accurate. In this, different beacons are transmitted with details of coverage area sector and coordinates. Then, the sensor estimates its location on the basis of received sector information. Another scheme (Sastry et al. 2003) was given which was using in-region verification problem (this problem states how to check if the target node, which is proving its location, is falling in the region of interest) for location-based access control and protection. A robust statistical method (Li et al. 2007) was presented for making localization classes tolerant to attacks and compromises like triangulation and RF-based fingerprinting class. There are various other schemes also which secure positioning and localization for preventing the attacks and other malicious actions from an adversary (Chen et al. 2007).

8.10.2 Secure Location Scheme without Beacons

WSNs are deployed in open and hostile environments and in practical areas of the network; the nodes may not be enabled with a beacon. So, in such situations the position estimation can be achieved by approaches (Liu et al. 2005) which utilize the probability and deployment knowledge of the node. A beacon-less location discovery scheme (Fang et al. 2005) assumes that sensors of the same group are placed at the same time and at the same point and their locations may have a probability distribution that can be investigated. Following this assumption, the actual position of the node can be guessed in static networks by analyzing group memberships of nodes and using the maximum likelihood estimation method. Furthermore, a general scheme called localization anomaly detection (LAD) (Du et al. 2005), was discoursed to notice localization anomalies which resulted from pre- and postdeployment location inconsistencies.

8.11 Secure Data Aggregation

From the randomly scattered nodes, information of interest is detected, collected and processed, and forwarded to the sink in coordination with the neighbor nodes. The process of data combining and aggregating may be called as data fusion and data aggregation. The data aggregation is efficient in reducing communication overhead, but security issues may accompany. In WSNs, there are aggregator nodes, which respond to queries and collect information. There can be two types of malicious operations performed on data aggregation. In the first type, aggregators may receive wrong information from the nodes where the intermediaries know the transmitted data material. In the second type, the BS may receive wrong information or false data from corrupted aggregator nodes, which may be hidden. Data aggregation can be done based on two schemes: plaintext-based scheme and ciphertext-based scheme.

8.11.1 Plaintext-Based Scheme

Plaintext-based schemes are methods in which data aggregation is done in a readable form, that is, plaintext. These schemes can be based on nodes, certificates, etc., and are given as follows.

8.11.1.1 Scheme Defending Against One Compromised Node

The data aggregation in this is done with some delay (μTESLA) for security purpose (Hu and Evans 2003). This scheme is resilient, but may become vulnerable if any of the base or subnodes in the hierarchy are compromised.

8.11.1.2 Bidirectional Authentication Schemes

A secure in-network routing algorithms (Deng et al. 2003) was introduced with processes downstream and upstream between aggregators and normal nodes. In the downstream, sensor nodes authenticate commands from parent aggregators, and in the upstream the sensor nodes are authenticated by aggregators.

8.11.1.3 Neighbor's Certificate Schemes

In this scheme (Du et al. 2003), the neighbors have to verify the data aggregation report before being sent to the sink. Based on this concept, a witness-based approach (Du et al. 2003) was proposed for assuring the data sent to the BS from data fusion nodes. For monitoring purposes, randomly, some nodes were selected as a witness from the vicinity of the fusion node. Then, there was a need to screen out the data, which was falsely injected into the original data stream travelling between the nodes and the BS. So, a hop-by-hop authentication scheme (Zhu et al. 2004) was proposed which required at least $t + 1$ number of sensor nodes to validate the information before it was sent to the BS. Here, t represented the threshold of compromised nodes. This technique is good for large WSNs also. There was another scheme proposed (Vogt 2004) which also used interleaved authentication where t was equal to 1, that is, it was capable of finding and screening out false data packet when the compromised node is one. To find the false data packets dynamically, another scheme was proposed (Yu and Guan 2005) in which multiple sensors on the route had to verify the report, by using their one-way hash generated authentication keys, being sent to the BS.

8.11.1.4 Statistical Method

In this method, a statistical analysis is done to secure aggregation. A statistical en route filtering (SEF) mechanism (Ye et al. 2005) was proposed to distinguish and drop fake reports while forwarding the reports. In this method, a MAC was attached to the data packet and that has to be verified by each node when that packet is transmitted to the BS. A strong statistic estimation model was developed (Wagner 2004) for collecting error-prone, malicious, or spoofed data.

8.11.2 Cipher-Based Scheme

In this cipher text-based scheme, the intermediary nodes along the path do not have any knowledge of the transmitted data packet contents. The scheme, concealed data aggregation (CDA) (Girao et al. 2004), obscures sensed data end-to-end without affecting efficient in-network data aggregation. This scheme used an encryption transformation algorithm called a privacy homomorphism (PH) (Rivest et al. 1978).

   All the above-discussed schemes and methods aim to provide sufficient security and services while maintaining the wholesome goal of the WSN. These schemes are good and achieve their purpose almost satisfactorily. Still, there are some or the other issue left here and there. These issues are nothing but new challenges and points to begin new research for better performance, efficiency, and security.

8.12 Open Issues and Challenges

There has been extensive research in the area of WSN security; however, there are still some open issues which need to be addressed.

- **BS protection:** The BS is considered to be secure and robust, but in some applications like environmental monitoring and surveillance they may be susceptible to various attacks. Hence, the BS security is a major concern which needs to be properly addressed.
- **Cryptography:** Cryptographic techniques are considered to be important techniques for providing encryption in WSNs. Complex cryptography techniques are not suitable for WSNs and less complex, inexpensive, energy-efficient techniques need to be developed for WSNs.
- **Key management:** It is another important aspect of cryptography, which demands research. The current literature review suggests that symmetric key cryptography techniques are better suited for WSN. Further, one-way hash functions and distributed key mechanisms should be used. To defend the cryptographic attacks as per the application domain, the knowledge of node deployment along with location information and adaptive rekey distribution should be integrated.
- **Attack detections and preventions:** Secure schemes limit the attack, but these schemes operate at the centralized level, which increases the traffic at the BS or the nodes. Some schemes are cooperative schemes that take the help of neighbors to detect attacks. Cooperative schemes requires a lot of computation and monitoring so better techniques need to be evolved which are computationally light and do not induce traffic in the network.
- **Secure routing:** The nature of routing in WSN is different and hence a robust routing protocol demand energy consumption and cost. Secured routing protocols in WSNs should provide authentication while broadcasting. Network topology should not be easily interpreted by the adversaries. The protocol should have the capability of identifying malicious nodes and defending the routing attacks. System security and performance can be improved by identifying malicious nodes and isolating them from routing path, using localized algorithm, and using hierarchical structures of WSN.
- **Location-based algorithm:** It reveals the node identity, thus making it vulnerable to attacks; hence, location algorithms need to hide the information of node from the adversary and at the same time provides the information to the neighbors. This is a tedious task of providing reliable, accurate information and at the same time defending the attacks.
- **Secure data fusion:** Data fusion is an important aspect of data reduction and thus energy conservation; hence, statistical techniques can be used to filter the fake data. Ciphertext may be used to prevent the adversary from finding the location of nodes and also interpreting the messages.
- **Security and quality of service:** These are the two important aspects which need to be balanced and it calls for an active research in this area.

Although there are several open research areas, security concerns are specific to the application domain. One needs to understand the application and their security concerns before proposing the security mechanisms.

References

Abuhelaleh, M.A. and Elleithy, K.M. 2010. Security in wireless sensor networks: Key management module in SOOAWSN. *Int J Netw Security Appl (IJNSA)*. 2(4):67–68.

Ahmed, M.H., Alam, S.W., Qureshi, N., and Baig, l. 2011. Security for WSN based on elliptic curve cryptography. In *Proceedings of the First International Conference on 2011 International Conference on Computer Networks and Information Technology (ICCNIT'11)*, Abbottabad, Pakistan, 11–13 July 2011, 75–79.

Alcaraz, C., Lopez, J., Roman, R., and Chen, H-H. 2012. Selecting key management schemes for WSN applications. *Int J Comput Security*. 31(8): 956–966.

Amin, F., Jahangir, A.H., and Rasifard, H. 2008. Analysis of public-key cryptography for wireless sensor networks security. *World Acad Sci Eng Technol.* , IJCEACIE, 2(5):529–534. scholar.waset.org/1999.4/1056.

An, D. and Cam, H. 2007. Route recovery with one-hop broadcast to bypass compromised nodes in wireless sensor networks. In *Proceedings of the IEEE Wireless Communications and Networking Conference (WCNC)* March 11–15, IEEE Computer Society, Kowloon, pp. 2495–2500.

Anderson, R., Chan, H., and Perrig, A. 2004. Key infection: Smart trust for smart dust. In *Proceedings of the 12th IEEE International Conference on Network Protocols*, 2004, ICNP 2004, Berlin, Germany, pp. 206–215.

Anderson, R. and Kuhn, M. 1996. Tamper resistance– A cautionary note. In *Proceedings of the 2nd USENIX Workshop Electronic Commerce*, November 18, Oakland, California, Vol. 2, 1–11.

Arikumar, K.S. and Thirumoorthy, K. 2011. Improved user authentication in wireless sensor networks. In *2011 International Conference on Emerging Trends in Electrical and Computer Technology (ICETECT)*, Tamil Nadu, India, pp. 1010–1015.

Bechkit, W., Challal, Y., and Bouabdallah, A. 2013. A new class of Hash-Chain based key pre-distribution schemes for WSN. *Comput Commun*. 36(3):243–255.

Bekara, C., Laurent-Maknavicius, M., and Bekara, K. 2008. Mitigating resource- draining DoS attacks on broadcast source authentication on wireless sensors networks. In *IEEE International Conference on Security Technology*, 2008. SECTECH'08, pp. 109–116.

Bellare, M., Canetti, R., and Krawczyk, H. 1996. Keying hash functions for message authentication. In *Annual International Cryptology Conference*, August 18 (pp. 1–15). Springer-Verlag: Berlin Heidelberg, vol. 1109.

Blom, R. 1985. An optimal class of symmetric key generation systems. In *Workshop on the Theory and Application of Cryptographic Techniques*, Paris, France, New York: Springer Berlin Heidelberg, 335–338.

Blundo, C., Santix, A.D., Boneh, D., and Franklin, M. 2001. Identity-based encryption from the weil pairing. In *Advance in Cryptology-crypto, Lecture Notes in Computer Science*, vol. 2139.

Boneh, D. and Franklin, M. 2001. Identity-based encryption from the Weil pairing. In: *Advances in Cryptology—CRYPTO 2001. CRYPTO 2001*, Kilian J. (ed.). Lecture Notes in Computer Science, vol 2139. Berlin, Heidelberg: Springer.

Boyen, X. 2003. Multipurpose identity-based signcryption, a Swiss army knife for identity-based cryptography. In *Annual International Cryptology Conference*, August 17, 2003. CRYPTO 2003. In: Boneh D. (eds) Advances in Cryptology – CRYPTO 2003. Lecture Notes in Computer Science, Springer Berlin Heidelberg. vol 2729, pp. 383–399.

Boyle, D. and Newe, T. 2008. Securing wireless sensor networks: Security architectures. *J Netw*. 3(1):65–77.

Brands, S. and Chaum, D. 1994. Distance-bounding protocols. In *Proceeding Advances in Cryptology – EUROCRYPT '93: Workshop Theory Application Cryptographic Techniques*, vol. 765/1994. Springer Berlin/Heidelberg.

Camtepe, S.A. and Yener, B. 2005. Key distribution mechanisms for wireless sensor networks: A survey. Report TR-05-07. Computer Science Department at RPI Tech Technical Report TR-05-07, Rensselaer Polytechnic Institute, Troy, New York, pp.05–07.

Camtepe, S.A. and Yener, B. 2007. Combinatorial design of key distribution mechanisms for wireless sensor networks. *IEEE/ACM Trans Netw (TON)*. 15(2):346–358.

Capkun, S., Cagalj, M., and Srivastava, M. 2006. Secure localization with hidden and mobile base stations. In *Proceedings of IEEE INFOCOM 25TH IEEE International Conference on Computer Communications*, April 23–29, Barcelona, Spain, pp. 1–10.

Capkun, S. and Hubaux, J.-P. 2005. Secure positioning of wireless devices with application to sensor networks. In *INFOCOM 2005. 24th Annual Joint Conference of the IEEE Computer and Communications Societies. Proceedings,* March 13–17, Miami, FL, USA, Vol. 3, pp. 1917–1928.

Castelluccia, C., Mykletun, E., and Tsudik, G. Efficient aggregation of encrypted data in wireless sensor networks. In *Proceedings of the 2nd International Conference on Mobile Ubiquitous Systems: Networking Services , 2005. MobiQuitous 2005,* IEEE San Diego, CA, USA, pp. 109–117.

Chan, H. and Perrig, A. 2003. Random key pre-distribution schemes for sensor networks. In *Proceedings of the 2003 IEEE Symposium on Security and Privacy,* May 2003, IEEE May 11–14, Washington DC, USA, 197–213.

Chan, H. and Perrig, A. 2005. PIKE: Peer intermediaries for key establishment in sensor networks. In *Proceedings of the 24th Annual Joint Conference of the IEEE Computer and Communications Societies (INFOCOM'05).* Miami, FL, March 13–17, 2005, Vol. 1, 524–535.

Chang, S-M., Shieh, S., Lin, W.W., and Hsieh, C-M. 2006. An efficient broadcast authentication scheme in wireless sensor networks. In *Proceedings of the 2006 ACM Symposium on Information, Computer and Communications Security (ASIACCS' 06).* Taipei, Taiwan, 21–24 March, 2006.

Chen, L. and Kudla, C. 2002. *Identity-based authenticated key agreement protocols from pairings. Proceedings of the 16th IEEE Computer Security Foundations Workshop,* June 30–July 02, 2003, IEEE Pacific Grove, CA, USA, pp. 219–233

Chen, X., Makki, K., Yen, K., and Pissinou, N. 2009. Sensor network security: A survey. *IEEE Commun Surveys Tuts.* 11(2), 2nd Quarter.

Chen, Y., Trappe, W., and Martin, R.P. 2007. Attack detection in wireless localization. In *INFOCOM 2007. 26th IEEE International Conference on Computer Communications.* May 06–12, Anchorage, AK, pp. 1964–1972.

Chorzempa, M., Park, J.M., and Eltoweissy, M. 2005. SECK: Survivable and efficient keying in wireless sensor networks. In *24th IEEE International Performance, Computing, and Communications Conference,* April 7–9, 2005. IPCCC 2005. Phoenix, AZ, USA, pp. 453–458.

Chowdhury, M., Kader, M.F., and Asaduzzaman. 2013. Security issues in wireless sensor networks: A survey. *Int J Future Generation Commun Netw SERSC* 6(5):97–116.

Colbourn, C.J. and Dinitz, J.H. 1996. *The C R Chand Book of Combinatorial Designs.* Boca Raton, FL: CRC Press.

Corin, R.D., Russello, G., and Salvadori, E. 2011. TinyKey: A light-weight architecture for wireless sensor networks securing real-world applications. In *IEEE Eighth International Conference on Wireless On-Demand Network Systems and Services (WONS),* January 26–28, Bardonecchia, pp. 68–75.

Daemen, J. and Rijmen, V. 2013. *AES Proposal: Rijndael. National Institute of Standards and Technology,* p. 1. Retrieved 21 February, 2013. Das, M.L. 2009. Two-factor user authentication in wireless sensor networks. *IEEE Trans Wirel Commun.* 8(3):1086–1090.

Deb, B., Bhatnagar, S., and Nath, B. 2003. Information assurance in sensor networks. In *Proceedings of the 2nd ACM International Conference on Wireless Sensor Networks and Applications (WSNA '03).* ACM, Sep. 2003, New York, NY, USA, 160–168.

Delgosha, F. and Fekri, F. 2006. Threshold key-establishment in distributed sensor networks using a multivariate scheme. In *Proceedings INFOCOM 2006. 25th IEEE International Conference on Computer Communications,* April 23–29, Barcelona, Spain, pp. 1–12.

Deng, J., Han, R., and Mishra, S. 2003a. Enhancing base station security in wireless sensor networks. Technical Report, University of Colorado, CU-CS-951-03, Department of Computer Science, University of Colorado.

Deng, J., Han, R., and Mishra, S. 2003b. Security support for in-network processing in wireless sensor networks. In *Proceedings of the 1st ACM Workshop Security Ad hoc Sensor Networks,* 83–93.

Deng, J., Han, R., and Mishra, S. 2005. Countermeasures against traffic analysis attacks in wireless sensor networks. In *IEEE Secure Communication.*

Deng, J., Han, R., and Mishra, S. 2006. INSENS: Intrusion-tolerant routing in wireless sensor networks. *Comput Commun.* 29:216–230.

Di Pietro, R., Mancini, L.V., Law, Y.W., Etalle, S., and Having, P. 2003. LKHW: A directed diffusion- based secure multicast scheme for wireless sensor networks. In *First International Work shop on Wireless Security and Privacy* (WiSPr03).

Diffie, W. and Hellman, M.E. 1976. New directions in cryptography. *IEEE Trans Inf Theory*. 22(6):644–654.

Ding, M., Chen, D., Xing, K., and Cheng, X. 2005. Localized fault-tolerant event boundary detection in sensor networks. In *Proceedings of IEEE INFOCOM*, 2:902–913.

Dinker, A.G. and Sharma, V. 2016. Attacks and challenges in wireless sensor networks. In *2016 IEEE 3rd International Conference on Computing for Sustainable Global Development (INDIACom)*. New Delhi, 3069–3074.

Djenouri, D., Khelladi, L., and Badache, N. 2005. A survey of security issues in mobile ad hoc and sensor networks. *IEEE Commun Surveys Tuts*. 7:2–28.

Du, W., Deng, J., Han, Y.S., and Varshney, P.K. 2003. A witness-based approach for data fusion assurance in wireless sensor networks. In *Proceedings of IEEE Global Telecommunication Conference*.

Du, W., Deng, J., Han, Y.S., and Varshney, P.K. 2005a. A pair-wise key pre-distribution scheme for wireless sensor networks. *ACM Trans Inf Syst Security* (TISSEC). 8(2):228–258.

Du, W., Fang, L., and Ning, P. 2005a. LAD: Localization anomaly detection for wireless sensor networks. In *Proceedings of the 19th IEEE International Parallel Distributed Processing Symposium*, 41a–41a.

Du, W., Han, Y.S., Chen, S., and Varshney, P.K. 2004. A key management scheme for wireless sensor networks using deployment knowledge. In *Proceedings of IEEE INFOCOM 04*. Hong Kong, 586–597.

Du, W., Wang, R., and Ning, P. 2005b. An efficient scheme for authenticating public keys in sensor networks. In *Proceedings of the 6th ACM International Symposium Mobile Ad Hoc Networking Computing (MobiHoc)*.

Eldefrawy, M.H., Khan, M.K., Alghathbar, K. 2010. A key agreement algorithm with rekeying for wireless sensor networks using public key cryptography. In *IEEE*.

Eltoweissy, M., Heydari, H., Morales, L., and Sudborough, H. 2004. Combinatorial optimization of Group key management. *J Netw Syst Manag*.

Eltoweissy, M., Moharrum, M., and Mukkamala, R. 2006. Dynamic key management in sensor networks. *IEEE Commun Mag*.

Eltoweissy, M., Wadaa, A., Olariu, S., and Wilson, L. 2005. Group key management scheme for large-scale sensor networks. *Ad Hoc Netw*. 3(5):668–688.

Eschenauer, L. and Gligor, B.D. 2002. A key-management scheme for distributed sensor networks. In *Proceedings of the 9th ACM Conference on Computer and Communication Security*, Washington, DC, 41–47.

Fang, L., Du, W., and Ning, P. 2005. A beacon-less location discovery scheme for wireless sensor networks. In *Proceedings of IEEE INFOCOM*.

FIPS PUB 46-2. 1993. *Data Encryption Standard (DES)*, December 1993.

Gaubatz, G., Kaps, J.P., and Sunar, B. 2004. Public key cryptography in sensor networks. In *1st European Workshop on Security in Ad-Hoc and Sensor Networks* (ESAS2004).

Geetha, R. and Kannan, E. 2015. A hybrid key management approach for secure communication in wireless sensor networks. *Indian J Sci Technol*. 8(15), July 2015, doi: 10.17485/ijst/2015/v8i15/57603.

Girao, J., Westhoff, D., and Schneider, M. 2004. CDA: Concealed data aggregation in wireless sensor networks. In *Proceedings of the ACM WiSe*.

Gu, J., Chen, S., and Zhuang, Y. 2009. Double guarantee for security localization in wireless sensor network. In *IEEE*.

Gura, N., Patel, A., Wander, A., Eberle, H., and Shantz, S. 2004. Comparing elliptic curve cryptography and RSA in 8-bit CPU. In *2004 Workshop on Cryptographic Hardware and Embedded Systems*, August 2004.

Gurudatt, K., Shelk, R., Gaikwad, K. et al. 2013. Wireless sensor network security threats. In *Communication and Computing, 5th International Conference on Advances in Recent Technologies in Bangalore*, IET.

Hankerson, D., Menezes, A., and Vanstone, S. 2004. *Guide to Elliptic Curve Cryptography*. New York, NY: Springer. ISBN 0-387-95273-X.

Healy, M., Newe, T., Lewis, E. 2009. Security for wireless sensor networks: A Review. In *SAS 2009 – IEEE Sensors Applications Symposium*. New Orleans, LA, 17–19 February, 2009.

Hu, L. and Evans, D. 2003. Secure aggregation for wireless networks. In *Proceedings of the Symposium Applications Internet Workshops*, 384–391.

Hu, Y., Perrig, A., and Johnson, D. 2003. Packet leashes: A defense against wormhole attacks in wireless ad hoc networks. In *Proceedings of IEEE INFOCOM*.

Huang, X., Ahmed, M., and Sharma, D. 2011. Protecting from inside attacks in wireless sensor networks. In *IEEE International Conference on Dependable, Autonomic and Secure Computing*.

Huang, Q., Cukier, J., Kobayashi, H., Liu, B., and Zhang, J. 2003. Fast authenticated key establishment protocols for self-organizing sensor networks. In *Proceedings of the 2nd ACM International Conference on Wireless Sensor Networks and Applications.* New York, NY: ACM Press, 141–150.

Huang, C. and Du, D. 2005. New constructions on broadcast encryption and key pre- distribution schemes. In *Proceedings of IEEE IN FOCOM05.* Miami: IEEE Press, 515–523.

Hung, K.S., Lui K.S., and Kwok, Y.K. 2007. A trust-based geographical routing scheme in sensor networks. In *2007 IEEE Wireless Communications and Networking Conference.* Kowloon, 2007, pp. 3123–3127. doi: 10.1109/WCNC.2007.577.

IETF RFC 4346. 2006. The Transport Layer Security (TLS) Protocol, Version 1.1, April 2006.

Intanagonwiwat, C., Govindan, R., and Estrin, D. 2000. Directed diffusion: A scalable and robust communication paradigm for sensor networks. In *Mobile Computing and Networking.*

Jian, Y., Chen, S., Zhang, Z., and Zhang, L. 2007. Protecting receiver-location privacy in wireless sensor networks. In *Proceedings of IEEE INFOCOM*, 1955–1963.

Karapistoli, E. and Economides, A.A. 2012. Wireless sensor network security visualization. In *IEEE 4th International Workshop on Mobile Computing and Networking Technologies.*

Karlof, C. and Wagner, D. 2003. Secure routing in sensor networks: Attacks and countermeasures. In *First IEEE International Workshop on Sensor Network Protocols and Applications.*

Karlof, C., Sastry, N., and Wagner, D. 2004. TinySec: A link layer security architecture for wireless sensor networks. In *Proceedings of the 2nd International Conference on Embedded Networked Sensor Systems,* 162–175.

Kavitha, T. and Sridharan, D. 2010. Security vulnerabilities in wireless sensor networks: A survey. *J Inf Assur Secur.* 5:31–44.

Koblitz, N. 1987. Elliptic curve cryptography. *Math Comput.* 48(177).

Kodali, R.K. 2013. Implementation of ECDSA in WSN. In *IEEE International Conference on Control Communication and Computing* (ICCC).

Kodali, R.K. and Chougule, S.K. 2013. Hierarchical key agreement protocol for wireless sensor networks. *Int J Recent Trends Eng Technol.* 9(1), July 2013.

Kodali, R., Sarma, N. 2013. Energy efficient ECC encryption using ECDH. In *Emerging Research in Electronics, Computer Science and Technology,* Lecture Notes in Electrical Engineering, vol. 248, Springer, pp. 471–478.

Kousalya, C.G. and Raja, J. 2009. An energy efficient traffic-based key management scheme for wireless sensor networks. In *International Conference on Networking and Digital Society.*

Krauss, C., Stumpf, F., and Eckert, C. 2007. Detecting node compromise in hybrid wireless sensor networks using attestation techniques. In *Proceedings of the Security and Privacy in Ad-Hoc and Sensor Networks,* vol. 4572.

Law, Y.W., Etalle, S., and Hartel, P.H. 2003. Assessing security in energy efficient sensor networks. In *Proceedings of the 18th IFIP TC11 International Conference on Information Security Privacy Age Uncertainty (SEC).*

Lazos, L. and Poovendran, R. 2002. Secure broadcast in energy-aware wireless sensor networks. In *IEEE International Symposium on Advances in Wireless Communications (ISWC 02).*

Lazos, L. and Poovendran, R. 2003. Energy-aware secure multicast communication in ad-hoc networks using geographic location information. In *Proceedings of International Conference on Acoustics, Speech and Signal Processing (ICASSP'03).* Vol. 6. 201–204.

Lazos, L. and Poovendran, R. 2004. SeRLoc: Secure range-independent localization for wireless sensor networks. In *Proceedings of the 3rd ACM Workshop Wireless Security,* 21–30.

Lee, H., Kim, Y.H., Lee, D.H., and Lim, J. 2007. Classification of key management schemes for wireless sensor networks. In *The 2007 International Workshop on Application and Security Service in Web and PervAsive eNvironments (ASWAN 07),* June 2007.

Lee, J. and Stinson, D.R. 2004. Deterministic key pre-distribution schemes for distributed sensor networks. In *Proceedings of ACM Symposium on Applied Computing 2004.* Waterloo, Canada, 294–307.

Lee, J. and Stinson, D.R. 2005. A combinatorial approach to key predistribution for distributed sensor networks. In *Proceedings of the IEEE Wireless Communication and Networking Conference,* 2:1200–1205.

Li, Z. and Gong, G. 2013. On the node clone detection in wireless sensor networks. *IEEE Trans Netw.* 21(6): 1799–1811. http://dx.doi.org/10.1109/tnet.2012.2233750.

Li, M., Koutsopoulos, I., and Poovendran, R. 2007. Optimal jamming attacks and network defense policies in wireless sensor networks. In *Proceedings of IEEE INFOCOM*, 1307–1315.

Li, Z., Trappe, W., Zhang, Y., and Nath, B. 2005. Robust statistical methods for securing wireless localization in sensor networks. In *Proceedings of the 4th International Symposium Information Processing in Sensor Networks*.

Liu, D. and Ning, P. 2003. Efficient distribution of key chain commitments for broadcast authentication in distributed sensor networks. In *Proceedings of the 10th Annual Network Distributed System Security Symposium (NDSS)*, 263–276.

Liu, D. and Ning, P. 2003. Establishing pairwise keys in distributed sensor networks. In *Proceedings of 10th ACM Conference on Computer and Communications Security (CCS03)*. Washington, DC: ACM Press, 41–47.

Liu, D. and Ning, P. 2003. Location-based pairwise key establishments for static sensor networks. In *Proceedings of the 1st ACM Workshop on Security of Adhoc and Sensor Networks*, 72–82.

Liu, D. and Ning, P. 2005. Improving key pre-distribution with deployment knowledge in static sensor networks. *ACM Trans Sensor Netw.* 1(2):204–239.

Liu, D., Ning, P., and Du, W. 2005. Attack-resistant location estimation in sensor networks. In *Proceedings of the 4th International Symposium Information Processing Sensor Networks*.

Liu, D., Ning, P., and Du, W. 2005. Detecting malicious beacon nodes for secure location discovery in wireless sensor networks. In *Proceedings of the 25th International Conference on Distributed Computing Systems (ICDCS)*.

Liu, F., Cheng, X., and Chen, D. 2007. Insider attacker detection in wireless sensor networks. In *Proceedings of the IEEE INFOCOM*, 1937–1945.

Liu, Q., Liu, L., Kuang, X., and Wen, Y. 2012. Secure service and management for security-critical wireless sensor network. In *Sixth International Conference on Innovative Mobile and Internet Services in Ubiquitous Computing*.

Lou, W. and Kwon, Y. 2006. H-SPREAD: a hybrid multipath scheme for secure and reliable data collection in wireless sensor networks. *IEEE Trans Vehicul Tech.* 55(4):1320–1330.

Malan, D. 2004. Crypto for tiny objects. Harvard University TR-04-04.

Manzo, M., Roosta, T., and Sastry, S. 2005. Time synchronization attacks in sensor networks. In *Proceedings of the 3rd ACM Workshop Security of Ad Hoc and Sensor Networks (SASN'05)*.

Mathias, B. and Wade, T. 2003. An authentication framework for hierarchical ad hoc sensor networks. In *Proceedings of the 2nd ACM Workshop on Wireless Security (WiSe '03)*, ACM, New York, NY, USA, 79–87. http://dx.doi.org/10.1145/941311.941324

Merkle, R. 1980. Protocols for public key cryptosystems. In *Proceedings of the IEEE Symposium on Research in Security and Privacy*, April 1980.

Moisés Salinas, R., Gina Gallegos, G., and Gonzalo Duchén, S. 2009. An authentication protocol for sensor networks using pairings. In *IEEE*.

Naccache, D. and Stern, J. 2000. Signing on a postcard. In *Proceedings of Financial Cryptography, Lecture Notes in Computer Science*, 1962:121–135.

Nanda, R., Tiwari, S., and Krishna, P.V. 2011. Secure and efficient key management scheme for wireless sensor networks. In *IEEE International Conference on Electronics Computer Technology (ICECT)*.

Newsome, J., Shi, E., Song, D., and Perrig, A. 2004. The Sybil attack in sensor networks: Analysis and defenses. In *Proceedings of the 3rd International Symposium on Information Processing in Sensor Networks*, 259–268.

Pandey, A. and Tripathi, R.C. 2010. A survey on wireless sensor networks security. *Int J Comput Appl.* 3(2):43–49. (0975–8887).

Perrig, A. et al. 2008. TESLA: Multicast source authentication transform introduction, IETF working draft. *IEEE Commun Surveys Tuts.* 3rd Quarter.

Perrig, A., Szewczyk, R., Tygar, J.D., Wen, V., and Culler, D.E. 2002. SPINS: Security protocols for sensor networks. *Wireless Netw.* 8:521–534.

Peter, S., Piotrowski, K., and Langendoerfer, P. 2007. On concealed data aggregation for WSNs. In *Proceedings of the 4th IEEE Consumer Communication Networking Conference (CCNC)*, 192–196.

Pietro, R.D., Mancini, L.V., Law, Y.W., Etalle, S., and Havinga, P.J.M. 2003. LKHW: A directed diffusion-based secure multicast scheme for wireless sensor networks. In *Proceedings of International Conference on Parallel Processing Workshops*, 397–406.

Przydatek, B., Song, D., and Perrig, A. 2003. SIA: Secure information aggregation in sensor networks. In *Proceedings of the 1st International Conference on Embedded Networked Sensor Systems*.

Rassam, M.A., Maarof, M.A., and Zainal, A. 2011. A novel intrusion detection framework for wireless sensor networks. In *IEEE International Conference on Information Assurance and Security*.

Ren, K., Lou, W., and Zhang, Y. 2006. LEDS: Providing location-aware end-to-end data security in wireless sensor networks. In *Proceedings of the IEEE INFOCOM*.

Ren, K., Lou, W., and Zhang, Y. 2007. Multi-user broadcast authentication in wireless sensor networks. In *4th Annual IEEE Communications Society Conference*, 223–232.

Ren, X. and Yu, H. 2006. Security mechanisms for wireless sensor networks. *IJCSNS Int J Comput Sci Netw Security*. 6(3):155–162.

Rivest, R.L., Shamir, A., and Adleman, L. 1978. A method for obtaining digital signatures and public-key cryptosystems. *Commun ACM*. 21(2):120–126.

Rivest, R.L., Adleman, L., and Dertouzos, M.L. 1978. On data banks and privacy homomorphisms. In *Proceedings of Foundations Secure Computation*, 169–179.

Rivest. 1996. IETF RFC 2040, The rc5, rc5-cbc, rc5-cbc-pad, and rc5-cts algorithms. October 1996.

Sastry, N, Shankar, U., and Wagner, D. 2003.Secure verification of location claims. In *Proceedings of the 2nd ACM workshop on Wireless security (WiSe '03)*. ACM, New York, USA, 1–10. doi. org/10.1145/941311.941313.

Sastry, A.S., Sulthana, S., and Vagdevi, S. 2013. Security threats in wireless sensor networks in each layer. *Int J Adv Netw Appl*. 4(4):1657–1661.

Schneier, B. 1996. *Applied Cryptography*, 2nd edition. New York, NY: Wiley.

Shah, P.G., Huang, X., and Sharma, D. 2010. Analytical study of implementation issues of elliptical curve cryptography for wireless sensor networks. In *IEEE 24th International Conference on Advanced Information Networking and Applications Workshops, Processing*, 2002, 403–410.

Shamir, A. 1979. How to share a secret. *Commun ACM*. 22(11):612–613, November 1979.

Shamir, A. 1984. Identity-based crytosystems and signature schemes. In *Advances in Cryptology-Crypto84, Lecture Notes in Computer Science*. Berlin: Springer, 196:47.

Shannon, C.E. 1949. Communication theory of secrecy systems. *Bell Sys Tech J*.

Shanyue, B. and Liqing, C. 2012. A new key management protocol for wireless sensor network. In *2012 International Conference on Computer Science and Service System,* Nanjing, 2012, pp. 991–994. doi: 10.1109/CSSS.2012.251.

Shao, M., Zhu, S., Zhang, W., and Cao, G. 2007. pDCS: Security and privacy support for data-centric sensor networks. In *Proceedings of IEEE INFOCOM*, 1298–1306.

Sharif, L. and Ahmed, M. 2010. The wormhole routing attack in wireless sensor networks (WSN). *J Inf Process Syst*. 6(2), June 2010.

Shi, E. and Perrig, A. 2004. Designing secure sensor networks. *IEEE Wirel Comm*. 11(6): 38–43. doi: 10.1109/MWC.2004.1368895.

Singh, V.P., Jain, S., and Singhai, J. 2010. Hello flood attack and its countermeasures in wireless sensor networks. *IJCSI Int J Comput Sci Issues*. 7(3).

Spencer, J. 2000. *The Strange Logic of Random Graphs. Algorithms and Combinatorics*. Berlin: Springer, vol. 22.

Stephens, D.L, Jr. and Peurrung, A.J. 2004. Detection of moving radioactive sources using sensor networks. *IEEE Trans Nucl Sci*. 51:2273–2278.

Tajeddine, A., Kayssi, A., Chehab, A., and Elhajj, I. 2014. Authentication schemes for wireless sensor networks. In *Mediterraneann Electrotechnical Conference (MELECON)*.

Tanenbaum, A.S., Wetherall, D.J. 2013. *Computer Networks,* 5th edition. India: Pearson Education.

Tubaishat, M., Yin, J., Panja, B., and Madria, S. 2004. A secure hierarchical model for sensor network. *SIGMOD Rec*. 33(1):7–13. doi.org/10.1145/974121.974123.

Vogt, H. 2004. Exploring message authentication in sensor networks. In *Proceedings of the 1st European Workshop Security Ad-Hoc Sensor Networks (ESAS)*.

Wagner, D. 2003. Cryptanalysis of an algebraic privacy homomorphism. In *Proceedings of the 6th Information Security Conference (ISC)*.

Wagner, D. 2004. Resilient aggregation in sensor networks. In *Proceedings of the 2nd ACM Workshop Security Ad hoc Sensor Networks*, 78–87.

Walters, J.P., Liang, Z., Shi, W., and Chaudhary, V. 2006. *Wireless Sensor Network Security: A Survey*. Auerbach Publications, CRC Press.

Wallner, D., Harder, E., and Agee, R. 1999. Key management for multicast: issues and architectures, June 1999, RFC 2627.

Wander, A., Gura, N., Eberle, H., Gupta, V., and Shantz, S. 2005. Energy analysis of public-key cryptography on small wireless devices. In *PERCOM '05 Proceedings of the Third IEEE International Conference on Pervasive Computing and Communications*, March 2005.

Wang, Y., Attebury, G., and Ramamurthy, B. 2006. A survey of security issues in wireless sensor networks. *IEEE Commun Surveys Tuts.* 8:2–23.

Watro, R., Kong, D., Cuti, S. et al. 2004. Tinypk: Securing sensor networks with public key technology. In *Proceedings of the 2nd ACM Workshop on Security of Ad Hoc and Sensor Networks* (SASN 04). New York, NY: ACM Press, 59–64.

Wei-hong, W., Yi-ling, C., and Tie-ming, C. 2009. Design and implementation of an ECDSA-based identity authentication protocol on WSN. In *Natural Science Fund Project* (Y106290), *IEEE*.

Wood, A.D., Fang, L., Stankovic, J.A., and He, T. 2006. SIGF: A family of configurable, secure routing protocols for wireless sensor networks. In *Proceedings of the 4th ACM Workshop Security of Ad hoc Sensor Networks*.

Wood, A.D. and Stankovic, J.A. 2002. Denial of service in sensor networks. *IEEE Comput.* 35:54–62.

Yang, G., Rong, C., Veigner, C., Wang, J., and Cheng, H. 2006. Identity-based key agreement and encryption for wireless sensor networks. *IJCSNS Int J Comput Sci Netw Security.* 6(5B):182–189.

Ye, F., Luo, H., Lu, S., and Zhang, L. 2005. Statistical en-route filtering of injected false data in sensor networks. *IEEE J Select Areas Commun.* 23:839–850.

Yin, J. and Madria, S. 2006. SecRout: A secure routing protocol for sensor networks. In *Proceedings of the 20th International Conference on Advanced Information Networking Applications (AINA)*, 1:393–398.

Younis, M.F., Ghumman, K., and Eltoweissy, M. 2006. Location-aware combinatorial key management scheme for clustered sensor networks. *IEEE Trans Parallel Distrib Syst.* 17:865–882.

Yu, Z. and Guan, Y. 2005. A robust group-based key management scheme for wireless sensor networks. In *Proceedings of IEEE Wireless Communications and Networking Conference (WCNC)*.

Yu, Z. and Guan, Y. 2006. A dynamic en-route scheme for filtering false data injection in wireless sensor networks. In *Proceedings of IEEE INFOCOM*.

Zhang, J. and Varadharajan, V. 2010. Wireless sensor network key management survey and taxonomy. *Int J Netw Comput Appl.* 33, Elsevier.

Zhang, J., Li, J., and Liu, X. 2009. An improved pairwise key management scheme for wireless sensor networks. In *International Symposium on Computer Network and Multimedia Technology*.

Zhang, W. and Cao, G. 2005. Group rekeying for filtering false data in sensor networks: A pre-distribution and local collaboration-based approach. In *Proceedings of IEEE INFOCOM*.

Zhou, Y. and Fang, Y. 2006. BABRA: Batch-based broadcast authentication in wireless sensor networks. In *Proceedings of the 49th Annual IEEE Global Telecommunication Conference (GLOBECOM'06)*.

Zhou, Y., Fang, Y., and Zhang, Y. 2008. Securing wireless sensor networks: A survey. *IEEE Commun Surveys Tuts.* 10, 3rd Quarter.

Zhu, S., Setia, S., Jajodia, S., and Ning, P. 2004. An interleaved hop-by-hop authentication scheme for filtering of injected false data in sensor networks. In *Proceedings of IEEE Symposium Security Privacy*, 259–271.

Chapter 9

Communication, Localization, Coverage, Error and Control, Time Synchronization, Naming and Addressing, and Cross-Layer Issues

Anuradha Pughat, Parul Tiwari, Vidushi Sharma, and Neeta Singh

Contents

9.1   Introduction 214
9.2   Communication in Wireless Sensor Networks 214
    9.2.1   Energy Saving Methods in Communication 215
        9.2.1.1   Duty Cycling Approach 215
        9.2.1.2   Data-Driven Approach 215
        9.2.1.3   Mobility-Based Approaches 216
    9.2.2   Aspects of the Physical Layer 216
    9.2.3   Communication Protocols 216
    9.2.4   Energy-Efficient Modulation Techniques in Physical Layer 217
9.3   Localization Issues 218
    9.3.1   Range-Based Localization 220
    9.3.2   Range Free Localization 220
    9.3.3   Challenges in Localization 221
        9.3.3.1   Computational and Energy Constraints 221
9.4   Error and Control Issues 222
    9.4.1   Aim of the Error Control 223
    9.4.2   Error Control Approaches 223

          9.4.2.1 Automatic Repeat Request 223
          9.4.2.2 Forward Error Correction 223
          9.4.2.3 Hybrid Automatic Repeat Request 224
      9.4.3 Error Correcting Codes 224
      9.4.4 Challenges in Error Control 224
  9.5 Time Synchronization Issues 225
      9.5.1 Time Synchronization Protocols 225
          9.5.1.1 Sender–Receiver-Based Protocols 226
          9.5.1.2 Receiver–Receiver-Based Protocols 226
          9.5.1.3 Receiver-Only–Based Protocols 227
  9.6 Naming and Addressing Issues 227
      9.6.1 Address Allocation and Assignment 228
      9.6.2 Types of Addressing 229
          9.6.2.1 Content-Based Addressing 229
          9.6.2.2 Geographic Addressing 229
      9.6.3 Research Issues and Challenges Related to Naming and Addressing 230
  9.7 Coverage Issues in WSNs 230
      9.7.1 Types of Coverage 230
          9.7.1.1 Target Coverage 231
          9.7.1.2 Area Coverage 231
          9.7.1.3 Barrier Coverage 231
      9.7.2 Deployment Strategies and Coverage 231
      9.7.3 Coverage Protocols 231
      9.7.4 Challenges in Coverage 234
  9.8 Cross-Layer Issues 234
      9.8.1 Cross-Layer Interaction between MAC Layer, Physical Layer, Network Layer,
          and Application Layer 234
      9.8.2 Open Research Challenges in Cross-Layer Design 235
References 235

9.1 Introduction

Sensor networks differ from traditional wired and wireless networks in terms of computation capabilities, energy, size, and memory. Due to this, the physical layer demands energy efficiency in modulation and coding schemes. There are several challenges in other operative mechanisms such as localization, time synchronization, naming and addressing of nodes, and network coverage. All the operative protocols in these areas have to be energy efficient and less complex. This chapter addresses these challenges and outlines the physical layer design concepts along with the other operative mechanisms of various protocols.

9.2 Communication in Wireless Sensor Networks

In the current scenario, the technical developments in terms of communication are changing rapidly, and low-cost energy-efficient sensor nodes are required in wireless sensor networks (WSNs). A frequency band is used for communication because of the inefficiency in using a single frequency to communicate. As far as the WSNs are concerned, they use some of the license free bands called industrial, scientific, medicine (ISM) bands. These bands do not need permission from any authorized body to transmit the data and can be used with other frequency bands. There exist many systems, which use an ISM band of 2.4 GHz such as the IEEE802.11b.

The demand of allocating and using the radio frequency (RF) spectra is rapidly growing due to the increasing number of wireless and mobile communication applications. The choice of the radio carrier frequency differs based on the applications, e.g., wall penetration, different atmospheric attenuation, mobile phones, satellite, military services, etc. The industry has reached the limits of the current static spectrum allocation and this points toward the open challenges of the dynamic spectrum allocation. This type of allocation provides sparsely used spectrums to the users. An efficient and dynamic spectrum is used with the help of spectrum utilization and modulation, which is adaptive. With the help of dynamic spectrum allocation, the main aim of the WSNs to transmit the data efficiently can be fulfilled. For example, in real-time communication and for sending large files of audio and video clicked by sensor nodes needs quite a high bandwidth as well as a dynamic spectrum.

9.2.1 Energy Saving Methods in Communication

There exists a number of energy saving methods for WSNs; some of the frequently used approaches are described in the next section.

9.2.1.1 Duty Cycling Approach

During communication, some of the nodes remain idle while others are active. Even when the nodes are idle, they consume energy; so, the better way is to use the radio in the best possible mode. The duty cycling approach plays an important role. The fraction of the active time of a node in one cycle is called its duty cycle. Thus, WSNs can be operated in different modes, such as transmit mode, receive mode, idle mode, and sleep mode. In cases where the nodes are idle, i.e., the system is not communicating, its circuit is still turned on. In this case, there is continuous energy consumption and it is approximately equal to that of the active state. The option to shut down the radio is a better choice in the idle mode. It is better to put the current inactive nodes in sleep mode, whereas active nodes can switch off the radio when there is no network job.

9.2.1.2 Data-Driven Approach

These approaches can be mainly classified as data reduction and energy-efficient data acquisition. Data reduction is the process in which the larger entity of the collected data from sensors is converted into smaller useful entity so that at a later stage the same data can be retrieved without any loss. This concept is very important in power management as the transmission of data from nodes uses up a lot of power, thus by reducing the size of the data, power consumption can be reduced. At the same time, the data reduction approach also concentrates on preventing the nodes from transmitting data to the sink. This reduces the transmission load on the node as well as the communication and processing overheads at the sink side. The other approach, i.e., energy-efficient data acquisition, is an approach where data acquisition is reduced based on energy-efficient algorithms. However, this is not exclusive to reducing energy consumption through sensing. It also reduces the number of communications along with reducing sensed data. Adaptive sampling takes advantage of correlated and gradually changing data, thus reducing the number of data sensings. Hierarchical sampling looks into the dynamics of the nodes such as accuracy and energy consumption, thus ensuring a balance between the aforementioned attributes. Another energy-saving approach is the model-based active sampling in which data is presented by building upon the sampled models. Due to this, less number of communications exists between the nodes.

9.2.1.3 Mobility-Based Approaches

As discussed earlier for the designing of WSNs, a sparse structure can be used in case of the same application requirements. In mobility-based approaches, no restriction on connectivity is required. Further, the communication between wireless sensor nodes needs a radio connection as a physical layer in which energy is consumed when the radio sends or receives data. Since the main aim of WSNs is the optimal use of their cost and energy, the design of the physical layer of a WSN becomes very important in this context.

9.2.2 Aspects of the Physical Layer

Modulation and demodulation of the data are associated with the physical layer. The transceiver has three modes: idle, sleep, and active. Thus, the key to effective energy management is to switch the radio off when the radio channel is idle. The open research area is to minimize the time and energy utilized to switch between different modes during the transmitting and receiving states. Heinzelman et al. (2002) suggested that there are two factors responsible for energy loss in a wireless transmission. First, one is the loss due to the channel and another is the fixed energy cost to run the transmission and reception circuitry. Both the factors have a relation with the hop distance. The increment of hop distance is responsible for channel loss and as the number of hops increases, the cost increases linearly. Thus, there should be a balance between optimal hop distance and the amount of energy consumed (Cui et al. 2005). Moreover, the authors discussed the different linear and nonlinear modulation schemes for power consumption related to circuit and transmission. Second, the scheme in which the transceiver is switched off periodically or after certain time schedules, the chances of packet loss are higher. Thus, the third option is to use efficient modulation techniques in the physical layer. The chances of the transmission's success depend upon the modulation of the system. The designing of the physical layer components change with the probability of the transmission success (Wang et al. 2001). Wang suggested that the amount of start-up energy consumed is related with the energy consumed in a transmission. In case of higher modulations such as M-ary modulations, the transmitted energy increases for a fixed bit error rate (BER), whereas the number of transmissions decreases. Since the "on" time of a transmitter is very short for higher modulation, these modulation schemes are preferred as energy-efficient schemes, although their cost is high. Wang et al. (2001) discussed how the start-up time of a physical layer was very important and how one could choose an optimal modulation scheme.

9.2.3 Communication Protocols

In the domain of WSNs, there are two main communication protocols: 6LowPAN and ZigBee. The 6LowPAN (released in 2007 by Internet Engineering Task Force [IETF]) is an open standard communication protocol. It consumes less power, minimum data rate, and needs low-cost personal area networks (PANs). It is a combination of two-Internet protocol (IPV6) and low-power PAN. It can also be used with the relationship of time variance among the nodes in WSNs.

ZigBee is popular for a low-cost, low-power, advanced communication protocol for small devices. It is used for low-rate wireless personal area networks (LR-WPANs). ZigBee is preferably used in body sensor networks (BSNs). BSNs are a sensor or group of sensors attached to a patient and a coordinator for collecting raw data. This data is sent, analyzed, and processed in the control devices through the network. The ZigBee coordinator as a controlling device works with interrupt to reduce power consumption in the network in gathering the raw data, for example, healthcare monitoring. In addition, ZigBee is applied in a mesh network of routers to relay data from different patients to the access point (AP). The AP is connected to the Internet to allow collaboration of doctors, medical centers, and other data centers that gather patient records, so that decisions can be made.

ZigBee consists of two layers. One of the layers is application support and another one is the network/security layer (see Figure 9.1). The data rates between 10 and 250 kbps over a 10–75 m range is easily communicable using ZigBee network devices.

9.2.4 Energy-Efficient Modulation Techniques in Physical Layer

An appropriate modulation scheme is required for effective communication systems and thus, the survivability and lifetime of WSNs. Choice of the correct scheme of modulation depends on the network traffic and reliable communication in a WSN. The various modulation schemes make the channel capable of sending maximum data over the unsecure channel with high security. Since all the modulation schemes are not energy efficient, it is very important for WSNs to use the optimum modulation scheme in terms of energy efficiency and minimum error. This would yield a reduction in the usage of battery and improvement in the system performance considerably. The researcher used different digital signal techniques (such as quadrature phase shift keying [QPSK], binary phase shift keying [BPSK], frequency shift keying [FSK], and many more).

Let us talk about the basics of the modulation techniques, which have been used in the WSNs design. In amplitude modulation (AM), when the incoming signal is a sequence of 0 and 1 value, the modulation process is called *amplitude shift keying* (*ASK*). In frequency modulation (FM), when the incoming signal is digital, the modulation process is called FSK. In FM, when 0.5 modulation index is used, it is called *minimum shift keying* (*MSK*). The MSK is capable of detecting coherent and noncoherent signals and can amplify the power efficiently. The incoming digital signal with phase modulation (PM) is called phase shift keying (PSK). One of the applications of BPSK is the *direct sequence spread spectrum* (*DSSS*) transceivers. Depending upon the type of signal (1 and 0), the phase of a constant amplitude carrier signal varies between 0° and 180°. The transitions, which occur due to the inversion of carrier, produce a very wide transmitted spectrum (Yang et al. 2005). Furthermore, the authors suggested a method for the appropriate selection of modulation schemes such as BPSK, 16 quadrature amplitude modulation (16QAM), and 8PSK. Other techniques such as QPSK, 8PSK, 16PSK, and offset QPSK have also been used for WSNs in the literature. Table 9.1 summarizes a quick view of different modulation schemes.

User	Application layer (APL)
ZigBee	Application support layer (APS)
	Network/security layer
IEEE802.15.4	MAC layer
	PHY layer

Figure 9.1   **IEEE802.15.4 stack.**

Table 9.1   Features of Various Modulation Techniques

Type of Modulation	Number of Phase Angles	Number of Bits	Advantage
Quadrature phase shift keying (QPSK)	4	2 bits at a time	High bit rate
8-phase shift keying (8-PSK)	8	3 bits at a time	Increased bit rate with same bandwidth
16-PSK	16	4 bits at a time	Higher bit rate
Offset QPSK (OQPSK)	Phase variations are slow	Q data stream is offset from I stream	Tolerance of OQPSK waveform is high
Quadrature AM (QAM)	Two carriers with a 90° phase shift used	Two amplitude modulation (AM) signals are used in single channel	Bandwidth is doubled

The authors in literature have shown that the adaptively chosen modulation and coding scheme can provide better system performance. Several authors have used BPSK, QPSK, 16QAM, and 64QAM to achieve a better overall system performance. Shih et al. (2001) suggested that the binary modulation scheme with an effective start-up power dominant condition is more energy efficient. Abouei et al. (2011) observed the concept of green modulation over Rayleigh flat-fading channels to ensure energy efficiency in WSNs. For better understanding, Table 9.2 presents the coding, performance parameters, and overall system performance of different modulation techniques such as BPSK, QPSK, 16QAM, 64QAM, M-ary phase shift keying (MPSK), M-ary quadrature amplitude modulation (MQAM), M-ary frequency shift keying (MFSK), and 8PSK.

The MSK and BPSK are preferable modulation techniques during data transmission from ordinary sensor nodes to their cluster heads and from cluster heads to border nodes of the cluster. This shows that as the transmission distance from border node to base station is more, PSK or QAM modulation techniques are preferred. Modulation techniques like QAM and PSK provided higher channel throughput, but at the cost of increased power consumption, whereas modulation techniques such as MSK and BPSK proved to be energy efficient but at the cost of channel throughput.

9.3 Localization Issues

Localization is a network management process for providing a physical location to each node during system initialization. It has three main components: distance estimation, position computation, and localization algorithms. The technique is more essential in the smart environment for scale of measurement, accuracy, and precision. Today, sensor nodes use in-built special module, i.e., location finding system. These nodes coordinate for the events of interest and the system uses either global positioning system (GPS) module or the localization software/hardware module. Since, the GPS modules increase the cost of the tiny and resource-constrained WSNs, one

Table 9.2 Comparison of Various Modulation Schemes

Modulation Techniques	Codes	Performance Parameters	Overall System Performance
BPSK QPSK 16QAM 64QAM	Convolutional code	Spectral efficiency (SE) Block error rate	Good system performance
	Reed–Solomon (RS) code	Bit error rate (BER)	Efficient physical layer design
QPSK 16QAM	Hybrid automatic repeat request (HARQ) code	Code rate Signal-to-interference plus noise ratio (SINR) gain	Efficient error control mechanism
MPSK MQAM MFSK	–	BER Energy lifetime	Suitable for short distance communication
16QAM 8PSK BPSK	Turbo codes	Average throughput, frame error rate (FER), signal-to-noise ratio (SNR)	The maximum average throughput achieved
8PSK 16QAM	Extended Bose–Chaudhuri–Hocquenghem (BCH) code	BER	Multistage decoding and modulation used

solution space to this uses the GPS module to few special nodes called *seed nodes*. Again, this can hide the useful information from the remaining nodes. Hence, research on the localization issues related to no-GPS approaches is required. Application problems such as: What is the current temperature at this location? What is pressure on this object? What anomaly is detected at this location? What is the animal count in any location? etc. can be solved using the localization itself (Whitehouse et al. 2004). The localization is classified into two categories: centralized and distributed. In the centralized scheme, the centralized node computes and sends the location to each node while, each node knows its location in the distributed scheme. Moreover, the nodes send information to the neighboring nodes or cluster heads with the help of their exact location and position. Some position-based routing protocols have potential to enhance energy efficiency with less control overheads. The distributed localization schemes can be range-based or range-free. In the range-based localization schemes, the range information is required while in range-free schemes, each node estimates its location information getting a broadcast message from the seed nodes. Other approaches to localization are cooperative and noncooperative target localization, cooperative infrastructure localization, blind source localization, passive target localization, and passive self-localization. The cooperative schemes give effective, accurate, and energy-efficient solution to sensor nodes.

The traditional localization uses range measurement techniques such as trilateration (Oguejiofor et al. 2013), triangulation (Savarese et al. 2016), and maximum likelihood (ML) multilateration (Xu et al. 2016), shown in Figure 9.2.

In the trilateration and triangulation techniques, the position of the node is calculated by the node distance with the intersection of three circles and with the angle of triangle, respectively. As shown in Figure 9.2a, the outer three nodes with known coordinates (x, y) and angles (θ) create

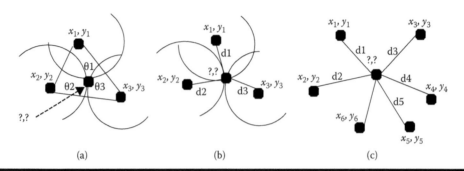

Figure 9.2 Traditional localization techniques. (a) Triangulation technique, (b) trilateration technique, and (c) multilateration technique.

a triangle and help in evaluating the coordinates of the central node. Figure 9.2b describes the trilateration technique in which based on the distance between known and unknown nodes, the location of unknown node is calculated. The multilateration gives more accuracy in the results and uses multiple numbers of nodes in finding the node location (Figure 9.2c).

A case study of a smart kindergarten project (UCLA) has been given in Raghavendra et al. (2004), which shows how children's development can be monitored using fine-grained localization. In this project, two sensor platforms have been used. One type of sensor node, i.e., "iBadge" is attached to the child and another is the "Medusa MK-2." The self-configuring Medusa MK-2 sensor node uses beacon signals to localize and track the iBadge nodes. Thus, the iBadge tags associated with the child gets their location coordinates and can be tracked by the central processing station.

9.3.1 Range-Based Localization

The range-based localization protocols use range measurements from the fixed location sensor nodes. Few distance measurement methods are received signal strength (RSS), time of arrival (ToA), time difference of arrival (TDoA), and angle of arrival (AoA). The measurements from the mobile nodes provide more accuracy. The range-based protocols are ad hoc localization system (AHLoS) (Savvides et al. 2001), mobile-assisted localization (MAL) (Priyantha et al. 2005), localization with noisy range measurements (LNRM) (Moore et al. 2004), and AHLoS and time-based positioning scheme (TPS) (Cheng et al. 2004). The AoA method is more accurate but requires hardware change.

9.3.2 Range Free Localization

These range-free protocols are more energy efficient than range-based protocols as they do not use range measurement methods. However, a lack of range measurement methods limits the amount of information and makes them less accurate. Some range-free localization protocols are convex position estimation (CPE) (Doherty et al. 2001) and approximate point-in-triangle (APIT) (He et al. 2005), centroid, high-resolution robust localization (HIRLOC), secure range-independent localization (SERLOC), and active distributed localization algorithm (ADLA). The angle of the antenna and communication range can improve the accuracy in HIRLOC protocol. The SERLOC protocol is used to find node location in untrusted environment.

9.3.3 Challenges in Localization

- The design of a localization-based sensor node requires a ranging mechanism for measuring the physical distance and angle of nodes. These ranging mechanisms use radio waves, visible light, or acoustic signals the range estimation. The calculation of range can be done by observing RSS or RF time-of-flight (ToF) of the received signal. Unfortunately, these simple techniques require exact knowledge of transmitting power, time synchronization, and clock accuracy. However, the time synchronization issue can be minimized using acoustic signals. The acoustic signals path loss near to ground is much lower than the loss in RF signals. Again, the solid obstructions in the path of acoustic signal and the large-size acoustic rangers become a challenge for WSN designers.

- Another important localization issue is related to the physical layer, as the noise in the environment can cause the measurement variations. The algorithm design becomes a challenge with the physical layer measurement error or if environment changes. Thus, the efficient optimization techniques are required for minimizing the noise and overheads.

- On the other hand, an understanding of network topology, node behavior, and node geometry affects these designs and the customization in hardware and software needs before the development of energy-efficient nodes.

- The choice of the range estimation method for any application is again a challenging task. Each method has its pros and cons. The selection is resource and application dependent.

- The distributed localization approach is more scalable while the centralized approach is more accurate. The beacon-based algorithms are suitable for localized coordinate system and the beaconless approach for global coordinate systems. So, the choice of these approaches depends on the scenario.

- There is diversity in the error characterization behavior of WSNs; the proposed solutions need to be addressed again. The real-time experimentation on the localization will explore the measurement modalities on the systems under consideration.

- As we know, the localization algorithms are application dependent. The lack of global solution for localization is another hot research topic for beginners. Thus, the designing of accurate, flexible, low-power consuming, and high-performance localization algorithms/ methodologies requires recent research efforts.

Finally, we can understand the localization techniques such as cost energy, hardware, software, installation, processing, communication, and price of a node. The above-mentioned points explore the possible localization approaches in WSNs.

9.3.3.1 Computational and Energy Constraints

The RSS-based localization approach is the most energy and bandwidth efficient as no additional hardware is required for setting up this localization on a node. Moreover, these protocols are more attractive if their estimation error can be minimized. The statistical methods can minimize the error in measurement but require more energy for results on a large number of samples. The authors in Moravek et al. (2011) use statistical methods minimizing the RSS uncertainty. The investigated results show that the energy consumption is not directly related with the RSS accuracy. During the initial phase (first five measurements), the accuracy increases at a faster rate and after that gradually with increase in number of samples. Thus, the estimation depends on the application requirement and interest of the receiver node.

The energy and communication cost of a three-anchor node RSS localization technique has been investigated in a 100-node scenario (Kotwal et al. 2012). The known anchor nodes transmit their location information to the neighboring nodes. Then, the location of the unknown node is estimated with localization error and distance error. It has been investigated that the mean error after three iterations is very small and the accuracy about a few centimeters, which is appropriate for indoor applications.

In another work (Zou and Chakrabarty 2003), significant energy saving has been done with the target localization approach. The cluster head gathers the limited information from the sensor nodes and executes the localization algorithm for best results. This energy aware, probabilistic localization technique consumes smaller energy at low-computational cost.

Trilateration, triangulation, and ML multilateration techniques are error prone and do not give practically feasible solution to WSNs. The highly accurate ML techniques used for finding the node location consumes large computational power. Apart from these approaches, probabilistic methods can also be used for node distance estimation (Ramadurai and Sichitiu 2003).

These optimization solutions for localization require large memory space too. Then, this becomes a problem for resource-limited sensor nodes. Therefore, global optimization gives a solution in which a sink node computes and communicates location to each node. This reduces the computational power of individual nodes at the cost of communication power. On the other hand, GPS solutions do not give cost-effective implementation to WSNs. Moreover, the centralized approaches provide high accuracy at the cost of computation. In addition, the reduced function sensor nodes pace the problem while implementing localization algorithms. These application-specific nodes are used for low-cost operations and have limited components and protocols on the board. Multimeasurements at different time slots can be combined for accurate location. As we know, minimum communication and collaboration inside the network can minimize the energy consumption. Now, we can say that the lightweight, reliable, and low-power consuming localized protocols are required for the energy-limited sensor nodes.

9.4 Error and Control Issues

In case of WSNs, it is required that the power consumption in communication should be low so that energy can be saved. Furthermore, there are lots of restrictions on energy results in error. Thus, error control is of prime importance for WSNs. Due to the low-power communication constraints, error-prone links occur in WSN channels. A number of techniques are used to reduce the redundancy in the traffic. To get the desired energy efficiency, the filtered packets must be reliable. Hence, energy-efficient error control is of extreme importance.

The length of the network lifetime can be increased by putting the sensor node radios to sleep as and when possible. Therefore, it is also important to use the node's on-time efficiently. Sampling intervals and transmission techniques must be chosen and used in such a way that this active time can be used effectively. Here, the design of energy and latency-efficient error control schemes play an important role. Thus, error-control ensures the correct transmission and has a control on possible errors.

To provide reliable data communication in WSNs, two main error control strategies are used. The first one is automatic repeat request (ARQ) and the second is forward error correction (FEC). There have been some studies that consider the energy efficiency of error control techniques in sensor networks. These strategies use retransmission of packets and various error correcting codes (ECCs) to overcome the harmful consequences of the channel.

9.4.1 Aim of the Error Control

The main aims of the error control is to ensure that data transport are

- Error-free and transmit exactly the sent bits
- In-sequence and to send them in the original order
- Duplicate-free and should be lossless

9.4.2 Error Control Approaches

Error control can generally be realized by backward error control (ARQ), FEC, or a combination of the two, i.e., hybrid automatic repeat request (HARQ).

9.4.2.1 Automatic Repeat Request

This strategy is based on retransmissions. It is preferably used in process control networks. In ARQ, the sender node adds error detection codes called parity bit to the data. Sink node checks the correctness of the received data. If there is an error in the received packet, the sink node rejects it and requests the sender node to retransmit the same packet. Although the strategy is very simple, low cost increases the throughput, and requires less overhead, it fails to improve the overall energy efficiency in WSNs. The ARQ strategy results in latency and excessive energy cost. These two drawbacks cannot be avoided in time-critical control applications. Some researchers have termed it as energy inefficient as it consumes more energy due to retransmitting every erroneous packet (Sankarasubramaniam et al. 2003). They considered different error control techniques and the investigation was focused on the question of optimal packet size for WSNs. The ARQ scheme uses positive acknowledgment (ACK) or negative acknowledgment (NACK) to send feedback to the transmitter. If ACKs are used, then the transmitter will retransmit if it has not received an ACK packet within a prespecified time-frame. The major drawback with ARQ is the high cost.

9.4.2.2 Forward Error Correction

To overcome the limitation of ARQ, a second technique is used for reliable data transmission called FEC (Pottie and Kaiser 2000). It is also called as channel coding. This improves the performance of error control. ECCs are utilized to add redundancy to the packet. This redundancy detects some bit errors and corrects it at the receiver end. Therefore, the problem of retransmission is resolved. The transmit power required for BER or frame error rate can be minimized due to coding also involved in ECCs. However, this results in a high cost of extra energy consumption in encoding, decoding, and transmitting redundant bits. In most cases, encoding energy is considered to be negligible while the decoding process consumes significant amount of energy. Therefore, FEC is used in those communications where retransmissions are relatively costly. There are many types of FEC codes, where the most commonly used are divided into two categories; block codes, and convolutional codes. A block code first divides the message to be transmitted into smaller blocks of a predefined length. These blocks are then encoded into code words. In the case of convolutional codes, each bit is encoded as a function of a predefined number of preceding bits. Lettieri et al. (1997) used a combination of several codes and introduce hybrid scheme using block and convolutional codes.

9.4.2.3 Hybrid Automatic Repeat Request

It is a challenge in WSNs to choose an optimum ECC keeping both the performance and energy consumption in mind. ARQ provides reliable communication through retransmissions, which will be costly in poor channels where retransmissions occur frequently. FEC performs better in poor channels, while the redundant bits become an undesired cost when channel conditions are good. Some researchers have studied HARQ schemes, which include the advantages of both error correcting schemes by combining ARQ and FEC. The FEC-based HARQ scheme with Bose–Chaudhuri–Hocquenghem (BCH) codes is proposed by Akyildiz and Vuran (2010). It is not good for all applications in WSNs (Eriksson 2011), but it is limited to only specific applications and consumes a large amount of energy (Agarwal et al.). The selection of an optimum ECC for WSN is studied by Abughalieh et al. (2010).

9.4.3 Error Correcting Codes

Several different types of ECCs exist, but we may categorize them into:

1. Block codes: Block codes are of a fixed length n_C, with n_C-k parity bits, and are decoded one block or codeword at a time, where k is the length of the information sequence.
2. Convolutional codes: For a rate k/n_C code, input is k bits, and output n_C bits at each time interval, but are decoded in a continuous stream of length $L > n_C$.

The Hamming code (HC) is known as one of the basic codes introduced by Richard Hamming in 1950. As long as the research is developed, more efficient and powerful codes have been generated such as Reed–Solomon (RS) and BCH. Thus, RS, BCH, and HCs are the most widely known block codes. Balakrishnan et al. (2007) reveal that BCH code outperforms over any other channel codes in terms of energy efficiency. This code requires much less encoding and decoding energy consumption. Goldsmith (2005) described a set of cyclic, linear BCH block codes. They are powerful codes for moderate to high signal-to-noise ratio (SNR), generally outperform all other block codes at high rates. Short block codes like HCs can be decoded by syndrome decoding. ML can also be used on a trellis with the Viterbi algorithm (Viterbi 1967) or maximum a posteriori (MAP) decoding with the Bahl–Cocke–Jelinek–Raviv (BCJR) algorithm (Bahl et al. 1974). The place of the existence of the error can be determined using RS and BCH codes. To determine the error location, these codes are decoded with a complex polynomial solver. In convolutional codes, encoding is performed in a continuous fashion rather than accumulating k data bits and then encoding into n-bit code word as in block codes. In case of convolutional codes, code word depends on both current k data bits and also on some earlier bits. The number of shifts a particular bit can influence output depends on constraint length. Convolutional codes are decoded on a trellis using either Viterbi decoding, MAP decoding, or sequential decoding. In addition to traditional block codes and convolutional codes, there exist yet more powerful codes such as turbo codes and low-density parity-check (LDPC) codes. All these codes have limited applications because of their computational complexity. Stronger codes are optimal to be used with end-to-end error control strategy while simple codes are best for node-to-node error-control strategy.

9.4.4 Challenges in Error Control

■ An error control design based on FEC or HARQ will have the advantage of being able to correct a certain amount of errors in a packet. To avoid retransmissions, the lower package error rate (PER) induced by FEC or HARQ could be used. This can further be explored.

- Lower PER could be managed by making longer hops in a multihop network. By using FEC or HARQ, one extra hop can be avoided.
- By making control on transmission power, one can get desired output in terms of energy efficiency as well as low cost.
- ECC is not always a practical and intelligent solution for increasing link reliability. In some applications, an uncoded system may actually be more energy efficient.
- Depending on the application and environment, analog decoders can be energy efficient in a WSN. A combination of low power consumption and moderately high to high throughput makes analog decoders quite practical and efficient for WSN use.

9.5 Time Synchronization Issues

Sensor nodes are self-organizing nodes and one of the important characteristics is that they synchronize among themselves to communicate with each other (energy-efficient radio schedule). Besides this, synchronization is also required for in-network processing, acoustic ranging, distributing an acoustic beam forming, and for developing other protocols designs/applications which require accurate time. Time synchronization in WSNs calls for special consideration due to their limited energy, computation power, memory, and size of sensor nodes. In traditional networks as there is no such limitations, we have more efficient solutions like GPS and network time protocol (NTP).

In case of WSNs, time synchronization handles clock synchronization of different sensor nodes in a single-hop and multihop environment. One of the challenges, which is of prime concern is the limited energy. Most of the applications of WSNs are powered by a battery having a limited life. Since the prime concern of these applications is increasing the lifetime of the network by energy conservation, the operations like time synchronization should be efficient. Limited bandwidth restricts the data rate and hence, one cannot exchange frequent messages among the sensor nodes, which is the main requirement of synchronization algorithms. Hardware limitations of sensor nodes further limit the processing capability and the memory required for the storage of synchronization algorithms. In applications employing a time-division multiple access (TDMA), to prevent collisions, the nodes should have synchronized clocks, so that the noncommunicating nodes can be switched to sleep mode to conserve energy (Bachir et al. 2010). In cross-layer network management, solutions such as intertwined medium access scheduling and in-network data aggregating synchronized clocks are required (Bagaa et al. 2013). Due to these resource constraints, the clock synchronization protocol should be lightweight and efficient in terms of communication overhead.

9.5.1 Time Synchronization Protocols

The time synchronization protocols can be classified into three categories: sender–receiver, receiver–receiver, or receiver-only. In sender–receiver protocols, data is exchanged between the sender and receiver. The receiver synchronizes its clock with the sender and a bidirectional link is required in such cases. The time sync protocol for sensor networks (TPSNs) (Ganeriwal et al. 2003), flooding time synchronization protocol (FTSP) (Maróti et al. 2004), and gradient time synchronization protocol (Sommer and Wattenhofer 2009) are examples of sender–receiver protocols. Here, the sender sends a beacon with the time stamp to the receiver and receiver adjusts its clock on the basis of the time stamp received. Receiver–receiver synchronization is achieved at a local level in contrast to synchronization achieved at network level by some of the protocols. Reference broadcast synchronization (RBS) protocol (Elson et al. 2002) is an example of receiver–receiver protocol. In this protocol, neighbors receive a synchronization message by the node and this is used as a reference time to adjust their clocks.

In receiver-only protocols, a group of nodes do not send or receive the synchronization message, instead they overhear the synchronization messages exchanged between a pair of sender–receiver nodes working on the principle of sender–receiver protocol, e.g., pairwise broadcast synchronization (PBS) (Kyoung-lae Noh et al. 2008).

9.5.1.1 Sender–Receiver-Based Protocols

Sender–receiver protocol follows the following steps:

- The transmitting node sends periodic messages with its time stamp containing the local time.
- The receiver synchronizes its clock with the sender's time stamp message.
- The delay message between sender and receiver is calculated by measuring the total time of sending and receiving the message.

The TPSN (Ganeriwal et al. 2003) provides synchronization at the network level. It has mainly two phases: initial level discovery phase and then the synchronization phase. In the discovery phase, the topology of nodes is identified by first identifying the root node and then building a spanning tree of the network. In the second phase, child nodes synchronize to their parent node in the tree by exchanging messages. The time stamp received by the child nodes helps in the calculation of transmission delay and the relative clock offset, but it does not compensate clock drift, which requires frequent resynchronization. The communications overhead is high as there is two-way message exchange but resynchronization overheads are not there. These limitations have been overcome in FTSP (Maróti et al. 2004). In FTSP, an ad hoc tree structure is formed. The root node is elected dynamically based on the smallest node identifier. After election, it periodically broadcasts its current time stamp into the network, which helps in forming the ad hoc tree structure. A linear regression table is used for conversion between the local hardware clock and the clock of the reference node. After the initialization phase, a node waits and listens for synchronization beacons sent by other nodes. Once the node has synchronized with the root node, it broadcasts its estimation of the global clock. If a node does not receive synchronization beacon for certain time duration then it declares itself the new root node otherwise it continue to adjust the time clock as per the root node. The gradient clock synchronization protocol synchronizes the node in decentralized way in contrast to centralized synchronization shown by TSPN and FTSP. Every node periodically broadcasts its time information. Synchronization messages received from direct neighbors are used to calibrate the logical clock.

9.5.1.2 Receiver–Receiver-Based Protocols

Physical channels broadcast the messages; this property is used by RBSs (Elson et al. 2002) for synchronization between the receivers. Here, a transmitter sends the message which is received by the receivers in the receiving range. Here, the variability in the time period of receiving the message is very less as the node records the time the moment it receives the message and then sends a message for adjusting the relative clock offset and the skew. RBS works well with single-hop time synchronization. However, in a clustered network the nodes which are participating in more than one cluster can be used to synchronize timing of these clusters. These common nodes thus help in network level synchronization. For better accuracy, GPS receivers can be used as a reference to bring about network-level synchronization and increase the efficiency of the network.

9.5.1.3 Receiver-Only–Based Protocols

Receiver-only is considered a better technique methodology for achieving clock synchronization with significant lesser message exchange, resulting in reduced overheads of bandwidth and energy. However, these protocols work on the concept that the nodes working on the receiver-only principle should lie within the communication range of the two nodes performing sender–receiver-based synchronization. In PBS, the cluster head is an example of receiver-only protocol. It synchronizes with the reference nodes through two-way message exchanges (i.e., using the sender–receiver methodology). The reference nodes are like leaders who synchronize among themselves and the neighboring nodes overhear their synchronization message and time stamps exchanged by these reference or leader nodes. In a multicluster system, these pair the receiver-only mechanism. The whole process can thus spread into a network wide level (Noh et al. 2008). This protocol has limitation in terms of the errors that arise due to number of hops and hence the synchronization accuracy decreases. Synchronization through piggybacked reference timestamps (SPiRT) (Benzaïd et al. 2016) overcomes these limitations. In this protocol, cluster members lying outside the communication range of the leader/reference node also overhear all timing messages sent by the cluster-head. Piggybacking the time, the timestamp received is used along with synchronization rounds to adjust the local clock and reduce the error arising out of multihop synchronization.

The distributed WSNs make heavy use of synchronized time, but often have unique requirements in terms of energy conservation and precision of the synchronization achieved along with reducing the complexity of the synchronization in terms of communication overheads.

9.6 Naming and Addressing Issues

Names are used to identify the objects like nodes and transaction. It may or may not be unique. Naming is used to improve energy efficiency and fault tolerance. On the other hand, the addresses are used to import information so that things can be found out. The naming and addressing issues are related to the network management in WSNs. Under this combined scheme, each node gets and identifies its and the neighboring node's name and location.

Two basic approaches to naming are low level and high level. Application independent naming is low-level whereas high-level naming is location independent. It is applied when communication between applications is required. Heidemann et al. (2001) used application oriented low-level naming. Others (Ali and Uzmi 2004) proposed an effective energy-efficient naming scheme. The unique node identifier (UID) provides a unique name to every node. This unique name consists of such components as name of the supplier, name of the item, its sequencing, etc. A name to a component can be assigned at manufacturing time or a temporary name can be assigned to increase the energy efficiency. The network identifier is used to distinguish the networks, which are working into similar environment and geographical area. The message authentication code (MAC) address is used to differentiate between neighbors of a node. A node goes into sleep mode after the decision of activation of the packets. Sometimes, a network address is essential to locate a node in multiple hop scenarios. It is related with the routing. The addressing helps routing in a multihop network. The network structure can be a base for taking decision on the type of routing to be implemented for any particular application in WSNs. The role of each sensor node in the *flat routing* is similar and these nodes collaborate with each other in forwarding the packets. They use attribute-based naming in order to specify the data properties. In the *hierarchical routing,* the cluster heads coordinate among their groups and others. The selection of cluster heads is based on their energy. The higher energy nodes are elected as heads and they perform processing while lower energy nodes perform the sensing task. The cluster head approach increases the scalability and lifetime of the entire network. The location-based routing or geo-routing keeps on updating the location and positional knowledge of the nodes. Further, this information is exchanged between neighboring nodes and the nodes update their relative coordinates. The high-level naming saves energy in the directed diffusion of flat network routing.

The uniqueness of addresses is classified as globally unique address, network-wide unique address, and locally unique address. The globally unique address occurs at most once all over the world. It uses 48-bits MAC address. The network-wide unique address is unique only within a network. A similar address can be used in a different network, while the local address can be used multiple times within a network.

Actually, the naming schemes save the energy when a node offers the data with high attributes. Through naming: the useful information is passed to the neighboring nodes and thus the entire network. This reduces the overhead and latency. A sink node sends a query to the nodes. The queries are compared with the knowledge of attributes in a node. Then, these nodes pass answers to the sink node queries.

9.6.1 Address Allocation and Assignment

A service to map between names and addresses is the domain name system (DNS). The DNS provides a bridge between the human-readable hostnames and machine-readable IP addresses. The DNS can provide two kinds of naming: domain name-based naming and an IP address-based naming. Applications such as Internet browsers are identified by the domain name, while the IP address within the routing protocol guarantees the packet forwarding.

In WSNs, the dynamic address assignment protocols are used to allocate the addresses as a prior or when needed (on demand). In the centralized address assignment scheme, a single authority node can control and monitor the address pool. It has some limitations; the central node is reachable only until the network is partitioned. If a node joins a group after the networks is partitioned, it cannot connect to the central node. This approach is not suitable and creates significant traffic. In distributed WSNs, all nodes within a network can provide and accept the same address assignment scheme. As told earlier, the node addresses need not to be unique all the time, in all the networks. There may be duplicity within the network and is called an address conflict. Therefore, research to resolve the address conflict detection and correction is required. A distinction between strong and weak duplicate address detection (DAD) is required. Doss et al. (2010) introduced the concept of dynamic addressing, in which a WSN can address the nodes without the knowledge of their location and position. This scheme ensures the dynamic reuse of addresses in event-based WSNs. This can be done at minimal latency and packet loss. The dynamic reuse of addresses increases the energy efficiency of network and eliminates the need for the DAD in the network. Further, the scheme ensures the uniqueness in the addresses of the sensor nodes and networks.

9.6.2 Types of Addressing

Application requirements of WSNs pose new challenges in node addressing. *Fixed and universal addressing* of sensor nodes is not a viable option, thus attribute-based naming or local identifications of sensor nodes have thus far been considered to address a specific sensor for various purposes such as node management, data querying, data aggregation, and routing. The *content-based and geographic addressing* can be used for addressing as shown in Figure 9.3.

9.6.2.1 Content-Based Addressing

This type of addressing is data-centric instead of id-centric. The data contents define the addressing instead of nodes. The middleware systems perform this type of addressing. The useful information or the useful content is only used in this type of addressing scheme. In WSNs, the sensors sense data continuously and then the data of interest is used to describe the addressing. The format of the content-based addressing describing interest is <attribute, value, operation>, e.g., <height, 30 m, NE>. Here NE is the operator which means matches if the actual value is <NOT EQUAL>. Cugola and Migliavacca (2009) introduced a routing protocol in which the contents are used for addressing in mobile sensor networks.

9.6.2.2 Geographic Addressing

The physical position of nodes defines the addresses in case of *geographic addressing* which is different from content-based addresses. In this type of address, coordinates of the *x*- and *y*-axes are used for attributes. The addresses may be a point on a circle or such type of objects such as a sphere or it may be a set of points of a polygon. Imieliński and Navas (1999) explained the idea of GPS-based geographic addressing to find out the resources. The idea of geographic addressing is to use the location information available about a node locally for routing, i.e., its own location and that of its neighbors without the knowledge of the entire network.

In WSNs, the success of any network operation depends upon the addressing scheme of the nodes. Ahmed (2012) applied a *software- and hardware-based* addressing scheme for WSNs.

A scheme for long thin WSNs was proposed by Pan and Tseng (2012). Their addressing and routing scheme was based on ZigBee protocol. A distributed address scheme is applied for the assignment of network address in case of ZigBee. Before the address assignment, the users collect some information related to router and network such as:

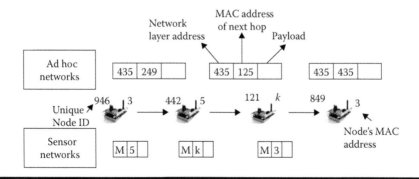

Figure 9.3   Address usages in WSNs.

- What is the maximum number of children of a router?
- What is the maximum number of child routers of a router? (It is limited to 5)
- How much is the depth of the network?

The addresses are assigned in a systematic fashion from top to bottom. To assign addresses to the coordinator, the whole address space is partitioned into blocks. The coordinator's child routers occupy the first blocks, whereas its own child end devices occupy last block. A detailed description of node addressing for ZigBee is explained in Pan and Tseng (2012).

9.6.3 Research Issues and Challenges Related to Naming and Addressing

- The two main aspects in addressing are address assignment and address representation. When the assignment is static, there is no true scaling issue. Dynamic assignment needs to be explored.
- The addresses and names in a sensor networks can be used for nodes, MAC address, network address, and network identifier. Research is ongoing on the energy-efficient naming and addressing schemes.
- The issue occurs when the unique node ID, which is allocated before deployment, is used as the MAC address.
- Research on geographic routing addresses two issues, one is routing packets successfully in a given topology and second is acquiring location information of nodes reflecting the given topology.
- Spatial reuse of addresses requires a dynamic address assignment protocol. Such a protocol can be centralized or distributed, but only distributed versions scale well. Network wide unique addresses scale poorly. Spatial reuse dramatically improves the scalability, as it is mainly the local node density and not the network size, which dictates the address size.
- The problem of the malfunctioning occurs in *software*-based addressing. Resolving this problem makes the sensor network energy inefficient.
- Although the unique and clear addressing is obtained with hardware-based scheme, only the efficient design of the addressing scheme can ensure the elimination of processing and transmission overheads.

9.7 Coverage Issues in WSNs

The quality of service provided by the various sensing nodes raises the concept of coverage, which depends on the sensor device and its applications. If the communication range of the sensor device is in relation with sensor range, the area of coverage and connectivity will be high. It must be at least double that of the sensing range. The quality of service depends on the type of coverage used.

9.7.1 Types of Coverage

The coverage problem can be taken at any level of network layer. It considers the parameters such as sensing direction, sensitivity, transfer function, accuracy, dynamic range, etc. There can be several coverage approaches. Different types of coverage are explained in the next section such as *target coverage*, *area coverage*, and *barrier coverage*.

9.7.1.1 Target Coverage

It may also be called point coverage in which the sensors are deployed to cover some points of interest or targets in the area. It observes a limited or fixed number of points. Examples of this coverage include military operations where energy consumption is the main motive with the best hit. Researchers (Zhang 2009; Zhang et al. 2009; Cardei et al. 2005) investigated target coverage with energy conservation.

9.7.1.2 Area Coverage

When the sensor nodes sense that all the points are in range, then area coverage is achieved. Area coverage is a set of targets or points. It is also known as full coverage or blanket coverage. In full coverage, at least one sensor node will cover every single point of the interested field.

9.7.1.3 Barrier Coverage

This type of coverage includes the bounded region of the interested field. This concept is the same as to obtain maximal breach path. A special type of barrier coverage is the sweep coverage. It is a moving barrier problem. The objective of the area and barrier is to keep an eye on each location in the specified area within the sensing range and detect any abnormality. Aliyu et al. (2016) developed a coverage enhancement problem for mobile sensor deployment.

9.7.2 Deployment Strategies and Coverage

Deployment strategy may be classified as dense or sparse. When a large number of sensor nodes are deployed, it is dense deployment while a sparse deployment would have less number of nodes. Several issues in WSNs like routing, connectivity, coverage, and its redundancy can be handled using appropriate node deployment strategy. With proper deployment, a high degree of coverage can be obtained. Poduri and Sukhatme (2004) suggested a deployment strategy in which each node will communicate with its neighbors and decide to move away until they get desired coverage with proper connectivity. Others (Umeki et al. 2009) proposed another approach in which the authors used a flying balloon for an ad hoc network system.

9.7.3 Coverage Protocols

Energy-efficient coverage protocols are clustering protocols and distributed protocols. Most of the clustering protocols require location information. Thus, most commonly used coverage protocols are location unaware distributed coverage protocols and location aware distributed coverage protocols. In location unaware distributed coverage protocols, the sensor nodes to find active nodes in their neighborhood use probes. The sleep–wake-up cycle is decided by probing nodes. On the other hand, in location aware distributed coverage protocols all the nodes are known to their location. Various types of distributed, location aware, location unaware, deterministic sensing model, computational geometry-based coverage protocols are summarized in Table 9.3

A detailed study of the above approaches is presented and analyzed in Cardei and Wu (2006). An energy-efficient problem related to heterogeneous WSNs was solved by Yu et al. (2016). They used centralized connected target k-coverage (CTC_k) and distributed CTC_k algorithm. Another

Table 9.3   Location Unaware and Location Aware Distributed Protocols

Location Unaware Distributed Protocols		Location Aware Distributed Protocols	
Probing environment and adaptive sleeping (PEAS)	• It has three operating modes. • Only the necessary nodes are active and others are in sleeping mode. Objective: Minimum number of nodes are waking up. Limitation: State overhead is low.	Coverage configuration protocol (CCP)	• The nodes can be in one of the three states: sleep, active, and listen. • It is a decentralized protocol Objective: To maintain full coverage. Limitation: It is unable to avoid sensing void
Probing environment and collaborating adaptive sleeping (PECAS)	• Node becomes active when no other nodes are active within its probing range. • PECAS has periodic scheduling. Objective: There is a balance of energy and the problem of network partitioning is also removed.	Enhanced configuration control protocol (ECCP)	• In ECCP, the outer region is also covered by neighboring nodes. • The number of active sensors is achieved and there is no sensing void. Limitation: Larger active nodes in comparison with CCP.
Controlled layer deployment (CLD)	• It uses deterministic node deployment, thus number of active nodes are used. It maintains 100% area coverage. Objective: To reduce the cascading effect. Limitation: Higher energy consumption because of the higher number of nodes used.	Optimal geographical density control (OGDC)	• Nodes can be in one of the three states: on, off, and undecided. • Minimum number of working nodes for 100% coverage is determined using density control algorithm. Limitation: OGDC has higher lifetime as compared to PEAS.
Discharge curve backoff sleep protocol (DCBSP)	• It is a probe-based protocol and works on battery discharge curve. • It avoids random and undesired regular wake-ups of sleeping nodes. Objective: Network lifetime high, coverage redundancy is minimized, and energy consumption is low.	Probabilistic coverage protocol (PCP)	• It is a distributed coverage protocol which forms hexagonal structures. Objective: Convergence rate is very high. PCP protocol requires on an average 10% energy consumption to maintain coverage similar to other protocols.

(Continued)

Table 9.3 (*Continued*)   Location Unaware and Location Aware Distributed Protocols

Location Unaware Distributed Protocols		Location Aware Distributed Protocols	
Distributed coverage calculation algorithm (DCCA)	• Coverage redundancy is minimized. • It avoids coverage voids. Limitation: Cascading effect cannot be avoided thus battery discharges rapidly.	Probabilistic coverage reserving protocol (PCPP)	It uses the limited number of active nodes for energy conservation and coverage area is full.
Coverage and energy strategy for wireless sensor networks (CESS)	• It finds the redundant working nodes. Coverage property is preserved. Limitation: Energy consumption is high due to large number of working nodes.	Balanced energy and coverage guaranteed protocol (BECG)	• Nodes work in three states: sleeping, checking, and working. Limitation: There may be coverage redundancy. Energy consumption is high.

work (Akhlaq et al. 2014) suggested a new algorithm for the C3 problem, i.e., coverage, connectivity, and communication. The periodic wake-up rate is discussed in the paper given by Gui and Mohapatra (2004). However, Cerpa and Estrin (2004) and Li et al. (2009) find out coverage metrics using conditional wake-up rate.

Ye et al. (2002) used simplified marker and cell (S-MAC) scheme, to conserve energy, the wake-up time of the sleeping sensors is periodic. A new mechanism, probing environment and adaptive sleeping (PEAS) was proposed by Ye et al. (2008). In PEAS, a sleeping node uses probes to check an active node within its probing range. The probing node goes into the active state only when it gets no reply from its neighbor node otherwise it goes back to sleep mode. The geographical adaptive fidelity (Xu et al. 2001) was suggested, which uses geographic location information to divide the area into fixed square grids. One and only one node staying awake to forward packets within each grid. Zhang and Hou (2005) determine that some of the sensor nodes should not be blind due to the turning off of nodes (which are inactive in data forwarding).

The optimal geographical density control (OGDC) approach defines that the density of the sensor is high so that a sensor can be found at any desirable point. However, it is not so easy practically. That is why, to increase the network lifetime with energy conservation, a novel searching algorithm, energy-efficient coverage control algorithm (ECCA) inspired by the multiobjective genetic algorithms (MOGAs) is proposed by Jia et al. (2009). In this paper, the authors maintained 100% coverage with minimum number of working nodes. A two-dimensional grid represents the sensor field. The mobility of the sink node is discussed in Kamat et al. (2007), and Sekhar et al. (2005) describes the dynamic coverage maintenance (DCM) scheme. Lin et al. (2013) and Yu et al. (2012) proposed the *k*-coverage problem, which was based the on game theory for hybrid sensor networks.

The performance measurement of all the coverage protocols and existing algorithms, which are based on active node count, sensing area coverage, and network lifetime is surveyed by More and Raisinghani (2016). In this paper, they showed the relative performance measurement of various coverage protocols.

9.7.4 Challenges in Coverage

- Energy-efficient coverage in WSNs can be achieved by appropriate sensor node deployment.
- Mobility in sensors is helpful for flexible designing of more efficient sensor deployment strategies for area coverage.
- Improper use of probing nodes may lead to a reduction in the network and node lifetime.
- The coverage degree and node failure probability are incorporated only in PEAS, whereas not all other protocols handle node failure probabilities.
- Geometrical pattern can be used to increase the coverage and energy efficiency.
- Sensor's collaboration can improve the coverage performance.
- Multiround sensor deployment for information coverage under the probabilistic sensing mode can be applied.

9.8 Cross-Layer Issues

After reviewing the analysis from experts and researchers in the field of energy-efficient WSNs, we have reached the state to say that efficient utilization of the sensor network energy is the most effective way to enhance its lifetime. There are several other cross-layer services such as synchronization, topology control, localization, task management, mobility management, power management, etc. The cross-layer techniques provide solutions to the load balancing, congestion, bandwidth allocation, routing, transmission power, modulation, reliability, data aggregation, packet overhead, and end-to-end delay. Although, cross-layer designs have been proved very effective in terms of the energy and performance improvement, the cross-layer design on the layers of the network stack reduces the chances of design improvements and level of modularity.

9.8.1 Cross-Layer Interaction between MAC Layer, Physical Layer, Network Layer, and Application Layer

The combination and interaction between layers such as physical (PHY), MAC, routing (ROUTE), and application (APP) gives the best solution to the problem of energy efficiency and performance control. These layers share the database with each other. The cross-layer designs can be distributed, centralized, manager-based, or nonmanager-based. The main cross-layer optimization techniques are classified according to the design on the layers of network stack in WSNs. They are the PHY-MAC-ROUTE (PMR) approach, PHY-MAC-APP (PMA) approach, MAC-ROUTE (MR) approach, PHY-MAC-ROUTE-APP (PMRA) approach, etc. The PMR cross-layer protocols (XLPs) are collision aware routing protocol (CARP), cross-layer routing protocol for multihop (CLMHR), cross-layer energy-efficient routing (XLE2R), cost link-based (CLB), and link distance link cost link error (LDCE). Among these protocols, CARP provides high-energy efficiency, throughput, and residual energy along with low packet loss, end-to-end delay, and overheads (Huang et al. 2011). The OPNET-based XLE2R protocol does not perform well for mobile node applications (Babulal and Tewari 2010).

The fair and delay-aware cross-layer (FDRX) data transmission protocol is the cross-layer approach for PMA. The QualNet simulator-based FDRX protocol estimates delay and data prioritization before transmission (Al-Anbagi et al. 2012). This increases the energy overheads. The MR protocols are: quality of service routing protocol (QSRP), hybrid access protocol based on clustering routing scheme (SCL), cross-layer congestion control (CLCC), balanced-routing protocol (BRP), cross-layer optimization approach (CLOA), and cross-layer approach for energy-efficient (CLAEE) MAC layer protocol. The SCL protocol bears high packet loss and overhead while this improves performance and energy under different traffic conditions (Zhang et al. 2012).

The C++-based QSRP protocol prioritizes the traffic before use and provides lower energy consumption, path loss, and end-to-end delay (Alwan and Agarwal 2013). The adaptive routing protocol (ARP) is a PMRA protocol. The NS-2-based ARP is designed for heterogeneous networks and provides energy efficiency at the cost of overheads (Khan et al. 2010).

The XLP is a PHY-MAC-ROUTE-TRANSPORT (PMRT) protocol. The cross-layer module (XLM) is a PHY-MAC-ROUTE-TRANSPORT-APP (PMRTA) protocol. The XLM and XLP protocols have a common communication block for every networking layer but increase implementation complexity (Akyildiz et al. 2006; Vuran and Akyildiz 2010).

9.8.2 Open Research Challenges in Cross-Layer Design

- Signal fading and path loss effects should be considered while designing the XLPs.
- Limited work has been done on the mobility effects on cross-layer design. Thus, the mobility effects such as topology reconfigurations should also be considered to fix the lower bound on energy.
- There is still demand for XLPs, which reduce the path loss, congestion, and end-to-end delay within network.
- These protocols share and interchange large amount of data between layers. These require large memory space, which is a limitation for the WSNs.
- A design change in one component can affect the entire system. This creates negative consequences such as the instability and modularity issues.
- The available discrete event simulators for WSNs, e.g., J-Sim, GloMoSim, QualNet, OPNET, etc., work on traditional layered architecture. Therefore, development of new software simulators for cross-layer implementation is a big research challenge today.
- The cross-layer design issues are not limited to only these above-mentioned points. A global optimum solution is required to minimize energy consumption and maximize network performance.

References

Abouei, J., David Brown, J., Plataniotis, K.N., and Pasupathy, S. 2011. On the energy efficiency of LT codes in proactive wireless sensor networks. *IEEE Trans Signal Process.* 59(3):1116–1127. doi:10.1109/TSP.2010.2094193.

Abughalieh, N., Steenhaut, K., and Nowe, A. 2010. Low power channel coding for wireless sensor networks. In 2010 17th IEEE *Symposium on Communications and Vehicular Technology in the Benelux (SCVT2010).* 24–25 November 2010, Enschede, Netherlands, 1–5. IEEE. doi:10.1109/SCVT.2010.5720472.

Ahmed, M. 2012. Handshaking problem associated with addressing scheme for the nodes of a wireless sensor network. *Int J Adv Res Technol.* 1(5):74–78.

Akhlaq, M., Sheltami, T.R., and Shakshuki, E.M. 2014. C3: An energy-efficient protocol for coverage, connectivity and communication in WSNs. *Pers Ubiquit Comput.* 18(5):1117–1133. doi:10.1007/s00779-013-0719-2.

Akyildiz, I.F. and Vuran, M.C. 2010. Error control. In *Wireless Sensor Networks*, pp. 117–137. Chichester: John Wiley & Sons. doi:10.1002/9780470515181.ch6.

Akyildiz, I., Vuran, M., and Akan, O. 2006. A cross-layer protocol for wireless sensor networks. In *2006 40th Annual Conference on Information Sciences and Systems*. Princeton, NJ, 22–24 March 2006, Princeton, NJ, USA, 1102–7. doi:10.1109/CISS.2006.286630.

Al-Anbagi, I.S., Erol-Kantarci, M., and Mouftah, H.T. 2012. Fairness in delay-aware cross layer data transmission scheme for wireless sensor networks. In *2012 26th Biennial Symposium on Communications, QBSC 2012*. 28–29 May 2012, Kingston, Ontario, Canada,146–149. doi:10.1109/QBSC.2012.6221370.

Ali, M. and Uzmi, Z.A. 2004. An energy-efficient node address naming scheme for wireless sensor networks. In *2004 International Networking and Communication Conference*. 11–13 June 2004, Lahore, Pakistan. doi:10.1109/INCC.2004.1366571.

Aliyu, M.S., Abdullah, A.H., Chizari, H., Sabbah, T., and Altameem, A. 2016. Coverage enhancement algorithms for distributed mobile sensors deployment in wireless sensor networks. *Int J Distrib Sens Netw.* 2016(1):1–9. doi:10.1155/2016/9169236.

Alwan, H. and Agarwal, A. 2013. A cross-layer-based routing with QoS-aware scheduling for wireless sensor networks. In *2013 ACS International Conference on Computer Systems and Applications (AICCSA)*, 27–30 May 2013, Ifrane, Morocco.

Babulal, K.S. and Tewari, R.R. 2010. Cross layer energy efficient routing (XLE2R) for prolonging lifetime of wireless sensor networks. In *2010 International Conference on Computer and Communication Technology (ICCCT)*. Allahabad, Uttar Pradesh, India, 17–19 September 2010, 70–74. IEEE. doi:10.1109/ICCCT.2010.5640385.

Bachir, A., Dohler, M., Watteyne, T., and Leung, K.K. 2010. MAC essentials for wireless sensor networks. *IEEE Commun Surv Tut.* 12(2):222–248. doi:10.1109/SURV.2010.020510.00058.

Bagaa, M., Younis, M., Ouadjaout, A., and Badache, N. 2013. Efficient multi-path data aggregation scheduling in wireless sensor networks. In *IEEE International Conference on Communications*, Budapest, Hungary, 1560–1564. doi:10.1109/ICC.2013.6654736.

Bahl, L.R., Cocke, J., Jelinek, F., and Raviv, J. 1974. Optimal decoding of linear codes for minimizing symbol error rate. *IEEE Trans Inf Theory.* 20(2):284–287. doi:10.1109/TIT.1974.1055186.

Balakrishnan, G., Yang, M., Jiang, Y., and Kim, Y. 2007. Performance analysis of error control codes for wireless sensor networks. In *Fourth International Conference on Information Technology (ITNG'07)*. 2–4 April 2007, Las Vegas, Nevada, 10:1–4. doi:10.1109/ITNG.2007.149.

Benzaïd, C., Bagaa, M., and Younis, M. 2016. Efficient clock synchronization for clustered wireless sensor networks. *Ad Hoc Netw.* 56:13–27. doi:10.1016/j.adhoc.2016.11.003.

Cardei, M., Thai, M.T., Li, Y., and Wu, W. 2005. Energy-efficient target coverage in wireless sensor networks. In *Proceedings IEEE 24th Annual Joint Conference of the IEEE Computer and Communications Societies*. 13–17 March 2005, Miami, FL, USA, 3(C):1976–1984. doi:10.1109/INFCOM.2005.1498475.

Cardei, M. and Wu, J. 2006. Energy-efficient coverage problems in wireless ad-hoc sensor networks. *Comput Commun.* 29(4):413–420. doi:10.1016/j.comcom.2004.12.025.

Cerpa, A. and Estrin, D. 2004. ASCENT: Adaptive self-configuring sensor networks topologies. *IEEE Trans Mobile Comput.* 3(3):272–285. doi:10.1109/TMC.2004.16.

Cheng, X., Thaeler, A., Xue, G., and Chen, D. 2004. TPS: A time-based positioning scheme for outdoor wireless sensor networks. In *Proceedings-IEEE INFOCOM*. 7–11 March 2004, Hong Kong, 4:2685–2696. doi:10.1109/INFCOM.2004.1354687.

Cugola, G. and Migliavacca, M. 2009. A context and content-based routing protocol for mobile sensor networks. *Wirel Sens Netw.* 5432:69–85. doi:10.1007/978-3-642-00224-3_5.

Cui, S., Goldsmith, A.J., and Bahai, A. 2005. Energy-constrained modulation optimization. *IEEE Trans Wirel Commun.* 4(5):2349–2360. doi:10.1109/TWC.2005.853882.

Doherty, L., Pister, K.S.J., and El Ghaoui, L. 2001. Convex position estimation in wireless sensor networks. In *Proceedings IEEE INFOCOM 2001. Conference on Computer Communications. Twentieth Annual Joint Conference of the IEEE Computer and Communications Society (Cat. No.01CH37213)*. 22–26 April 2001, Anchorage, AK, USA, 3:1655–1663. doi:10.1109/INFCOM.2001.916662.

Doss, R.C., Chandra, D., Pan, L., Zhou, W., and Chowdhury, M.U. 2010. Dynamic addressing in wireless sensor networks without location awareness. *J Inf Sci Eng.* 26:443–460.

Elson, J., Girod, L., and Estrin, D. 2002. Fine-grained network time synchronization using reference broadcasts. *ACM SIGOPS Oper Syst Rev.* 36(SI):147. doi:10.1145/844128.844143.

Eriksson, O. 2011. Error control in wireless sensor networks: A process control perspective. Ph.D. thesis. Uppsala University, Uppsala, Sweden.

Ganeriwal, S., Kumar, R., and Srivastava, M.B. 2003. Timing-sync protocol for sensor networks. In *Proceeding of 1st International Conference on Embedded Networked Sensor System (SenSys)*. Los Angeles, CA, 5–7 November 2003, 47:34–40. doi:10.1111/j.1469-8137.2009.02811.x.

Goldsmith, A. 2005. *Wireless Communications*, p. 250. Cambridge, New York, NY: Cambridge University Press. doi:10.1017/CBO9780511841224.

Gui, C. and Mohapatra, P. 2004. Power conservation and quality of surveillance in target tracking sensor networks. In *Proceedings of the 10th Annual International Conference on Mobile Computing and Networking*. Philadelphia, PA, 26 September–1 October, 2004, 129–143. doi:10.1145/1023720.1023734.

He, T., Huang, C., Blum, B.M., Stankovic, J.A., and Abdelzaher, T.F. 2005. Range-free localization and its impact on large scale sensor networks. *ACM Trans Embed Comput Syst.* 4(4):877–906. doi: 10.1145/1113830.1113837.

Heidemann, J., Silva, F., Intanagonwiwat, C., Govindan, R., Estrin, D., and Ganesan, D. 2001. Building efficient wireless sensor networks with low-level naming. *ACM SIGOPS Oper Sys Rev.* 35(5):146. doi: 10.1145/502059.502049.

Heinzelman, W.B., Chandrakasan, A.P., and Balakrishnan, H. 2002. An application-specific protocol architecture for wireless microsensor networks. *IEEE Trans Wirel Commun.* 1(4):660–670. doi:10.1109/TWC.2002.804190.

Huang, J., Zhang, K., and Yu, M. 2011. A cross-layer collision-aware routing protocol in wireless sensor networks. In *2011 International Conference on Computer Science and Service System (CSSS)*, Nanjing, China, 1909–1913. doi:10.1109/CSSS.2011.5974721.

Imieliński, T. and Navas, J.C. 1999. GPS-based geographic addressing, routing, and resource discovery. *Commun ACM.* 42(4):86–92. doi:10.1145/299157.299176.

Jia, J., Chen, J., Chang, G., and Tan, Z. 2009. Energy efficient coverage control in wireless sensor networks based on multi-objective genetic algorithm. *Comput Math Appl.* 57(11–12):1756–1766. doi:10.1016/j.camwa.2008.10.036.

Kamat, M., Ismail, A.S, and Olariu, S. 2007. Optimized-hilbert for mobility in wireless sensor networks. In *2007 International Conference on Computational Science and Its Applications (ICCSA 2007)*. 26–29 August 2007, Kuala Lumpur, Malaysia, 554–560. IEEE. doi:10.1109/ICCSA.2007.85.

Khan, Z.A., Auguin, M., and Belleudy, C. 2010. Cross layer design for QoS aware energy efficient data reporting in WSN. In *7th International Symposium on Wireless Communications Systems (ISWCS 2010)*. York, UK, 19–22 September 2010, TBD York, United Kingdom, 526–530. doi:10.1109/ISWCS.2010.5624535.

Kotwal, S.B., Verma, S., and Abrol, R.K. 2012. RSSI WSN localization with analysis of energy consumption and communication. *Int J Comput Appl.* 50(11):975–8887.

Kyoung-lae Noh, E.S. and Qaraqe, K. 2008. A new approach for time synchronization in wireless sensor networks: Pairwise broadcast synchronization. *IEEE Trans Wirel Commun.* 7(9):3318–3322. doi:10.1109/TWC.2008.070343.

Lettieri, P., Fragouli, C., and Srivastava, M.B. 1997. Low power error control for wireless links. In *Proceedings of the 3rd Annual ACM/IEEE International Conference on Mobile Computing and Networking*. Budapest, Hungary, 26–30 September 1997, 139–150. doi:10.1145/262116.262142.

Li, J., Andrew, L.L.H., Foh, C.H., Zukerman, M., and Chen, H.H. 2009. Connectivity, coverage and placement in wireless sensor networks. *Sensors.* 9(10):7664–7693. doi:10.3390/s91007664.

Lin, K., Wang, X., Peng, L., and Zhu, X. 2013. Energy-efficient K-cover problem in hybrid sensor networks. *Comput J.* 56(8):957–967. Oxford University Press. doi:10.1093/comjnl/bxt020.

Maróti, M., Kusy, B., Simon, G., and Lédeczi, A. 2004. The flooding time synchronization protocol. In *SenSys'04-Proceedings of the Second International Conference on Embedded Networked Sensor Systems*. Baltimore, MD, 03–05 November 2004, 39–49. doi:10.1145/1031495.1031501.

Moore, D., Leonard, J., Rus, D., and Teller, S. 2004. Baltimore, MD, USA, Robust distributed network localization with noisy range measurements. In *Proceedings of the 2nd International Conference on Embedded Networked Sensor Systems*. Baltimore, MD, 03–05 November 2004, 50–61. doi:10.1145/1031495.1031502.

Moravek, P., Komosny, D., Simek, M., Girbau, D., and Lazaro, A. 2011. Energy analysis of received signal strength localization in wireless sensor networks. *Radioengineering.* 20(4):937–945.

More, A. and Raisinghani, V. 2016. A survey on energy efficient coverage protocols in wireless sensor networks. *J King Saud Univ Comput Inf Sci.* doi:10.1016/j.jksuci.2016.08.001.

Noh, K.L., Wu, Y.C., Qaraqe, K., and Suter, B.W. 2008. Extension of pairwise broadcast clock synchronization for multicluster sensor networks. *Eurasip J Adv Signal Process.* 2008, 1–10: 286168. doi: 10.1155/2008/286168.

Oguejiofor, O.S., Aniedu, A.N., Ejiofor, H.C., and Okolibe, A.U. 2013. Trilateration based localization algorithm for wireless sensor network. *Int J Sci Mod Eng (IJISME).* 1(10):21–27. http://citeseerx.ist. psu.edu/viewdoc/download?doi=10.1.1.684.6710&rep=rep1&type=pdf.

Pan, M.-S. and Tseng, Y-C. 2012. ZigBee-based long-thin wireless sensor networks: Address assignment and routing schemes. *Int J Ad Hoc Ubiquitous Comput.* 12:147–156.

Poduri, S. and Sukhatme, G.S. 2004. Constrained coverage for mobile sensor networks. In *IEEE International Conference on Robotics and Automation, 2004. Proceedings. ICRA '04. 2004.* 26 April–1 May 2004, New Orleans, Louisiana, USA, 1:165–171. doi:10.1109/ROBOT.2004.1307146.

Agarwal, R., Popovici, E. and O'Flynn, B. 2006. Adaptive wireless sensor networks: A system design perspective to adaptive reliability. In *International Conference on Wireless Communications and Sensor Networks,* vol. 6, 216–225.

Pottie, G.J. and Kaiser, W.J. 2000. Wireless integrated network sensors. *Commun ACM.* 43(5):51–58. doi:10.1145/332833.332838.

Priyantha, N. B., Balakrishnan, H., Demaine, E.D., and Teller, S. 2005. Mobile-assisted localization in wireless sensor networks. In *Proceedings IEEE 24th Annual Joint Conference of the IEEE Computer and Communications Societies.* 13–17 March 2005, Miami, FL, USA, 1(C):172–183. doi:10.1109/ INFCOM.2005.1497889.

Raghavendra, C.S., Sivalingam, K.M., Znati, T.F., and Estrin, D. 2004. *Wireless Sensor Networks.* New York, NY: Kluwer Academic.

Ramadurai, V. and Sichitiu, M.L. 2003. Localization in wireless sensor networks: A probabilistic approach. In *Proceedings of the 2003 International Conference on Wireless Networks (ICWN 2003),* Las Vegas, NV, June 2003, 275–281.

Sankarasubramaniam, Y, Akyildiz, I.F., and McLaughlin, S.W. 2003. Energy efficiency based packet size optimization in wireless sensor networks. In *Proceedings of IEEE International Workshop on Sensor Network Protocols and Applications.* 11 May 2003, Anchorage, AK, USA, 1–8. doi:10.1109/SNPA.2003.1203351.

Savarese, C., Rabaey, J.M., and Beutel, J. 2016. Location in distributed ad-hoc wireless sensor networks. In *2001 IEEE International Conference on Acoustics, Speech, and Signal Processing. Proceedings (Cat. No.01CH37221).* 7–11 May 2001, Salt Lake City, Utah, 2037–2040. IEEE. Accessed 28 November. doi:10.1109/ICASSP.2001.940391.

Savvides, A., Han, C.-C., and Strivastava, M. 2001. Dynamic fine-grained localization in ad-hoc networks of sensors. In *Proceeding MobiCom '01 Proceedings of the 7th Annual International Conference on Mobile Computing and Networking.* Rome, Italy, 166–179. doi:10.1145/381677.381693.

Sekhar, A., Manoj, B.S., and Siva Ram Murthy, C. 2005. Dynamic coverage maintenance algorithms for sensor networks with limited mobility. In *Proceedings-Third IEEE International Conference on Pervasive Computing and Communications, PerCom 2005.* 8–12 March 2005, Kauai Island, Hawaii, 2005:51–60. doi:10.1109/PERCOM.2005.15.

Shih, E., Cho, S.-H., Ickes, N. et al. 2001. Physical layer driven protocol and algorithm design for energy-efficient wireless sensor networks. In *Proceedings of the 7th Annual International Conference on Mobile Computing and Networking-MobiCom '01.* Rome, Italy, 16–21 July 2001, 272–287. doi:10.1145/381677.381703.

Sommer, P. and Wattenhofer, R. 2009. Gradient clock synchronization in wireless sensor networks. In *The 8th ACM/IEEE International Conference on Information Processing in Sensor Networks (IPSN 2009).* San Francisco, CA, 13–16 April 2009, 37–48.

Umeki, T., Okada, H., and Mase, K. 2009. Evaluation of wireless channel quality for an ad hoc network in the sky, SKYMESH. In *2009 6th International Symposium on Wireless Communication Systems.* Siena, Italy, 7–10 September 2009, 585–589. IEEE. doi:10.1109/ISWCS.2009.5285366.

Viterbi, A.J. 1967. Error bounds for convolutional codes and an asymptotically optimum decoding algorithm. *IEEE Trans Inf Theory.* 13(2):260–269. doi:10.1109/TIT.1967.1054010.

Vuran, M.C. and Akyildiz, I.F. 2010. XLP: A cross-layer protocol for efficient communication in wireless sensor networks. *IEEE Trans Mobile Comput.* 9(11):1578–1591. doi:10.1109/TMC.2010.125.

Wang, A., Cho, S.H., Sodini, C., and Chandrakasan, A. 2001. Energy efficient modulation and MAC for asymmetric RF microsensor systems. In *Proceedings of the 2001 International Symposium on Low Power Electronics and Design-ISLPED '01.* Huntington Beach, CA, 6–7 August 2001, 106–111. doi:10.1145/383082.383105.

Whitehouse, K., Jiang, F., Karlof, C., Woo, A., and Culler, D. 2004. *Sensor Field Localization: A Deployment and Empirical Analysis.* Technical Report. Berkeley, CA: University of California Berkeley.

Xu, J., He, J., Zhang, Y., Xu, F., and Cai, F. 2016. A distance-based maximum likelihood estimation method for sensor localization in wireless sensor networks. *Int J Distrib Sens Netw.* 2016(4):1–8. SAGE Publications. doi:10.1155/2016/2080536.

Xu, Y., Heidemann, J., and Estrin, D. 2001. Geography-informed energy conservation for ad hoc routing. In *Proceedings of the 7th Annual International Conference on Mobile Computing and Networking—MobiCom '01.* Rome, Italy, 16–21 July 2001, 70–84. doi:10.1145/381677.381685.

Yang, J., Khandani, A.K., and Tin, N. 2005. Statistical decision making in adaptive modulation and coding for 3G wireless systems. *IEEE Trans Veh Technol.* 54(6):2066–2073. doi:10.1109/TVT.2005.853445.

Ye, F., Zhong, G., Lu, S., and Zhang, L. 2008. PEAS: A robust energy conserving protocol for long-lived sensor networks. In *Proceedings of the International Conference on Network Protocols, ICNP,* Orlando, Florida, USA, 200–201. doi:10.1109/ICNP.2002.1181406.

Ye, W, Heidemann, J., and Estrin, D. 2002. An energy-efficient MAC protocol for wireless sensor networks. In *Proceedings of the Twenty-First Annual Joint Conference of the IEEE Computer and Communications Societies.* 23–27 June 2002, New York, NY, USA, 3:1567–1576. doi:10.1109/INFCOM.2002.1019408.

Yu, J., Chen, Y., and Huang, B. 2016. On connected target K-coverage in heterogeneous wireless sensor networks. In *Proceedings of the 2015 International Conference on Identification, Information, and Knowledge in the Internet of Things, IIKI 2015.* 22–23 October 2015, Beijing, China, 262–265. doi:10.1109/IIKI.2015.63.

Yu, J., Ren, S., Wan, S., Yu, D., and Wang, G. 2012. A stochastic K-coverage scheduling algorithm in wireless sensor networks. *Int J Distrib Sens Netw.* 2012(11):1–11. SAGE Publications. doi:10.1155/2012/746501.

Zhang, H. and Hou, J.C. 2005. Maintaining sensing coverage and connectivity in large sensor networks. *Ad Hoc Sens Wirel Netw.* 1:89–124. doi:10.1.1.119.1155.

Zhang, H. 2009. Energy-balance heuristic distributed algorithm for target coverage in wireless sensor networks with adjustable sensing ranges. In *2009 Asia-Pacific Conference on Information Processing.* 18–19 July 2009, Shenzhen University, Hong Kong, China, 452–455. IEEE. doi:10.1109/APCIP.2009.247.

Zhang, H., Wang, H., and Feng, H. 2009. A distributed optimum algorithm for target coverage in wireless sensor networks. In *2009 Asia-Pacific Conference on Information Processing.* 18–19 July 2009, Shenzhen University, Hong Kong, China, 144–147. IEEE. doi:10.1109/APCIP.2009.172.

Zhang, M., Babaei, A., and Agrawal, P. 2012. SCL: A cross-layer protocol for wireless sensor networks. In *Proceedings of the 2012 44th Southeastern Symposium on System Theory (SSST).* Jacksonville, FL, 11–13 March 2012, 179–184. IEEE. doi:10.1109/SSST.2012.6195152.

Zou, Y. and Chakrabarty, K. 2003. Energy-aware target localization in wireless sensor networks. In *Proceedings of the First IEEE International Conference on Pervasive Computing and Communications, 2003. (PerCom 2003)* March 23–26, 2003, Fort Worth, Texas, USA. doi:10.1109/PERCOM.2003.1192727.

Chapter 10

Advanced Applications and Challenges

Vidushi Sharma

Contents

10.1 Introduction...241
10.2 Applications Based on Type of Sensing Behavior.. 242
10.3 Periodic Sensing Applications.. 243
    10.3.1 Environmental Monitoring.. 243
    10.3.2 Facility Management 243
    10.3.3 Home Automation 243
    10.3.4 Military Surveillance 244
    10.3.5 Healthcare Monitoring Systems 244
10.4 Nonperiodic Applications... 244
    10.4.1 Event Detection Systems... 244
    10.4.2 Query-Based Systems..245
    10.4.3 Mobility-Based Systems..245
10.5 Applications Based on Deployment Environment 245
    10.5.1 Terrestrial WSNs 245
    10.5.2 Underground WSNs.. 246
    10.5.3 Underwater WSNs 246
10.6 Other Application Areas..247
References 247

10.1 Introduction

Wireless sensor networks (WSNs) offer a wide range of applications due to their easy deployment, scalability, and auto configuration of nodes. In fact, the applications are so diverse that they have permeated in every aspect of human life and all the global applications. Today WSNs are a part of almost every application area. For the purpose of understanding varied applications, we need to segregate and classify them on the basis of similar characteristics, which is again dependent on the type of WSN. Sensors can be classified on the basis of behavior and deployment environment. Sensor nodes sense the environmental parameters and communicate to gateway via sink. The sensing pattern of the sensor nodes may be periodical or nonperiodical. Periodical sensing behavior is also called time-dependent sensing. Nonperiodical sensing is not time based but is dependent on some criteria such as event occurrence, etc. Different deployment environments have their specific architectures. Hence, the applications can be classified on the basis of environment in which the sensor is deployed. For example, a terrestrial environment may differ from an aquatic environment and hence the architectures of both of them are different. They have specific types of sensor nodes, communication media, and specific sets of protocols to work in accordance with the system to overcome the limitations of that environment. Hence, the applications can also be classified on the basis of deployment environment.

10.2 Applications Based on Type of Sensing Behavior

Periodical sensing is applied in areas where monitoring of the environment parameters is required. In this the sensor nodes save energy by alternating between sleep and wake-up cycle. In the wake-up mode, they sense the environment and communicate and then move to the sleep cycle. This mechanism is less complex but requires a lot of synchronization between the sleep and active periods of sensor nodes of a particular area. In nonperiodical sensing, the sensing is event based, query based, and mobility based. In event-based sensing, the sensor nodes sleep when there is no event and as soon as event occurs the nodes become active. Several applications are based on this sensing like disaster control, theft control, and other environment control applications. In query-based sensing, the nodes are inactive till the time sink raises the query or demands information of the concerned area. Here the sink or the control center initiates the request for sensing, which activates the node of the remote location to perform their designated operations. Mobility-based sensing is another important area of nonperiodic sensing wherein the nodes or the sink are mobile. There are several applications like wildlife tracking and defense tracking systems for surveillance that use mobility-based sensing. A classification of applications based on the sensing behavior is presented in Figure 10.1.

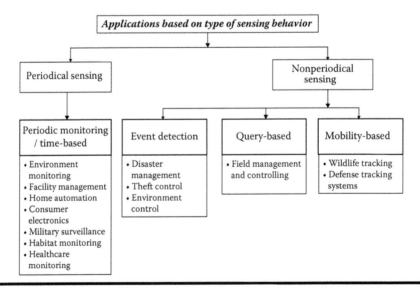

Figure 10.1 Classification of periodic and nonperiodic sensing applications.

10.3 Periodic Sensing Applications

Periodic sensing applications are mainly used for monitoring purposes and they provide information about the environment in which the sensor nodes are deployed.

10.3.1 Environmental Monitoring

Environment monitoring includes sensing of various environmental parameters such as temperature, pressure, humidity, light, pollutants in different environmental conditions like marine, floor, sea shore, terrestrial, and hilly terrains. Monitoring of environmental systems helps in providing information about the conditions of prevailing environmental parameters. Human beings have realized that environmental instability affects many aspects of their life and existence, but it is difficult to measure it due to the tough conditions prevailing in these environments such as extreme temperature conditions, less oxygen on high altitude, inaccessibility to deep valleys and dense forest areas, etc. Environmental monitoring comprises of various applications such as agriculture, field and greenhouse monitoring; monitoring of various habitats like marine, freshwater, terrestrial, forest, etc.; and climate monitoring (Othman and Shazali 2012).

Agriculture and field monitoring primarily focus on field conditions like temperature, moisture, pesticide content, and growth of crops and plants. Greenhouse monitoring measures carbon dioxide content, temperature, and moisture. Li et al. (2011) used a ZigBee-based WSN to monitor greenhouse parameters; later web-based greenhouse monitoring systems were also developed. Climate monitoring is also an important concern for environmentalists. Due to global warming, glaciers are melting, the sea level is rising, and the flora and fauna are being affected. Environmentalists record the data of climate monitoring system to anticipate any calamites due to climatic changes and propose remedial actions. Forest monitoring is another area of study for analyzing the biodiversity and ecological imbalance in forest areas. Habitat monitoring is the broader umbrella which studies the ecological imbalance of various habitats. Habitat is the ecological environment or area where different species coexist and maintain the ecological balance. Ecological imbalance may be caused due to pollutants and change in climatic conditions. This needs to be checked and corrected.

10.3.2 Facility Management

In private and public organizations, one needs to coordinate with various facilities. Facilities are tangible assets that support any organization such as buildings, infrastructure of transport, bridges, lightning system, traffic management system, etc. (Malatras et al. 2008). In these areas, WSN helps in monitoring the assets to ensure improved working conditions by reporting faults so that corrective measures can be taken. It helps in optimizing resources like reducing power consumption, switching between systems based on certain conditions, and controlling temperature and other environmental parameters. Surveillance and security are another aspect of facility management. WSN is preferred over wired solution because of its ease of deployment, scalability, and cost effectiveness.

10.3.3 Home Automation

Home automation systems have led to the concept of smart homes called e-homes. These systems provide interoperability between power, electronics, and electrical-based devices. The objective is to reduce power consumption and monitor whole household facilities. Nowadays, these systems based on WSN are also being used for helping elderly people and persons with disability.

ZigBee-based home automation system was also proposed (Gill et al. 2009). Wifi-based home automation systems are also becoming popular these days. Consumer electronics is another area where WSNs have made the electronic goods intelligent. Different types of sensors have helped in managing and controlling various electronics devices like bio sensing, temperature, pressure sensing, etc.

10.3.4 Military Surveillance

In military surveillance, WSNs are designed to measure electromagnetic signals, light, pressure, and sound signals for detecting explosions. Special chemical and biological sensors are developed to find the presence of people objects or explosive vapors. WSNs provide cost-effective solutions for gathering information. Mobile robots are designed with infrared (IR) imaging sensors, cameras, and provide estimates of the location of intruders. Research challenge in the military surveillance is to incorporate energy-harvesting modules in sensor networks to increase the lifetime and endurance of the network. Another challenge is to improve algorithms of localization, coverage, and connectivity with the sink.

10.3.5 Healthcare Monitoring Systems

Healthcare monitoring systems are of two types. One is remote patient monitoring and the other hospital monitoring systems. In remote patient monitoring, the patient's vital characteristics are measured like blood pressure, pulse rate, location and movement statistics of patient, etc. The patient is monitored from a remote location and in case of any emergency medical help is provided to them. In a hospital monitoring system, the critical patients in intensive care units are monitored and any change in vital characteristics is informed to doctors. A threshold for all vital characteristics is provided; in case the threshold is crossed, an alarm is raised to ensure timely medical care. Sensors are being extensively used in operative mechanisms for patients and postoperative care.

10.4 Nonperiodic Applications

Nonperiodic applications are not dependent on time factors. The nodes become active when any event occurs or a query is generated by the sink. Mobility-based applications are also not time dependent. Stochastic modeling is generally used to forecast and study the behaviors of such systems.

10.4.1 Event Detection Systems

In event detection systems, a trigger is required to activate the sensor node. Wake-up radio concept is used in some systems for activating the sensor node. Disaster management and various control systems are the examples of event detection-based systems.

The detection of natural calamities is very important and at times human intervention for detection is not possible. WSNs provide information regarding flood, volcanic activity, and forest fire. Seismic sensors are able to detect earthquakes. Event-driven working of sensors has less power consumption and has reduced communication overheads. Mechanisms are devised to avoid multiple nodes sending same event information. The overall cost of the system also decreases with event-based sensor networks.

10.4.2 Query-Based Systems

Periodic monitoring systems are energy-consuming systems as compared to nonperiodic systems. In a few applications, it is seen that at times user is presented with lot of data that may be redundant to the user. Environmentalists at times may seek sensor data as per their requirement. They send the request to the sink which in turn activates the sensor nodes of a particular region and then provide data as demanded. This is a pull technique of data gathering. Periodic monitoring follows the push technique whereby the nodes push data to the sink. Sometimes, queries are generated periodically; then the system behaves like periodic monitoring system.

10.4.3 Mobility-Based Systems

Mobility-based systems have their own set of requirements. Here, the event-generating system is mobile and the nodes nearby sense the movement trajectory and report it to the system. Wildlife-tracking systems are classical example of the mobility-based sensing systems. Endangered and precious animals' behaviors and movements are monitored by mobility-based sensor networks. Sensors are placed as collars around the neck of the tiger or lion and are used to track its movement. A vehicular sensor network is another example where the sensor nodes placed on the vehicles' systems are mobile. Sensor nodes provide warning and feedback to the driver regarding lane changing, left-hand drive, if the door of the vehicle is open, obstacle notification while reversing a car, and—very important—collision detection warnings.

10.5 Applications Based on Deployment Environment

A network deployment environment plays an important role in developing sensor network architecture. Applications may be classified on the basis of the environment in which they are deployed (Figure 10.2). They may include terrestrial WSN, underground WSN, underwater WSN, and multimedia WSN.

10.5.1 Terrestrial WSNs

Terrestrial WSNs are structured and have less constraints. Here, the nodes communicate with the sink which may be connected to the base station. Though energy is a limiting factor, energy-harvesting techniques can be applied to ensure the perpetual lifetime of the network. Terrestrial WSNs may be deployed in random manner or in structured manner. Human intervention is possible in most of the terrestrial applications and hence they are easier to handle. Several techniques have been devised for energy conservation like low-duty cycle, optimal routing, and energy efficient

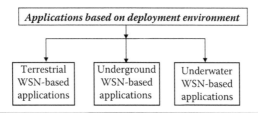

Figure 10.2 Classification of deployment environment-based applications.

MAC protocols. Healthcare monitoring systems, ICU management systems, facility management systems, and traffic monitoring and management systems are few examples of such systems.

10.5.2 Underground WSNs

Cost of deploying, maintaining, and operating underground WSNs is very high as the equipment cost is more as compared to terrestrial systems. The architecture of underground WSNs consists of the number of sensor nodes below the ground and an interface is created with the terrestrial system. Figure 10.3 depicts the architecture of underground mine WSN.

The underground mine network is connected to central monitoring station and in turn is connected to the terrestrial network. Some of the open issues of such networks are limited battery power, attenuation, and signal loss. Deployment of sensors and recharging of batteries are also a cumbersome task and at certain place difficult to replace.

10.5.3 Underwater WSNs

Water occupies almost one-third of the earth's surface, so the major portion of our earth environment is water. It is a rich place for biodiversity with different species living in communion. Ocean sensing and monitoring, coral reef monitoring, water quality monitoring, and marine fish farm monitoring are a few examples of underwater WSNs. All these applications are dependent on special sensors which measure the salinity of the water, turbidity, pH value, and oxygen density. For measuring the health of marine fauna, chlorophyll sensors and other chemical sensors are used. The challenges of

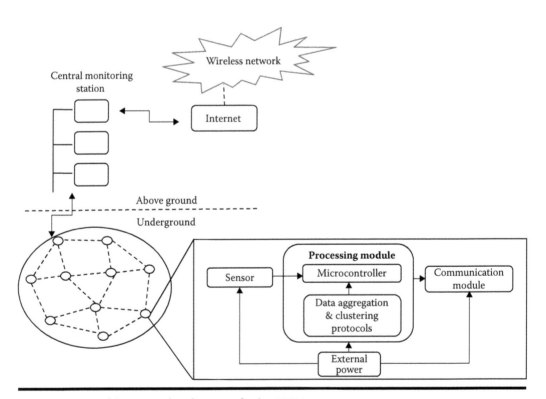

Figure 10.3   Architecture of underground mine WSNs.

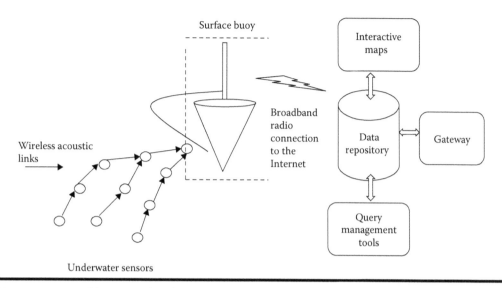

Figure 10.4   Architecture of underwater sensor network.

these networks are that the sensor nodes need to be water resistant and robust to bear stress of waves, typhoons, tides, etc. Maintenance and the deployment of nodes are another challenge to be overcome by the sensor nodes. Figure 10.4 depicts a general architecture of underwater sensor network.

The architecture comprises of a complex flotation device called buoy that is used for marine monitoring. A buoy consists of wireless sensor node for sensing and communication, energy harvesting component, and underwater sensors. Underwater sensors send information to a surface buoy, which is connected to gateway and database repository through radio connections. Interactive maps provide the location of the sensor node to manage the control. The failure rate of sensor nodes is high and the network faces propagation delay and bandwidth constraints. Underwater sensor deployment and maintenance are challenging tasks, but looking at the applications it is an important area of research (Xu et al. 2014).

10.6 Other Application Areas

Multimedia WSN has cropped up for exchanging and transferring images, audio, and video files. These are used for various purposes: monitoring traffic on roads, capturing images of vehicles violating the rules of traffic, and sending multimedia files of remote locations for remote monitoring. Development of these sensor networks has opened new avenues of research in terms of data compression, data processing, and bandwidth management. Sensor nodes are equipped with microphones, infrared cameras, and video cameras.

References

Gill, K., Yang, S. H., Yao, F., and Lu, X. 2009. A ZigBee-based home automation system. *IEEE Transactions on Consumer Electronics*. 55 (2):422–430. doi:10.1109/TCE.2009.5174403.

Li, L. L., Yang, S. F., Wang, L. Y., and Gao, X. M. 2011. The greenhouse environment monitoring system based on wireless sensor network technology. *2011 IEEE International Conference on Cyber Technology in Automation, Control, and Intelligent Systems*, Kunming, China, 265–268. doi:10.1109/CYBER.2011.6011806.

Malatras, A., Asgari, A., and Baugé, T. 2008. Web enabled wireless sensor networks for facilities management. *IEEE Syst J.* 2(4):500–512. doi:10.1109/JSYST.2008.2007815.

Othman, M. F., and Shazali, K. 2012. Wireless sensor network applications: A study in environment monitoring system. *Procedia Eng.* 41:1204–1210. doi:10.1016/j.proeng.2012.07.302.

Xu, G., Shen, W., and Wang, X. 2014. Applications of wireless sensor networks in marine environment monitoring: A survey. *Sensors.* 14(9):16932–16954. doi:10.3390/s140916932.

Chapter 11

The Future for Sensor Networks—Cloud and IoT

Arjun K. Sirohi

Contents

11.1 Introduction...249
11.2 Sensor Networks—From Big Data and Cloud Computing to Internet of Things..........250
11.3 Web 3.0..253
11.4 IoT and IoE..255
11.5 Fog Computing..256
11.6 Infrastructure and Data Challenges—Collection, Integration, and Analysis.................257
11.7 Data Security and Privacy 260
11.8 Distributed Architecture to Enable Local Actionable Intelligence and Insights.............261
11.9 Performance and Scalability Challenges... 262
11.10 Intersection of AI, Data Science, and Machine Learning with Sensor Networks'
      Data in IoT.. 263
11.11 Success Factors for Mass Adoption and Commercialization of IoT............................ 263
References 264

11.1 Introduction

In the previous chapters, we described the basic architecture of wireless sensor networks (WSNs). The protocol stack structure for resource-constrained WSN was also described. The sensing, processing, and communication processes in the context of energy use were explained. We also introduced the reader to sensor node hardware and software use along with the features of commercially available node components. Factors affecting the operation of WSNs such as hardware design, scalability, transmission medium, topology and control, power consumption in sensing, processing, and communication were then presented. To minimize energy consumption, routing techniques were proposed including tactics specific to WSN, such as data aggregation and in-network processing, clustering, different node role assignment, and data-centric methods. Several aspects of energy-harvesting techniques, their features, advantages, and limitations were then discussed. In Chapter 10, we saw the wide range of applications in WSNs such as monitoring, tracking and surveillance of borders; manufacturing instrumentation; traffic monitoring; building structure monitoring; and environmental monitoring including forests, oceans, etc. (Yick et al. 2008; Zhao and Guibas 2004).

The challenges in medical, military, industrial, home, environmental, and traffic applications were discussed to further broaden the scope of WSNs. As may be apparent by now, ultimately it all comes down to collecting massive amounts of data through these WSNs. For any successful application, WSNs are only the means to an end, which is to make smart and intelligent decisions based on the data collected (Chaâri et al. 2016). This brings us to this exciting chapter of what lies ahead—the current and future trends.

11.2 Sensor Networks—From Big Data and Cloud Computing to Internet of Things

A picture is worth a thousand words, they say. So let us start with a 10,000 feet view of the role of sensor networks in the fast-emerging networked world of big data, cloud computing, and Internet of Things (IoT). Figure 11.1 provides a high-level overview of how sensors fit into the larger scheme of things and how they will drive the next generation of applications.

We currently live in a world where lots of exciting things are happening around a collection of massive amounts of structured and unstructured data, collectively termed big data. At the same time, there are many technology advancements taking place that are of historic proportion. For example, the computing power of today's devices like smartphones is many times the power of large mainframe computers of just a few decades ago. It is becoming easier to put together very large numbers of commodity computer servers as well as storage servers to be able to store massive amounts of data and apply emerging technologies like data science and machine learning to

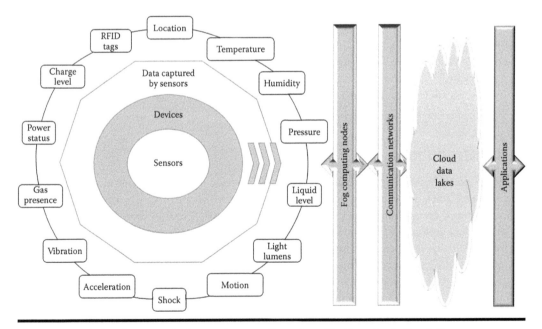

Figure 11.1   Role of sensors in the next generation of applications including IoT.

extract meaningful, actionable intelligence from this data that is transforming the way daily lives of people across countries, boundaries, and peoples are affected. A combination of these factors—the availability of massive amounts of data; the availability of cheaper, expandable, and, elastic storage media to store this data; scalable computing power to process this data; and, last but not the least, the ever-growing, cheaper, and faster network connectivity to make this data available to a vast majority of world citizens—has placed mankind at the threshold of a massive revolution that is changing the way we live. And mind you, these changes are no longer restricted to the information technology industry. Changes are happening across a swath of industries and changing the business models of how various industries run their businesses. New industries are emerging that were unimaginable just a few years ago. The breadth of this revolution and the velocity at which it is progressing make it worthwhile for us to pause and assimilate what is happening, what may lie ahead, and what are the challenges and opportunities this revolution brings. While no one can predict the future, in this chapter we make an attempt to present to the reader the current state and some of the emerging trends and technologies. Our hope is that the readers will not only become more aware of these technologies but also start thinking of ways of how they can apply and make use of these technologies, make improvements to them, and as a result bring about lasting changes in their societies to benefit the common person and make lives better and more enjoyable and at the same time bring about sustainable growth.

So let us start with an understanding of these terms. What is big data, one may ask. There are many different definitions and thoughts around this term, while most agree that it started as a term used in distributed computing but the meaning and purpose of it is evolving. In the simplest terms, big data refers to very large data sets that cannot be processed using traditional methods to draw intelligence from them. One may ask, so what is the definition of very large.? This is a good question to ask because the boundaries defining large data sets are ever expanding. A few years ago, tens of terabytes were considered very large; today we talk in terms of petabytes of data. Big data can be characterized by not just the size but also by the other two major characteristics that make it challenging to process big data—first, the speed at which data is being generated keeps increasing with the proliferation of devices like wireless sensors, and second, the type and variety of data being generated and captured are large and at many times with gaps in between. The digitization of commerce and communications along with the adoption of social media by the ever-increasing number of people has fueled the growth of big data (Yaqoob et al. 2016).

Some daily-life examples are photographs and messages being captured and sent by common people, digital commerce, scientific trials, as well as various kinds of sensors that collect raw data. The technological advances in the capture, transmission, and storage of such big data made it easy to start storing these vast amounts of data by adding cheap, new storage units to an elastic storage system. Such storage systems have built-in redundancies that preserve the data even if these are some failures in parts of the storage system. The challenge, though, is to find new ways to make use of this vast amount of big data to extract valuable inferences, insights, and anomalies. The availability of cheap commodity servers that can add horizontally scalable computing power is what is making it possible to process big data in a relatively compressed timeframe. What used to take a few days to compute a few years ago, can now be computed in seconds. It helps create newer models of computation and analysis where analytical tasks can be broken down into smaller tasks to make use of the massively parallel computing power available in modern data centers. With these advances, it is becoming easier to keep historical data and the context in which the data was generated. As this historical data grows, gaining insights into trends over the time periods becomes easier and so does behavioral analytics or market basket analysis. The main takeaway from trying to understand big data is that it affects every industry—governments, businesses, industries, medicine, social services, etc.—and that finding new ways of capturing, transmitting, storing, and analyzing this avalanche of a wide variety of data will be the key to solving some of the world's most challenging problems, ultimately leading to better lives for world citizens, whose lives are moving in dramatic ways from the physical world to the digital world.

Next, let us look at cloud computing and what it means. To start with, cloud computing can be used in two separate contexts—the personal computing world of individual consumers and the use by businesses to enable their employees and customers to conduct business operations. However, the underlying theme in both contexts is that it involves providing information technology services as a paid subscription service. Imagine a scenario when, a few decades ago, a family in a remote village would run their own power generator to create and use electricity for running the home appliances. Then when a power grid becomes available, the family simply becomes a subscriber to this power grid and consumes however much electricity it needs, pays for that amount only, and is relieved that it does not have to invest in buying, maintaining, and running a power generator. This is what we term as a subscription-based service. Cloud computing is quite exactly similar—it is a subscription-based service to which individual consumers, as well as businesses, can subscribe to. So what does cloud computing provide? Providers of cloud computing services offer services ranging from storage, computing power, as well as applications, including databases. For individual consumers, being able to store their files, documents, photographs, etc., in the cloud provides an easy solution to store once and access it anywhere where they are connected to the Internet. Examples of such services include e-mail services like those from Google, Yahoo, and others. Unlike using an e-mail application on their computer, users simply use the e-mail service by using a browser connected to the Internet. Similarly, services like Google Docs or Live 365 or Dropbox allow consumers to store their files in the cloud and access them from anywhere as well collaborate and share these with others. Until a few years ago, without cloud computing, such collaborative processes where multiple users could be updating the same document were unimaginable. However, now, users can share a document and make edits simultaneously, thereby increasing productivity. On the business side, cloud computing goes a little farther. Let's take an example from before the year 2000. Imagine a new startup business that wished to automate their marketing, sales, and service operations. Before the advent of cloud computing, the startup would have had to buy servers, buy applications, hire some information technology professionals, and then try to make the application available to their employees. This would involve upfront capital expense to buy hardware and software, some guesswork to estimate required computing resources for the present and future growth, and also running expenses to keep the data center, servers, and applications running along with the need to keep paid information technology staff on their payroll. However, today, with cloud computing, a similarly placed startup company has a wide variety of choices available to it via the many vendors of cloud computing. Without having to invest any upfront capital, a startup business today can subscribe to cloud computing services and have their business operations running in a relatively very compressed timeframe. This is the biggest appeal of cloud computing. So what kinds of subscriptions can cloud computing providers provide? While the field is still evolving, most industry professionals divide cloud computing into a few broad categories. The three terms—IaaS, PaaS, and SaaS—are the primary categories of the cloud computing model. IaaS refers to Infrastructure as a Service, which is essentially nothing but the provision of hardware equipment like servers, virtual machines, storage, and networking components over the Internet network. PaaS refers to Platform as a Service, which involves the delivery of hardware along with the operating systems that run the hardware. Last, but not least, SaaS is Software as a Service and refers to Web-hosted applications, such as e-mail and other communications, business applications like sales, service, marketing, digital commerce, and analytics. Each of these services has different capabilities as well as pros and cons. These three cloud computing services are often called the SPI model (SaaS, PaaS, and IaaS). There are some other terms that have begun to be used and we want you, the reader, to be aware of these. In general, there is a concept of "Anything or Everything" as a Service, termed as XaaS, short for "X" as a Service, meaning anything or everything as a service. For example, database vendors provide Database as a Service (DaaS), Storage as a Service (SaaS), Network as a Service (NaaS), Monitoring as a Service (MaaS), Communications as a Service (CaaS), and Identity as a Service (IDaaS). Another classification of cloud computing involves the amount of privacy and controls surrounding a cloud solution. In this context, there are two types of cloud services—public cloud and private cloud—and also a combination of these two, commonly referred to as a hybrid cloud. Since clouds can reside on any hardware, in any data center, run by any vendor, it is easy to think of them as an abstraction layer over the physical hardware. The access to all of the cloud resources is virtual. So what makes a cloud public or private? The distinguishment between the two is made by asking a simple question—are we the only ones to have access to the physical set of resources? Or are we sharing the physical resources with others. The private cloud could very well be hosted in a shared data center but as long as we have exclusive access, it is a private cloud. A public cloud, on the other hand, refers to hardware and software resources that are shared by many subscribers. However, even in a public cloud, it does not mean that subscribers can have access to other subscriber's applications or data. The public cloud service providers put in place enough access and security checks that separate the applications and data from one subscriber to another, keeping them private. A hybrid cloud, thus, refers to a cloud service wherein parts of the cloud are public while some parts are private or exclusive and more secure.

In the next few sections, we present some of the evolving terminologies representing changing paradigms in the fast-changing world of technology related to data computing.

11.3 Web 3.0

Ultimately, all of the big data and IoT feed into driving the development of new Web-based applications where not only humans interact with such applications but also devices communicate with other devices, including those with sensors embedded into them. This is the new era of Web 3.0. Before embarking on understanding what Web 3.0 means, let us take a step back and recap what Web 1.0 and Web 2.0 meant. When the Internet and the World Wide Web started (Web 1.0), the main theme was to provide static information, where people could go and read about the many topics affecting their worlds. For example, there were books, company information, and medical content websites. The phase was characterized by the initial building of the Web, making it accessible to large numbers of people, and trying to commercialize it. The key technologies of interest were Hyper Text Transport Protocol (HTTP), open standard markup languages such as Hyper Text Markup Language (HTML), and Extended Markup Language (XML), rapid and faster Internet access through Internet service providers (ISPs), the first Web browsers, Web development platforms and tools, Web-centric software languages such as Java and JavaScript, the creation of new websites using newer technologies, the commercialization through the next generation of business applications like customer relationship management (CRM), and newer business models to drive electronic commerce. In a nutshell, the main characteristics of this era were static read-only content, one-way communication, individual publishing, and personal websites.

This lasted until the early 2000s when gradually the theme of the Internet started changing to that of collaboration, group participation, multiway communication, and closer involvement of users who started contributing to content on the websites through tools like blogging. This was termed Web 2.0 era, which as we know today is an interactive and social web facilitating collaboration between people. This was different from Web 1.0, which provided static information on websites, where people read websites but did not have much interaction. Newer tools lead to activities such as podcasting, blogging, tagging, social bookmarking, and Web content voting. The users were given a new level of power to interact and drive the Web content that they and their peers wanted to view or collaborate on. The key themes were user interaction and participation and collaboration. The emergence of the mobile Internet and mobile devices like smartphones, tablets, and such provided a new platform to increase the adoption and growth of the Web, particularly outside of the United States. The emergence of open source operating systems and free applications drove these mobile devices to be used by an ever-increasing number of users around the developing world.

So what is Web 3.0? Well, in the simplest terms, we refer to Web 3.0 as driving the next big fundamental change in how websites are created and how people interact with them, the people interaction being the dominant theme. At another level, Web 3.0 is also leading to not just human interaction but also device interaction among the devices themselves. So how is it different from Web 2.0 one may ask? The answer lies in recognizing concepts like connective intelligence that leads to connecting data, concepts, applications, and ultimately people. It is the recognition of some of the technologies that are approaching maturity models, and the adoption of these technologies into the Internet is what will distinguish Web 3.0 from its predecessors. So what are these technologies and trends driving us toward Web 3.0? First and foremost, ubiquitous connectivity that includes broadband adoption, mobile Internet access, and mobile devices is a key technology trend. The more the number of users that connect to the Internet over broadband, the more people interaction there is. So adoption and development of broadband and wireless technologies such as 4G and long-term evolution (LTE) spectrum are making it possible to create Web content that can enable more people interaction. The next technology shift is the democratization and adoption of open technologies. Unlike before, where businesses created proprietary protocols and proprietary software platforms, there is recognition that involving a larger community of all interested parties accelerates the development and adoption of such platforms and software. This has led to the availability of open application programming interfaces (APIs) and protocols, open data formats, open source software platforms, open data via the Creative Commons, Open Data License, etc. These open source platforms and software are helping drive distributed computing to new levels. As more data is generated and collected, there is a need for newer processing models to consume these large amounts of data. A big technology driver for Web 3.0 is the maturing of artificial intelligence (AI). The evolving data science and machine learning techniques have become possible due to the availability of the large amounts of data needed to train machines as well as the availability of computing power that have made it possible to compress the timeframe in which machine learning and AI can be applied on large data sets to produce intelligence. Another promising technology is natural language processing that is being integrated with machine learning to help users ask questions in a natural language and get relevant precise answers. Contrast this with users going to a Web browser search engine and trying to type precise search terms to be able to get answers that they were looking for. Asking questions in a natural language and getting precise answers is becoming more and more possible as we start enabling machines to capture, store, read, understand, and reason with data. Machine learning and data science technologies are leading to the creation of prediction engines and anomaly detection engines. Another term that you may come across when talking about Web 3.0 is Semantic Web. According to the World Wide Web Consortium (W3C), "The Semantic Web provides a common framework that allows data to be shared and reused across application, enterprise, and community boundaries." It is about developing standards like common formats for integration of data originating from many different sources as well as capturing the semantics of how such data is related to real-world objects so that data does not get restricted to the applications that produce and consume it. So in brief, Web 3.0 can be expected to be more connected, open, and intelligent, with semantic Web technologies, distributed databases, distributed computing, natural language processing, machine learning, machine reasoning, and autonomous agents. We can expect an ever-growing number and types of devices to be connected to the Web. Smartphones, autonomous cars, household appliances like refrigerators, heating, and cooling controls, wireless cameras, etc., with all these devices generating data for consumption on the Internet as well as by other devices. Devices will be able to exchange data with each other and even generate new information based on machine learning techniques. We can expect the Internet to perform tasks much faster and more efficiently. We can also expect machines to generate data, transmit data, as well as consume data generated by other machines in a whole new way. WSNs generating and consuming data is a very relevant example in this context. This concept of devices and machines generating data, consuming data, and intelligently processing it to provide real-time intelligence and quick answers leads to the next section, where we discuss the IoT and Internet of Everything (IoE).

11.4 IoT and IoE

The IoT is essentially a coming together of data generated by sensors and actuators in many of today's physical devices like cars, thermostats, water meter readers, and other gadgets, transmitted wirelessly or over wired networks to storage systems connected to the Internet and the application of newer computation techniques like distributed processing, data science, and machine learning on this collection of data to provide meaningful insights and intelligence, including predictive and recommendatory answers. It has been estimated that there are already over 10 billion such devices generating ever-growing amounts of data. The IoT is not only affecting and providing opportunities to the private sector organizations but also to public organizations like cities, transportation systems, and energy grids. While the IoT helps in connecting physical objects via networks to capture, transmit, store, and process data, the IoE adds people to the mix of objects, data, and process. When we talk about the rapidly increasing amounts of data resulting in big data, a majority of such data is being fueled by the IoT initiatives. Even though the physical objects like cars, planes, manufacturing robots, plants, etc., generate large amounts of data, it is the small amounts of data generated by sensor networks at increasing velocities that is expected to become the major source of big data. Even though it seems that the major component of the IoT is the physical objects or things, it is actually the data generated by these things that are of the key value. Processing this large amount of data arriving at rapid speeds needs storage systems and data processing capabilities that can keep pace to generate intelligence and insights in near real time. So the devices and things are only a means to an end—they generate data individually in small amounts, but it is the collection of data from all of such objects can provide insights to drive innovation and efficiencies in the system these objects belong to within a given geographical area. A high-level diagram of the IoT architecture layers is shown in Figure 11.2.

A key characteristic of IoT environments is that the devices and objects can be quite heterogeneous, work under different physical conditions, generate data in different formats, and be able to transmit this data using heterogeneous networks. Another key characteristic of the IoT is that devices and objects can interact with each other without human intervention, thereby relieving people to be available for other more valuable work. The second component of IoTs, beyond the physical devices

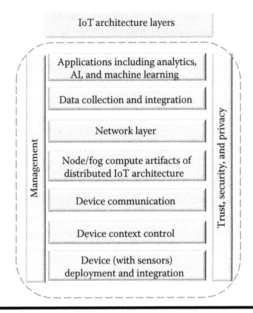

Figure 11.2 IoT architecture layers.

and sensors, is the applications and processes that are needed to turn the collected data into insights that can drive decision-making with or without human intervention or interaction. The applications include data science and machine learning algorithms that can enable decision-making and automation of subsequent actions and processes or be able to connect with people to provide intelligence and predictions in order for people to make decisions, thereby completing the circle for IoE. The ultimate goal of the IoT and IoE is to make life better for human beings by providing data-driven decision-making and actions without the need for human interference or with minimal human intervention. The central theme of IoT and IoE is not just the interconnected devices and the large amounts of data they produce but what can the data tell us and what kind of decision-making can become possible if we apply the right data analytics tools and applications to the data. Instead of relying on inference, we can now know for sure based on the data. Instead of estimating or guessing, we can now predictably determine. Some of the industries already benefiting from he IoT and IoE are manufacturing, transportation, supply chain management, weather, and environment monitoring.

As the number of connected things grows exponentially and the amounts of data they generate and transmit increases, the applications processing such data may find it difficult to keep pace with the velocity and volume of incoming data. This brings some of its own challenges in the concept and architecture of the IoT and IoE whereby data is generated and transmitted over networks to centralized storage systems where specialist applications would process the data. That leads us to the next section where this centralized architecture is challenged, giving to a new paradigm termed fog computing.

11.5 Fog Computing

The term fog computing essentially means some decentralization of data analysis and decision-making from the cloud computing architecture. While most IoT and IoE applications can benefit from the centralized architecture of cloud computing where data is collected and transmitted to a central place where specialized software applications can analyze it and provide intelligence out it, there are some IoT applications like connected autonomous vehicles, smart power grids, smart cities, and WSNs that have needs that cannot be met with the centralized IoT architecture. One may ask why a centralized approach would not work for all IoT applications. The answer lies in understanding that in many such applications, the network bandwidth, low latency, heterogeneity of data, and scalability become a challenge. Let us take smart autonomous cars for example. Such cars have many different devices and sensors that either generate data or consume data. However, the need for autonomous cars is to be able to use such data being generated and transmitted from other cars as well as their own data, apply data processing on the data in real time, and take decisions in a split second. Obviously, the architecture of cloud computing to first generate data, transmit all of the data over networks, store the data, and apply centralized processing on it to send back decisions to the smart cars would take so much time that it would defeat the purpose of autonomous cars. In such use cases, wouldn't it be better to make use of the smart cars systems themselves to generate, collect, collate, and process data that can enable the split-second decision-making needed for their use? At the same time, some amounts of data generated can still be transmitted to the cloud computing system for later use. This type of architecture where data processing and analytics are applied closer to the devices or objects is sometimes termed as fog computing or edge computing. One way to look at it is cloud computing is up there in the cloud and fog computing is closer to the ground or close to the objects. In such an architecture, we make use of the storage and computing power of the devices themselves to make real-time decisions. The fog computing architecture uses the same principles of cloud computing but pushes down the actual computing closer to the edge of networks where the devices are located. Thus, for critical applications, fog computing provides the low latency and also avoids the network bandwidth limitations thus making the architecture more scalable, robust, secure, and reliable. So while cloud computing is here to stay with all its advantages, fog computing is emerging as a crucial supplement to the IoT architecture due to its advantages of low latency, low network bandwidth requirements, security, and reliability. Thus, going forward, we can expect to have IoT and IoE include a mix of cloud computing and fog computing.

11.6 Infrastructure and Data Challenges—Collection, Integration, and Analysis

One of the biggest challenges for the IoT is determining the right architecture. The amount and type of data being generated, collected, transmitted, stored, and processed are big consideration in the design of an IoT system. For example, mobile network operators are already finding that the systems they designed few years ago are becoming incapable of handling the data volume, velocity, and variety. The analytics applications that were designed are now becoming bottlenecks due to the processing capabilities designed in the architecture of a few years ago. Thus, scalability and performance are big challenges even when the data is structured. When the data combines structured as well as unstructured data, the problems of integrating this data becomes a challenge. Distributed processing and distributed databases have provided architectural solutions to enable organizations deal with the large amounts and variety of data. Cloud computing with a centralized storage and processing may be appropriate for some requirements and industries while for other requirements low latency and/or network bandwidth would be key considerations. As we explained in the previous section, most IoT systems may need a balance of both fog computing architecture as well as cloud computing architecture. Another consideration is whether there is a requirement for any preprocessing of data that needs to happen before sending the data to the cloud. For example, some applications may work better if the collected data were to be aggregated or compressed before sending it over the network. The traditional models of first collecting data and then analyzing it no longer holds good for most real-time applications. Some analysis must be done as the data arrives to be able to generate meaningful insights with actionable intelligence. Some of the newer technologies like Hadoop are now able to provide economical ways to improve scalability and parallel processing. There are efforts under way to rearchitect data centers in ways that were considered risky a few years ago. For example, since gigabit networks are now available in data centers, it is no longer necessary to standardize commodity servers with x amount of processing capability, y amount of memory, and z amount of storage. In the newer architecture, separating computing power from storage capacity connected through these faster networks is making it possible to optimize not only the capacity utilization but also makes it much easier to add computing power or storage as the need be without any problems. In these newer architectures, redundancies can be built in for both computing as well as storage which provides the necessary protection from failures. Another challenge for enterprises collecting and storing big data is the building up of enough capacity to be able to back up all this data. As data volumes grow, organizations will need to define retention policies and automate selective backup of the data that must be archived. This selection process of data to keep versus data to delete will add to the workload increasing the need for processing, storage, and network resources that may already be short. At some tipping point, organizations will have to weigh in on the cost–benefit analysis of keeping large amounts of data for analysis versus the analytical benefits gained from analyzing all the data. Integration of data collected by the various types of devices is a big challenge. Most IoT devices and sensors send data in different formats and use different interfaces. Device and sensor manufacturers as well as IoT platform providers make available APIs. Some of the APIs are discussed here but this list is neither comprehensive nor a recommendation or approval of any API. The BloomSky API allows developers to link their personal weather stations to a variety of different outputs. The Fencer API is a geofence Web service for mobile apps, Web apps, and IoT. It offers "geofencing as a service." Mmuzzley allows developers to integrate applications with an IoT platform which features products related to lighting, thermostats, automotive, and health. AT&T M2X Keys API allows you to manage master, device, collection, and distribution keys for all resources or a device or stream while using the AT&T M2X REST API. Manage M2X device distributions using this API. It will allow you to represent a group of devices that begin with the attributes of the original device template when created. AT&T M2X MQTT API uses the Message Queuing Telemetry Transport (MQTT) protocol to streamline the connection between devices and M2X. This allows you to build applications and services as well as gather and translate data in real time. Telecoms cloud is a REST-like HTTP API used to connect data in the cloud. Access features include voice, SMS, fax, data processing, and more. Used with Web, mobile, or Internet-connected device, Telecoms.SNAP PAC REST is a RESTful API to Opto 22 industrial programmable automation controllers (PACs) to enable rapid industrial IoT application development and reduced time to market in machine and system. Predix Time Series API offers sensor data management, distribution, and storage. Predix Traffic Planning API offers metadata obtained from lighting sensors along public roadways. Information such as speed and direction is available for historical and real-time development purposes. The IOStash IoT PaaS API allows developers to integrate an IoT platform into their own applications. Predix Asset Data API can create and store instanced asset models for machine types. For example, an asset model that includes all pumps in an organization. This REST API returns data in JSON format. Roq.

ad Cross-Device User Identification API lets users submit a list of device identifiers and get back a list of identifiers that belong to the same users as those on the input list. The OGC SensorThings REST API by SensorUp allows developers to access and integrate the functionality of OGC SensorThings with other applications. Some example API methods include retrieving a list. The Netbeast API is a builder platform for automating the deployment of IoT applications. The API uses automated dashboard controls to synchronize the functioning of IoT devices and appliances. Space Bunny is an IoT platform that can be used to monitor live streams, and to remotely control devices. This platform features safe message queues, user management, protocol bridging. The xMatters API provides methods for accessing devices, device types, device names, groups, group calendars, people, shifts, and sites programmatically. The Kaa Admin API allows developers to integrate their own applications with its IoT platform. The open source Kaa IoT platform provides an unobscured cloud for IoT development. The InstaUnite API allows customers to integrate remote fleet, asset and equipment management functions into their own systems. The API includes methods for querying metadata about provisioning. Ncryptify provides a RESTful API that maintains the security of data while abstracting the details of the encryption. Bttn is a big physical button powered by AA batteries that can be programmed to perform certain tasks. The Bttn intends to offer developers a REST API to program a "press" to trigger actions. Kontakt.io is a beacon hardware and software platform. The Kontakt.io API allows developers to get information on and update their beacon devices. Beacons are small Bluetooth radio transmitters. MoBagel is a real-time analytics platform designed for IoT devices. It features machine learning, predictive analytics, and triple encryption. This API uses JSON data format. The METAQRCODE API provides developers a way to create their own XML metadata and put it inside a QRCode. This API is Rest based and can be consumed using JSON or XML. Samsung ARTIK cloud is an open data exchange platform designed to bring order to the chaos, break down the silos, and empower you to bring about the promise of IoT. Instacount is a cloud-based counting service that features sharded counters. A sharded counter is a counter whose count is stored in independent shards, and its purpose is increasing performance. The Weaver API provides tools for developers using the Weaver IoT services and frameworks. Use of the Weaver API enables handling many devices with no relation to brand or vendor. mnubo is an analytics platform that aims to improve different kinds of hardware by extracting value from sensor data. mnubo accomplishes this by collecting object data which is processed within mnubo. The Pimatic API allows developers to perform external scripting on Pimatic or create frontends for its home automation framework. Pimatic is designed to provide an extensible platform for home automation. The UnificationEngine API can be used to access UnificationEngineAPI endpoints, which can communicate with various connectors, such as Facebook, Twitter, LinkedIn, Xing, Weibo, and more. The Sensorist API allows developers to access information from environmental sensor systems programmatically. The DeviceHub API allows developers to integrate the solutions available from DeviceHub platform into their applications, to develop IoT projects for different devices. DeviceHive is a framework for machine-to-machine communication that can be used to bring connected devices into the IoT. DeviceHive provides control software and is platform specific. The Web MIDI API is a browser-based JavaScript web API allowing low-level interactions with music or lighting devices synced with a host computer.

However, the lack of standardization of such APIs as well as specifications that provide a common way to describe the data is a big challenge needed to be addressed if the IoT is to be successful. There are many efforts under way to address this standardization problem. A successful IoT architecture would, therefore, stress and focus on making data integration an integral part of the strategy.

11.7 Data Security and Privacy

The importance of data security and privacy stems from the fact that as per forecasts by Gartner, there would be 26 million connected devices by the year 2020. There has been a lot of research on protocols and standards to ensure interoperability of these connected devices, each with sensors that collect and transmit small amounts of data that accumulates over time. While this data may help drive the IoT to make smarter decisions and provide value to consumers, there are challenges surrounding privacy and security concerns. The distributed nature of the IoT and the fact that a large number of such embedded devices are installed in public places with very limited resource capacity makes them very vulnerable to threats. These threats exist at every architecture layer of the IoT. These include the sensor devices themselves, but also in the networks like wireless and cellular networks, in the cloud infrastructure, in the data storage layers, as well as the applications connecting to the IoT. Thus, as depicted in Figure 11.2, the management, as well as privacy and security aspects, must be considered at all the architecture layers of IoT. Solutions being currently considered are encrypted data communications from sensors to base station, encrypted and authenticated data transmission from base stations to the transceivers and from the transceivers to the cloud application and database, as well as employing many layers of security to keep the data safe in the cloud. There are additional considerations when data is transmitted from devices to other devices in the vicinity. For example, for the autonomous and driverless cars, it is extremely critical that the data generated in real time from one car is not exposed to any hacker's malintentions and manipulation before it is received by cars in its vicinity as that would cause complete havoc with the transportation system. Imagine the destruction that could be caused by one rogue car having its sensor and related software hijacked and then sending out incorrect signals to other cars. Similarly, imagine a hacker who could hack into a home device that is not secure and then start capturing not only private data but also be able to manipulate the home network to their advantage in many forms. This is the primary reason that many consumers still do not feel safe to connect their home devices to any networks.

With this knowledge, the IoT cloud providers today are very serious about data privacy and security threats. They carry out what is termed a threat model analysis using data flow diagrams, which detail out the various places where data originates, transforms, transmits, or gets stored for further use. An example of such a data flow diagram in the context of sensors and IoT cloud is depicted in Figure 11.3.

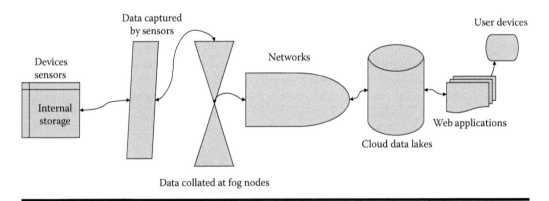

Figure 11.3 Data flow diagram depicting flows of data from devices/sensors to IoT applications on user devices.

Mitigation techniques and technologies are then applied at each point of the data flow diagram as well as to each data flow process with the aim of establishing trust boundaries. Various threat categories like spoofing, tampering, repudiation, denial of service, and elevation of privilege are considered at each data flow node along with a matrix of parameters like damage, reproducibility, exploitability, user impact, and discoverability to then decide what kind of technology and technique is applied to keep data safe and secure. There are many industry-standard techniques including cryptography, multifactor authentication, identity management, trust management, access control lists and permissions, as well as digital signatures and identifiers. Even though the initial devices with sensors were not secure, most new devices and sensor nodes are being manufactured to new security standards. However, as with all things digital, there is always a threat that needs to be met and security of data remains a concern.

11.8 Distributed Architecture to Enable Local Actionable Intelligence and Insights

Some of the initial IoT systems were designed keeping the cloud infrastructure and architecture in mind. It was perceived that all data could be collected, transmitted, and stored in the cloud where it could be processed using analytics to derive actionable business insights. However, it was quickly realized that not all types of IoT devices and use cases could lend themselves to this centralized architecture. In many use cases like parking lot management and road traffic control in a smart city, it would be better to have geographically localized data processing. In yet other cases like smart cars, it would just not be practical to collect, transmit, and process data in a centralized database because of the near real-time requirements of cars interacting with each other based on the data generated and transmitted by sensors in these smart cars. This brought into focus different distributed architectures for IoT where finding intelligence and actionable real-time inputs was necessary. Such distributed architectures provide many benefits while they have some limitations. For example, a distributed architecture for an IoT for smart household meters or a large manufacturing business would make it horizontally scalable whereby adding new compute nodes locally would be sufficient to support increased number of sensors. Such architectures also provide the benefit of preserving network resources and not unnecessarily transmit all the data to a central could server. A distributed IoT architecture for smart cars, thus, would be restricted to specified geographic limits to be able to provide the required closed-loop control as well as robustness needed for smart cars interoperability. Having such distributed architecture to support the IoT naturally makes real-time data analytics a possibility. If local nodes could process specific data in real time, all devices and sensors within that geographic zone could potentially benefit from real-time actionable intelligence. Such local nodes could simply be virtual machines supported by appropriate data processing applications that could process raw data to provide actionable intelligence. These localized data processing nodes would not just function stand-alone but rather become a part of the larger IoT architecture. Such distributed local nodes can be used to transmit aggregated data to the central IoT cloud databases. The data aggregation at these local nodes not only enables local, real-time intelligence but it also helps reducing the network bandwidth requirements of transmitting all the raw data over networks. Last but not the least, by not transmitting raw data, the distributed architecture's local data aggregation along with sufficient processing capacity adds a level of security that would not be available if the data from devices and sensors were to be transmitted over the networks to the central cloud data storage.

11.9 Performance and Scalability Challenges

The very nature of IoT systems makes performance, scalability, availability, and resiliency a challenge. With the ever-growing number of devices and sensors connected to any IoT system, it becomes a challenge to support heterogeneous devices and data formats while at the same time being scalable. Thus, newer hardware and software architectures as well as standards like EPCglobal (Electronic Product Code), which defines how radio-frequency identification (RFID) data is collected, filtered, aggregated, and transmitted, are being developed to address these concerns. In addition, there are standards being created for the different layers of IoT architecture: for example, for the infrastructure layer like 6LowPAN, IPv6, RPL; for identification layer like EPC, uCode, IPv6, URIs; for the communication layer like WiFi, Bluetooth, low-power wide-area network (LPWAN); for the discovery layer like Physical Web, multicast Domain Name System (mDNS), DNS service discovery (DNS-SD); for the data exchange layer like MQTT, Constrained Application Protocol (CoAP), AMQP, Websocket, Node; for device management layer like TR-069, Open Mobile Alliance-Device Management (OMA-DM); for semantic layer like JSON-LD, Web Thing Model; and for the multilayer framework layer like Alljoyn, IoTivity, Weave, Homekit, etc. The current heterogeneity of current IoT nodes or gateways as well as devices with embedded sensors increases the challenges. To achieve higher performance and scalability in IoT architectures, more and more processing capability needs to be pushed down to the nodes as well as devices that are not limited by the constraints of power, computing, and storage resources. Even within constrained environments, it is possible to embed smart devices that can take on some of the data aggregation and processing needs. This way, horizontal scalability can be achieved without relying on overly large centralized systems. The management and control of devices also poses scalability challenges and there is research being done to simplify the provisioning, configuration, and management of devices without any manual interventions. The solution for IoT architectures that can rely on centralized computing is to use cloud computing, which inherently offers the elasticity and scalability of adding computing as well as storage capacity on demand. These centralized cloud infrastructure environments are built with sufficient redundancies to be able to survive storage and system component failures. However, for IoT systems that need to rely on node computing, the resolution could be to build more redundancy and capacity in order to deal with performance spikes as well as scalability and failures. Using load balancers in the distributed node architecture is another option that can help even out demand for compute capacity by making sure that spare nodes take up processing when other nodes may be running out of capacity. Another challenge is to produce low power consuming and memory-efficient networking protocols and architectures. Information-centric networking (ICN) is one such future architecture that is being researched which can make the IoT network layer more efficient and scalable. ICN is a new architecture, especially suited for IoT and it places data at the center and replaces the client–server model with a new publish–subscribe model. In publish–subscribe model, requests for data can be made to the closest node and if the request cannot be met, it is passed on to the next available node and so on until the request is met. This is a lot simpler than existing architectures. The IoT network, when moved to an ICN, will bring about efficiencies and performance improvements. Data in ICN does need not be retrieved from a constantly-on network medium. Instead, replication and caching across network nodes make data movement easier in the IoT, and constant network connectivity is not needed. In short, as IoT systems start to take off, performance and scalability should not be an afterthought but rather built into each of the architecture layers of IoT.

11.10 Intersection of AI, Data Science, and Machine Learning with Sensor Networks' Data in IoT

IoT is making traction in many fields from manufacturing to smart cities to smart cars and healthcare systems. The one big promise of IoT systems is that is that they will make life better for people, conserve resources by making resource consumption more optimal, and relieve humans of many mundane and time-consuming chores. However, not many people realize that for such intelligent systems to deliver on their promises there is a lot of computing that needs to happen behind the scenes. AI is not a new area of research and has been around for many decades. However, it was mostly restricted to academic research with not many consumer use cases except specialized uses like robotics in manufacturing plants. In recent years, though, there are two major factors that are shaping the advancement and adoption of AI. One, the Internet and its myriad applications have started generating massive amounts of data that is being stored in data centers around the world. Two, the low cost as well as more power of computers and storage has significantly increased the amount of storage capacity as well as computing power by orders of magnitude. One may ask, then, how does it affect IoT systems. This is where we want to point out that we have now reached an inflection point where there is a lot of data and computing power available to be able to use the many old as well as new AI techniques. The deployment of the vast amount of sensor devices makes the collection, transmission, and storing of data easy for an IoT system. Also, the availability of computing resources at the nodes or gateways as well as the vast, elastic computing power available with cloud computing makes it now possible to use machine learning, deep learning, data science, and algorithms on this data to create intelligence that would not have been possible a few years ago. The more the data, the better the machine learning. The more the computation power, the shorter the time period to run machine learning. Thus, what used to take years in machine learning outcomes is now possible in hours. All this brings up exciting opportunities for IoT systems. There are many new areas that are benefiting from AI in IoT systems. Healthcare monitoring and delivery, smart cities and intelligent buildings, smart traffic management systems, smart cars, smart homes are all IoT applications that are reaching maturity levels very rapidly. It is now possible for vendors like Amazon to use this technology and sense what items in a home refrigerator need replenishment, order these on behalf of the homeowner, and have these delivered based on the owner's preferences. It is now possible to optimize energy consumption not only in commercial buildings but also consumer households by using machine learning and predictive intelligence by applying machine learning and predictive intelligence algorithms over the data collected over time the data collected over time and then using the IoT network to set controls in the building's energy-consuming appliances. Predictive scheduling ensure that when a building is not occupied for some time, the energy-consuming appliances slow or shut down and are switched back based on the predictive time the occupants are expected to arrive. It is also possible to monitor homes for providing security based on data collected by sensors and by using predictive intelligence. The possibilities are therefore endless for ways in which machine learning and AI can be used in all IoT systems to the benefit of individual consumers as well as for the larger good of mankind.

11.11 Success Factors for Mass Adoption and Commercialization of IoT

IoT systems are here to stay. However, there are certain factors that need to be considered for making IoT systems a success. We can categorize these success factors into two. The first category comprises of technology- and infrastructure-related factors. The second category comprises of business- and consumer-related factors. Let us briefly consider each of these.

On the technology side, the first big challenge is the interoperability of devices and standards for many things like networks, data transmission, etc. IoT systems need to have the devices and sensors communicating with each other in the most efficient manner. If there is a lack of standards, this interdevice communication will become hard and create barriers. The next technology challenge is connectivity and networks. For many existing businesses, it is a challenge to connect their existing stand-alone devices and sensors to a network. With innovations like fog/node computing, this can be addressed to some extent by using protocols like Wi-Fi to connect the local devices/sensors to newly installed fog nodes for collation and processing of data. This way, existing sensors only need to be connected to geographically localized nodes which are further connected to the IoT network. Keeping data secure within IoT systems is a huge challenge due to the nature of devices being in public areas. Thus, the security of the devices as well as of data being generated and collected can be an inhibitor. Solutions will be needed for both hardware and software to keep data secure. Other technology factors are automation, development of better AI applications to provide actionable intelligence in real time.

On the business and consumer side, the success factors include the creation of an ecosystem whereby manufacturers and consumers identify standard products that could be used for IoT systems. From a provider perspective, cost control while deploying a global-scale IoT system can be a big challenge before specific consumer-oriented applications could be offered to monetize it. The innovation from companies creating IoT systems will need to focus on customer needs and how better services can be offered to them in order to be successful. To provide comprehensive solutions, exploring new business models and engaging with a broader ecosystem are critical factors for success. In the new shared economy, there is the democratization of resources, as no business can afford to work in isolation because it alienates the consumers. So, businesses need to think about transforming their existing business models to adapt to the world of billions of connected things.

References

Chaâri, R., Ellouze, f., Koubâa, A. et al. 2016. Cyber-physical systems clouds: A survey. *Comput Netw.* 108:260–278. doi:10.1016/j.comnet.2016.08.017.

Yaqoob, I., Hashem, I.A.T., Gani, A. et al. 2016. Big data: From beginning to future. *Int J Inf Manag.* 36(6):1231–1247. doi:10.1016/j.ijinfomgt.2016.07.009.

Yick, J., Mukherjee, B., and Ghosal, D. 2008. Wireless sensor network survey. *Comput Netw.* 52(12):2292–2330. doi:10.1016/j.comnet.2008.04.002.

Zhao, F. and Guibas, L.J. 2004. *Wireless Sensor Networks: An Information Processing Approach.* Morgan Kaufmann. ftp://uiarchive.cso.uiuc.edu/pub/etext/gutenberg/

Index

A

Access control, security goal, 181
Access point (AP), 217
Acoustic energy harvesting system, 152–155
Acoustic microenergy harvester (AMEH), 155
Active attack, 184, 185
Adaptive random backoff protocol (ARBP), 33
Adaptive routing protocol (ARP), 235
Adaptive sampling, 215
ADCs. *See* Analog-to-digital converters
Address-centric approach, 170
Addressing issues, 227–230
Advanced Encryption Standard (AES), 192
Adversary's capability-based attacks, 185
AES. *See* Advanced Encryption Standard
Aggregation, 174
AI. *See* Artificial intelligence
Air quality index (AQI)
    representation of, 169
    sensor network of, 168
Alive nodes, 77–82, 116
Altered attack, 188
AM. *See* Amplitude modulation
Ambient energy model, 145
Ambient RF energy harvesting system, 156–158
AmbientRT, 10
AMEH. *See* Acoustic microenergy harvester
Amplitude modulation (AM), 217
Amplitude shift keying (ASK), 217
μAMPS, 119
Analog-to-digital converters (ADCs), 4
AP. *See* Access point
API. *See* Application programming interface
Application independent naming, 227
Application layer
    attacks, 189
    protocol stack for WSNs, 6
Application programming interface (API), 11
AQI. *See* Air quality index
ARBP. *See* Adaptive random backoff protocol
Architecture layers, IoT, 255–256
Area coverage, 231
ARM cortex, 144

ARP. *See* Adaptive routing protocol
ARQ. *See* Automatic repeat request
Artificial intelligence (AI), 254
    intersection of, 263
ASK. *See* Amplitude shift keying
Assurance, performance metrics, 182
Asymmetric key-based authentication scheme, 202
Asymmetric key management schemes, 198–200
Asymmetric/public key cryptography, 192–194
Asynchronous tree-based aggregation, 172
ATEMU, 15
Atmel ATmega128L, 7
Attacks-based security schemes, 189–191
Attacks in WSNs
    adversary's capability-based attacks, 185
    detections and preventions, 205
    host-based attacks, 186
    information in transit-based attacks, 185–186
    network-based attacks, 186–189
Authentication in WSNs, 200–202
Authorization, security goal, 181
Automatic repeat request (ARQ), 223
Availability, security goal, 181
Average flux density, 151
Average power consumption of WSN node, 140
Avrora, 11, 16

B

BABRA, 201
Backward secrecy, 182
Bahl–Cocke–Jelinek–Raviv (BCJR) algorithm, 224
Bandwidth utilization, 30
Barrier coverage, 231
Battery-operated device, 138
BCJR algorithm. *See* Bahl–Cocke–Jelinek–Raviv
    algorithm
Beacon-less location discovery scheme, 203
Beacons
    secure location scheme with, 202–203
    secure location scheme without, 203
Berkeley Software Distribution (BSD) license, 12
Bernoulli equation, 151
Bidirectional authentication schemes, 204

Big data, 250–251
Binary modulation scheme, 218
Binary phase shift keying (BPSK), 217, 218
Black hole attack, 188
Blanket coverage, 231
Block codes, 224
Blom's matrix-based key scheme, 197
BloomSky API, 258
B-MAC protocol, 34
Body sensor networks (BSNs), 217
Boost topology-based DC–DC converter, 149
BPSK. *See* Binary phase shift keying
Broadcast authentication, 201–202
Broadcast communication, 3
BSD license. *See* Berkeley Software Distribution license
BSNs. *See* Body sensor networks
BS participation scheme, 195
BS security, 183, 205
BTnut/NutOS, 12
Buck–boost converter, 143, 149
Buoy, 247

C

CARP. *See* Collision aware routing protocol
CAS. *See* Certificate-based authentication
C++-based QSRP protocol, 235
CBR. *See* Constant bit rate
CC2420 current consumption profile, 141, 144
CC-MAC protocol. *See* Correlation-based collaborative MAC protocol
CDA. *See* Concealed data aggregation
CDMA. *See* Code division multiple access
Centralized approach, 170
Centralized cloud infrastructure environments, 262
Centralized processing facility (CFP), 170
Centralized scheme, 219
Certificate-based authentication (CAS), 199
Certificate-based public key cryptographic system, 200
CFP. *See* Centralized processing facility
Channel coding, 223
Child nodes, 226
CHR. *See* Cluster-head relay routing
CHs. *See* Cluster heads
Cipher text-based scheme, 204–205
Clear-to-send (CTS), 33
Climate monitoring, 243
CLINK. *See* Complete LINKage
Closest polynomial scheme, 196
Cloud computing, 252, 257
Cluster-based data aggregation, 173–174
Cluster-head relay routing (CHR), 48
Cluster heads (CHs), 70, 77, 85
Clustering
    motivation and basic ideas, 70
    multiple criteria-based algorithms. *See* Multiple criteria-based algorithms

single criteria-based algorithms. *See* Single criteria-based algorithms
    use of, 70–75
Clustering-based MAC protocols, 34
Code division multiple access (CDMA), 173
Collision aware routing protocol (CARP), 234
Collision-based attacks, 190
Collision, data link layer attacks, 187
Combinatorial design-based key predistribution schemes, 198
Commercially available node components, 6–8
Communication
    energy-efficient modulation techniques in physical layer, 217–218
    energy saving methods in, 215–216
    of keys, 194
    physical layer aspects, 216
    protocols, 216–217
Communication path, nodes in, 173
Communication protocols, 70
Complete LINKage (CLINK), 72
q-Composite keys scheme, 196
Concealed data aggregation (CDA), 204
Confidence bounds, 65–66
Confidentiality, security goal, 181
Constant bit rate (CBR), 37
Constant jamming attack, 186
Content-based addressing, 229
Contention-based protocols, 33
Contiki, 11
Continuous channel, data link layer attacks, 187
Continuous exhaustion, 190
Continuous-time Markov decision process (CTMDP), 128
Control issues, 222–225
Conventional Helmholtz resonator (HR), 153
Converging-diverging flow channel, 152
Convolutional codes, 224
Cooja Simulator (Contiki OS), 17
Coordinated power management, 130–131
Correlation-based collaborative MAC (CC-MAC) protocol, 34
Coverage issues
    challenges, 234
    deployment strategies, 231
    energy-efficient coverage protocols, 231–233
    types, 230–231
CRM. *See* Customer relationship management
Cross-layer based MAC protocols, 35–36
Cross-layer designs, 234
    open research challenges in, 235–239
Cross-layer issues, 234–239
Cross-layer module (XLM), 235
Cross-layer network management, 225
Cross-layer protocols, 31
Cryptography, 191–194, 205
    asymmetric/public key cryptography, 192–194
    symmetric key cryptography, 192

CSMA-MPS, 34
CTMDP. *See* Continuous-time Markov decision process
CTS. *See* Clear-to-send
Customer relationship management (CRM), 253

D

DAD. *See* Duplicate address detection
Darwinian theory, 51
Data aggregation
    elements of, 167–169
    energy-efficient data aggregation techniques
        centralized approach, 170
        cluster-based data aggregation, 173–174
        in-network aggregation, 170–171
        tree-based approach, 171–172
    privacy preserving data aggregation, 175–176
    security in, 174–175
    summary and challenges in, 177
Data-centric approach, 169
Data-centric network architecture, 167
Data centric protocols (flat-based routing), 44–45
Data-driven approach, 215–216
Data link layer attacks, 187–188
Data link layer, protocol stack for WSNs, 6
Data origin and entity authentication, security goal, 181
Data privacy, aggregation, 174, 175
Data reduction process, 22, 215
Data security, 260–261
DBST algorithm. *See* Dynamic balanced spanning tree algorithm
DC–DC converter, 143
DD. *See* Directed diffusion
Deceptive jamming attack, 186
DEEC. *See* Distributed energy-efficient clustering algorithm
Deployment strategies, coverage, 231
Destruction attack, 186
Desynchronization attack, 188
DeviceHive, 259
DeviceHub API, 259
DHAC. *See* Distributed hierarchical agglomerative clustering
DHKE, 199
DigiMesh, 38
Digital signal techniques, 217
Directed diffusion (DD), 45
Direct-sequence spread spectrum (DSSS), 35, 217
Discrete-time Markov decision process (DTMDP), 128
Distributed architecture, 261
Distributed energy-efficient clustering algorithm (DEEC), 79
Distributed hierarchical agglomerative clustering (DHAC), 81, 98
Distributed IoT architecture, 261
Distributed localization approach, 221
Distributed local nodes, 261

Distributed protocols, location unaware and location aware, 232
Distributed scheme, 219
DLM. *See* Dynamically loadable module
DNS. *See* Domain name system
Domain name system (DNS), 228
DOS attack, 184, 189–191, 201–202
DPM. *See* Dynamic power management
DSMAC, 35
DSR protocol. *See* Dynamic source routing protocol
DSSS. *See* Direct-sequence spread spectrum
DTMDP. *See* Discrete-time Markov decision process
Duplicate address detection (DAD), 228
Duty cycle, 29, 138, 215
DVS techniques, 129–130
Dynamic address assignment protocols, 228
Dynamically loadable module (DLM), 10
Dynamic balanced spanning tree (DBST)algorithm, 171–172
Dynamic fluid energy harvesting system, 159
Dynamic key management scheme, 198
Dynamic power management (DPM), 116, 120, 121
Dynamic random access memory (DRAM), 122
Dynamic source routing (DSR) protocol, 37
Dynamic spectrum, 215
"Dynamic voltage scaling", 129

E

Earliest deadline first (EDF) scheduling, 10
EBB. *See* Euler-Bernoulli beam
EBS. *See* Exclusion basis system
ECC. *See* Elliptic curve cryptography
ECCA. *See* Energy-efficient coverage control algorithm
ECC-based asymmetric encryption system, 199
ECDSA, 199
Ecological imbalance, 243
EDF scheduling. *See* Earliest deadline first scheduling
EECS. *See* Energy-efficient clustering schemes
EEHC. *See* Energy-efficient heterogeneous clustered scheme
Efficiency, 182
Efficient modulation techniques, 216
Electrical energy, pressure converted into, 153
Electromagnetic torque, 150
Electromechanical Helmholtz resonator (EMHR), 153
Elitist GA-based routing algorithm, 51–53
Elitist GA, energy-efficient routing algorithm for WSN using, 49–55
Elliptic curve cryptography (ECC), 193
EMERALDS, 13
EMHR. *See* Electromechanical Helmholtz resonator
EmStar, 19
Encryption techniques, 191
Energy-aware stochastic model for WSNs, 128
Energy conservation, 70
    mechanisms in WSN, 22–23

Energy-constrained WSNs, challenges for, 23
Energy constraints, 221–222
Energy density, 141, 143
Energy-efficient clustering approach, 173
Energy-efficient clustering schemes (EECS), 70,
    79–81
  lifetime of, 82
  performance evaluation of, 81–83
Energy-efficient coverage control algorithm (ECCA),
    233
Energy-efficient coverage protocols, 231
Energy-efficient data aggregation protocol, 172
Energy-efficient data aggregation techniques, 169
  centralized approach, 170
  cluster-based data aggregation, 173–174
  in-network aggregation, 170–171
  tree-based approach, 171–172
Energy-efficient heterogeneous clustered scheme
    (EEHC), 78–79
Energy-efficient modulation techniques, in physical
    layer, 217–218
Energy-efficient privacy preservation technique, 176
Energy-efficient routing algorithm for WSN using Elitist
    GA, 49–55
Energy-efficient sensor nodes and networks, factors
    affecting, 19–22
Energy-efficient wireless sensor networks
  architecture of wireless sensor node and network, 4
  challenges for energy-constrained WSNs, 23
  commercially available node components, features
    of, 6–8
  design of energy-efficient sensor nodes and networks,
    factors affecting, 19–22
  energy conservation mechanisms in WSN, 22–23
  operating systems and networking environment for
    WSNs, 9–19
  protocol stack, 4–6
  WSN, 2, 3
Energy harvesting circuit, 142
Energy harvesting issues
  lifetime issue in, 138
  storage capacity, 139–140
Energy harvesting mechanisms, 155, 159
Energy harvesting techniques, 141, 249
  ambient RF energy harvesting system, 156–158
  flow of liquid, 149–152
  mechanical energy harvesting system, 152–155
  solar energy harvesting system, 141–145
  thermal energy harvesting system, 147–149
  wind energy harvesting system, 145–147
Energy saving methods, in communication, 215–216
Energy storage technologies, 143
Entity-based schemes, 194–195
Environmental monitoring, 243
EPCglobal, 262
Error and control issues
  aim of, 223
  approaches, 223–224
  challenges in, 224–225
  error correcting codes, 224
Euler-Bernoulli beam (EBB), 155
Event-based sensing behavior, 242
Event detection systems, 244
Event-driven sensor node, 128
Event-generating system, 245
Exclusion basis system (EBS), 198
Exclusion basis system-based key predistribution
    schemes, 198
Exclusive clustering, 74
Exhaustion, data link layer attacks, 187
External attack, 184
Extracted power, 138
EYES OS, 13

F

Facility management, 243
FAHP. *See* Fuzzy analytic hierarchy process
Fair and delay-aware cross-layer (FDRX) data
    transmission protocol, 234
Fault tolerance (reliability), 20
Fault-tolerant protocols, 190
FEC. *See* Forward error correction
Fencer API, 258
First-in–first-out (FIFO) mechanism, 10, 125
Fitness function, 59, 63
Fixed and universal addressing, of sensor nodes, 229
Flat routing, 228
Flexibility, performance metrics, 182
Flooding attack, 189
Flow of liquid, energy harvesting through, 149–152
FM. *See* Frequency modulation
Fog computing, 256–257
Fog/node computing, 264
Forest monitoring, 243
Forward error correction (FEC), 223
Forward secrecy, 182
Free space model, 56, 76
Frequency band, 214
Frequency modulation (FM), 217
Frequency range, 153, 154
Frequency shift keying (FSK), 217
Freshness, security goal, 181
FSK. *See* Frequency shift keying
Full coverage, 231
Fuzzy analytic hierarchy process (FAHP), 108
Fuzzy attribute weights, 104
Fuzzy clustering, 74
Fuzzy c-means (FCM) algorithm, 74, 75
Fuzzy decision matrix, 104
Fuzzy TOPSIS
  analysis, 106
  approach, 98–101
Fuzzy-weighted decision matrix, 105

G

GAF. *See* Geographic adaptive fidelity
Game-theoretic PM scheme, 130
GANGS protocol, 35
GASA. *See* Genetic algorithm simulated annealing
Genetic algorithm (GA) approach, 49
Genetic algorithm simulated annealing (GASA), 56
Geographic adaptive fidelity (GAF), 46, 47
Geographic addressing, 229–230
Geographical adaptive fidelity, 233
Geo-routing protocols, 191
Globally unique address, 228
Global Mobile Information System Simulator, 15
Global optimization, 222
Global positioning system (GPS), 46, 83, 218
Global power efficiency, of RF power harvester, 158
Global-scale IoT system, 264
GloMoSim. *See* Global Mobile Information System Simulator
GPS. *See* Global positioning system
GRAdient Broadcast (GRAB), 45
Gradient clock synchronization protocol, 226
Greedy aggregation, 171
Greedy scheme, 126–127
Greenhouse monitoring, 243
Group-based key management scheme, 197
Group key management algorithm, 197

H

Habitat monitoring, 243
HAC. *See* Hierarchical agglomerative clustering
Hamming code (HC), 224
Hardware constraints, 19
Hardware limitations of sensor nodes, 225
HARQ. *See* Hybrid automatic repeat request
Harvesting circuit design, 143
HBT. *See* Hierarchical binary tree
HC. *See* Hamming code
Health-care monitoring systems, 244
HEED. *See* Hybrid energy efficient distributed
Heliomote, 142
Hello flood attack, 188, 191
Heterogeneity-based protocols, 47–48
Hidden Markov model (HMM), 128
Hierarchical agglomerative clustering (HAC), 71
    algorithms with quantitative data, 73
Hierarchical-based routing, 29–30, 45–46
Hierarchical binary tree (HBT), 197
Hierarchical clustering, 71–73
Hierarchical routing, 228
Hierarchical sampling, 215
High-resolution robust localization (HIRLOC) protocol, 220
HMM. *See* Hidden Markov model
Home automation systems, 243–244

Hop-by-hop authentication scheme, 204
Horizontal scalability, 262
Host-based attacks, 186
Hybrid automatic repeat request (HARQ), 224
Hybrid cloud, 253
Hybrid energy efficient distributed (HEED), 46
Hybrid GA for routing, 55–60
Hybrid schemes, 200
Hybrid storage solar energy harvesting system, 145

I

IaaS. *See* Infrastructure as a Service
IBE-based scheme, 199
ICN. *See* Information-centric networking
ID-based key agreement schemes, 199–200
ID-based key management scheme, 200
ID-based one-way function scheme, 197
Ideal cell, voltage and current characteristics of, 143
Identity-based encryption (IBE) idea, 199
IDSQ protocol. *See* Information-driven sensor query protocol
IEEE802.15.4 stack, 217
IETF. *See* Internet Engineering Task Force
Industrial, scientific, medicine (ISM) bands, 215
Industry-standard techniques, 261
Information-centric networking (ICN), 262
Information-driven sensor query (IDSQ) protocol, 47
Information, in transit-based attacks, 185–186
Infrastructure as a Service (IaaS), 252
Initial IoT systems, 261
In-network aggregation, 170–171
In-network TAG approach, 170
INSENS, 202
Instacount, 259
Instant Contiki package, 17
InstaUnite API, 259
Integrity in WSNs, 200–202
Integrity, security goal, 181
Intel's iMote2, 129
Interactive maps, 247
Interfacing unit, 119
Internet, 253
Internet Engineering Task Force (IETF), 216
Internet of Everything (IoE), 255–256
Internet of Things (IoT), 180, 250–253, 255–256
    mass adoption and commercialization of, 263–264
IoE. *See* Internet of Everything
IOStash IoT PaaS API, 258
IoT. *See* Internet of Things
ISM bands. *See* Industrial, scientific, medicine bands

J

Jamming attack, 186
Java-Simulator (J-Sim), 18

K

Kaa Admin API, 259
KEK. *See* Key encryption key
Key distribution method, 194
Key encryption key (KEK), 197
Key management schemes, 205
    asymmetric key management schemes, 198–200
    symmetric key management schemes, 194–198
Kinetic energy, 152
Kontakt.io API, 259

L

LAD. *See* Localization anomaly detection
LAMA. *See* Link activation multiple access
Large-scale wind energy harvesting technique, 146
Latency and delay, 30
LDPC codes. *See* Low-density parity-check codes
LEACH. *See* Low-energy adaptive clustering hierarchy
LEACH protocol, 34
LEM. *See* Lumped element modeling
Li-battery, 145
Limited bandwidth, 225
Limited resources, security limitation, 183
Linear regression table, 226
Link activation multiple access (LAMA), 32
LiteOS, 11–12
LKHW. *See* Logical key hierarchy for WSNs
Localization anomaly detection (LAD), 203
Localization-based sensor node, 221
Localization issues, 218–222
Localized combinatorial keying (LOCK), 198
Localized data processing nodes, 261
Location-based algorithm, 205
Location-based routing, 46
LOCK. *See* Localized combinatorial keying
Logical key hierarchy for WSNs (LKHW), 198
Logical tree-based key predistribution schemes, 197
Long-term evolution (LTE) spectrum, 254
Low-density parity-check (LDPC) codes, 224
Low-energy adaptive clustering hierarchy (LEACH), 45, 46, 75–78, 81
Low-power design techniques, 19
Low-power modulation scheme, 5
Low-rate wireless personal area networks (LR-WPANs), 37, 216
LR-WPANs. *See* Low-rate wireless personal area networks
LTE spectrum. *See* Long-term evolution spectrum
Lumped element modeling (LEM), 155

M

MAC address. *See* Message authentication code address
MAC protocols. *See* Medium access control protocols
MADM techniques. *See* Multiattribute decision-making techniques

MagnetOS, 12
Mannasim, 16
MANTIS, 11
MAP decoding. *See* Maximum a posteriori decoding
Master key-based predistribution scheme, 194
MATLAB, 15–16
Matrix-based group key predistribution scheme, 196–197
MAUT. *See* Multiattribute utility theory
Maximal power point (MPP) system, for solar energy, 144
Maximum a posteriori (MAP) decoding, 224
Maximum power point tracking (MPPT) system, 143, 149
Maximum residual energy-based clustering algorithm, 86
Maximum residual energy-based clustering scheme, 83–90
MCDM techniques. *See* Multiple criteria decision-making techniques
Mechanical energy harvesting system, 152–155, 159
Medium access control (MAC) protocols, 5, 27
    classification of, 30–36
    features for, 28–29
    performance parameters of, 29–30
    simulations of, 36–39
Medusa II Nodes, 119
Medusa MK-2 sensor node, 220
Memory management, 10
Message authentication code (MAC) address, 227
Message Queuing Telemetry Transport (MQTT) protocol, 258
METAQRCODE API, 259
Microcontroller unit, 118
Middleware systems, 229
Military surveillance, 244
Minimum shift keying (MSK), 217, 218
Mission-critical applications, 33
Mitigation techniques, 261
Mobile robots, 244
Mobility-based approach, 216
Mobility-based protocols, 47
Mobility-based sensing behavior, 242
Mobility-based systems, 245
Model-based active sampling, 216
Modified genetic algorithm, flowchart of, 52
MODM. *See* Multiobjective decision-making
Modulation schemes, comparison of, 219
Modulation techniques, features of, 218
MOGAs. *See* Multiobjective genetic algorithms
MPPT system. *See* Maximum power point tracking system
MQTT protocol. *See* Message Queuing Telemetry Transport (MQTT) protocol
MSK. *See* Minimum shift keying
MSP430 microcontroller, 7
Multiattribute decision-making (MADM) techniques, 90

Multiattribute utility theory (MAUT), 91
Multicluster system, 227
Multihop
  authentication, 201
  communication, 6
  routing protocol, 12
Multimedia WSN, 247
Multiobjective decision-making (MODM), 90
Multiobjective genetic algorithms (MOGAs), 233
Multipath fading model, 56, 76
Multipath key reinforcement scheme, 196
Multipath protocols, 47
Multipath routing, 202
Multiple criteria-based algorithms
  Pareto-optimal theory, 91–92
  ranking using fuzzy TOPSIS, 98–102
  TOPSIS approach, 92–98
  VIKOR, ranking using, 102–111
Multiple criteria decision-making (MCDM) techniques, 90
Multiple transceivers, 33
Multithreaded model, 9

N

NAMA. *See* Node activation multiple access
Naming issues, 227–230
Naming schemes, 228
NanoQplus, 13
Nano-RK, 12–13
Naval Research Laboratory (NRL) group, 18
NCAEL. *See* Novel clustering approach for extending the lifetime
Negative ideal solution (NIS), 92
Neighbor's certificate schemes, 204
NetSim (Network Simulator), 18
Network-based attacks, 186
  application layer attacks, 189
  data link layer attacks, 187–188
  network layer attacks, 188
  physical layer attacks, 186–187
  transport layer attacks, 188–189
Network density, 20
Network deployment environment, applications based on, 245–247
Networking environment for WSNs, 9–10, 9–19
Network keys, 194
Network layer, 188, 191
  protocol stack for WSNs, 6
Network lifetime, 65
Network MAC, 34
Network security, 21
Network Simulator v2 (ns2), 16
Network Simulator v3 (ns3), 16–17
Network topology, 21, 205
Network traffic, 20
Network-wide unique address, 228

NIS. *See* Negative ideal solution
Node activation multiple access (NAMA), 32
Node capture attack, 188
Node heterogeneity, 173
Noncompensatory-based models, 91
Nonconventional algorithms, 61
Nonperiodical sensing behavior, 242
Nonperiodic sensing applications, 242, 244–245
Nonrepudiation, security goal, 181
Novel clustering approach for extending the lifetime (NCAEL), 79, 81
Novel searching algorithm, 233
NP-complete program, 49
NRL group. *See* Naval Research Laboratory group
NS-2-based ARP, 235
n-type semiconductor, 148

O

Objective Module Network Testbed in C++ (OMNeT++), 17
OGC SensorThings REST API, 259
OGDC approach. *See* Optimal geographical density control approach
OHC-based approach, 201
OMA-DM. *See* Open Mobile Alliance-Device Management
One-hop unicast authentication, 201
Open Mobile Alliance-Device Management (OMA-DM), 262
Open source Kaa IoT platform, 259
Open wireless sensor (Wi-Se) node, 144
Operating environment (applications), 21
Operational-level PM techniques, 126–129
OPNET, 17
Optimal geographical density control (OGDC) approach, 233
Optimal modulation scheme, 216
Outsider attack, 184
Overlapping clustering, 74
Overwhelm attack, 189

P

PaaS. *See* Platform as a Service
Packet delivery ratio, 30
Pairwise key predistribution scheme, 195
PAMAS. *See* Power-aware multiaccess with signaling
Parent node, 171, 172, 174
Pareto-optimal bar chart for CH selection, 95
Pareto-optimal cluster heads (CHs), 94
Pareto-optimal nodes, 101
Pareto-optimal plot for CH selection, 108
Pareto-optimal theory, 91–92, 105
Parity bit, 223
Partial clustering, 75
Partitional clustering, 71

Passive attack, 184, 185
Path-based DOS attack, 189
Pattern MAC protocol (P-MAC), 32
PEAS. *See* Probing environment and adaptive sleeping
Peer intermediary's participation for key establishment (PIKE) scheme, 195
PEEROS operating system, 13
PEGASIS. *See* Power-efficient gathering in sensor information systems
Performance metrics, 182
Periodical sensing behavior, 242
Periodic monitoring, 245
Periodic sensing applications, 242, 243–244
Perturbation, privacy preserving data aggregation, 176
Perturbed sensor network, 176
PH. *See* Privacy homomorphism
Phase modulation (PM), 217
Phase shift keying (PSK), 217
Photovoltaic (PV) cell, 141, 142
Physical layer
    aspects of, 216
    attacks, 186–187
    energy-efficient modulation techniques in, 217–218
    protocol stack for WSNs, 5
Piezoelectric crystal, 152–153
Piezoelectric energy harvesting method, 152
Piezoelectric generator, 154
Piezoelectricity effect, 153
PIKE scheme. *See* Peer intermediary's participation for key establishment scheme
Pilot Cell, 144
Pimatic API, 259
PIS. *See* Positive ideal solution
Plaintext-based scheme, 203–204
Platform as a Service (PaaS), 252
PM. *See* Phase modulation; Power management
P-MAC. *See* Pattern MAC protocol (P-MAC)
PMR cross-layer protocols (XLPs), 234
Point coverage, 231
Polynomial-based key predistribution schemes, 196
Positive ideal solution (PIS), 92
Power-aware multiaccess with signaling (PAMAS), 33
Power-aware WSNs
    communication, 125
    computing, 124–125
    hardware, 122–123
    sensing, 123–124
    software, 123
Power consumption (lifetime), 22
Power efficiency, of voltage multiplier, 158
Power-efficient gathering in sensor information systems (PEGASIS), 46
Power grid, 252
Power management circuit, 144
Power management (PM) in sensor node, 116
    challenges in, 132
    design techniques for low power, 120

power-aware WSNs, 121–125
power management, 125–131
sources of power consumption, 117–120
standard OSs for WSNs, 121
PowerTOSSIM, 14
Power unit, 119–120
PPDA. *See* Privacy preserving data aggregation
Predictive scheme, 127
Predix Asset Data API, 258
Predix Time Series API, 258
Predix Traffic Planning API, 258
Preemptive multitasking scheduling, 12
Primary energy storage, 145
Priority-based scheduling protocols, 32
Privacy, 260–261
    security goal, 181
Privacy homomorphism (PH), 204
Privacy preserving data aggregation (PPDA), 175–176
Private cloud, 253
Probing environment and adaptive sleeping (PEAS), 233
Probing node, 233
Production costs, 21
Proposed elitism-based GA routing, 50
Protocols
    communication, 216–217
    coverage, 231–233
Protocol stack for WSNs, 4–6
PRT vector, 62
PSK. *See* Phase shift keying
p-type semiconductor, 148
Public cloud, 253
Public key cryptography, 192–194
Public key cryptosystem, 198
Publish–subscribe model, 262
Pulse width modulation (PWM), 144
Pure probabilistic key predistribution schemes, 196
PV cell. *See* Photovoltaic cell
PV harvesters, 143
PWM. *See* Pulse width modulation

Q

QoS-based protocols, 48
Quality Networking (QualNet), 18
Quality of service (QoS), 20, 205, 230
QualNet. *See* Quality Networking
QualNet simulator-based FDRX protocol, 234
Query-based sensing behavior, 242
Query-based systems, 245
Query diffusion, 174

R

Radio energy dissipation model, 79
Radio energy model
    for transmitting, 76
    used in wireless sensor network, 80

Radio frequency (RF)
  energy harvester, 157
  spectra, 215
Radio-frequency identification (RFID), 262
Radio interference, 187
Radio unit, 118–119
Random jamming attack, 186
Random key predistribution scheme, 196, 197
Range adaptive random backoff protocol (RARBP), 33
Range-based localization protocols, 220
Range-based localization schemes, 219
Range-free protocols, 220
RARBP. *See* Range adaptive random backoff protocol
Rate-harmonized scheduling, 13
Rate monotonic (RM) scheduling, 13
Rayleigh flat-fading channels, 218
RBS protocol. *See* Reference broadcast synchronization protocol
Reactive jamming attack, 186
READA algorithm. *See* Redundancy elimination for accurate data aggregation algorithm
Real solar cell, voltage and current characteristics of, 143
Real-time operating system (RTOS), 121
Receiver-initiated cycled receiver (RICER), 34
Receiver-only-based protocols, 227
Receiver-only protocols, 226
Receiver-receiver-based protocols, 226–227
Receiver–receiver-based protocols, 226
Receiver–receiver synchronization, 225
Rectifiers, in voltage multiplier, 158
Redundancy elimination for accurate data aggregation (READA) algorithm, 172
Reference broadcast synchronization (RBS) protocol, 225
Reference nodes, 227
Replayed routing information, 188
Request-to-send (RTS), 33
Residual energy
  constant bit rate *vs.,* 37
  of nodes, 54
  of various schemes, 55
Resilience, performance metrics, 182
Resistance, performance metrics, 182
Resource-constrained WSN, 249
REST API, 258
RESTful API, 258
RF–DC converter, 156, 157
RFID. *See* Radio-frequency identification
RICER. *See* Receiver-initiated cycled receiver
RIOT, 13
Rivest–Shamir–Adleman (RSA), 192, 198
Robustness, performance metrics, 182
Robust statistical method, 203
Rockwell's Wins, 119
Root node, 226
Routing, 44
  energy-efficient routing algorithm for WSN, 49–55

genetic algorithms for, 48–49
  hybrid GA for, 55–60
  protocols for WSNs, types of, 44–48
  self-organizing migrating algorithm for, 60–66
Routing protocols, 202
RSA. *See* Rivest–Shamir–Adleman
RSA-based asymmetric encryption system, 199
RTOS. *See* Real-time operating system
RTS. *See* Request-to-send

S

SaaS. *See* Software as a Service
SaaS, PaaS, and IaaS (SPI) model, 253
Samsung ARTIK cloud, 259
SAR protocol. *See* Sequential assignment routing protocol
Scalability, 20, 182
Scheduled-based MAC protocols, 32–33
Scheduled MAC protocols, 32–33
Secret key system, 192
Secure data aggregation, 203–205
  cipher text-based scheme, 204–205
  plaintext-based scheme, 203–204
Secure data fusion, 205
Secure in-network routing algorithms, 204
Secure location, 202, 202–203
Secure range-independent localization (SERLOC) protocol, 220
Secure routing, 202, 205
Security, 205
  attacks in WSNs, 187
  goals, 181–182
Security, in data aggregation, 174–175
Seebeck effect, 147
Seed nodes, 219
SEF mechanism. *See* Statistical en route filtering mechanism
Self-destruction technique, 190
Self-management, 20
Self-organizing application, 182
Self-organizing migrating algorithm (SOMA), 60–66
Semantic Web, 254
Semi-Markov process (SMP), 128–129
Sender–receiver-based protocols, 226
SenOS, 14
SENSE. *See* Sensor Network Simulator and Emulator
Sensing behavior, applications based on, 242
Sensing unit, 120
Sensor management protocols (SMPs), 6
Sensor networks, 3, 9, 22, 250–253
Sensor network security
  aspects in
    goals, 181–182
    limitations, 182–183
    performance metrics, 182
  attacks in
    adversary's capability-based attacks, 185

host-based attacks, 186
   information in transit-based attacks, 185–186
   network-based attacks, 186–189
authentication and integrity in, 200–202
cryptography
   asymmetric/public key cryptography, 192–194
   symmetric key cryptography, 192
issues and challenges, 205
key management schemes
   asymmetric key management schemes, 198–200
   symmetric key management schemes, 194–198, 195
mechanisms, 189–191
secure data aggregation
   cipher text-based scheme, 204–205
   plaintext-based scheme, 203–204
secure location, 202–203
secure routing, 202
vulnerable components
   BS security, 183
   sensor node security, 184–185
Sensor Network Simulator and Emulator (SENSE), 19
Sensor nodes, 3, 138, 139, 242, 245
   comparative analyses of, 6, 7, 8
   components, features of, 6–7
   hardware components of, 19
   security, 184–185
Sensor protocols for information via negotiation (SPIN), 44
Sensors, 242
SensorSim, 18
SEP. See Stability election protocol
Sequential assignment routing (SAR) protocol, 48
SERLOC protocol. See Secure range-independent localization (SERLOC) protocol
SHELL, 198
Short circuit current, 143
Shuffling technique, privacy preserving data aggregation, 176
Signal processing scaling (SPS), 131
Signal-to-noise ratio (SNR), 224
Simple eight-node network, 73
Simple-power modulation scheme, 5
Simple random backoff protocol (SRBP), 33
Simplified marker and cell (S-MAC) scheme, 35, 39, 232
Simulink, 15–16
Single criteria-based algorithms
   distributed energy-efficient clustering algorithm, 79
   energy-efficient clustering scheme, 79–83
   energy-efficient heterogeneous clustered scheme (EEHC), 78–79
   low-energy adaptive clustering hierarchy, 75–77
   maximum residual energy-based clustering scheme, 83–90
   novel clustering approach for extending the lifetime, 79
Single-hop communication, 6
Single LINKage (SLINK), 72

Single path protocols, 47
Sinkhole attack, 188
Sink node, 228
Sleep–wake-up cycle, 231
SLINK. See Single LINKage
6LowPAN, 216
S-MAC scheme. See Simplified marker and cell scheme
SMP. See Semi-Markov process
SMPs. See Sensor management protocols
s-MREC. See Static maximum residual energy-based clustering
SNR. See Signal-to-noise ratio
Software-and hardware-based addressing scheme, 229
Software as a Service (SaaS), 252
Solar cell characteristics, 143
Solar energy, harvesting system, 141–145
Solar panel, 142
Solar radiation, 143
Sol-gel fabrication technique, 153
SOS, 12
Sound pressure level (SPL), 153, 154
Source and sink nodes, data aggregator between, 167
Space Bunny, 259
Sparse topology and energy management (STEM) protocol, 34
SPEED, 48
SPI model. See SaaS, PaaS, and IaaS model
SPIN. See Sensor protocols for information via negotiation
SPiRT. See Synchronization through piggybacked reference timestamps
SPL. See Sound pressure level
SPM. See Static power management
Spoofed attack, 188
SPS. See Signal processing scaling
SRBP. See Simple random backoff protocol
Stable election protocol (SEP), 79
Star-like tree-based key predistribution schemes, 197
Static maximum residual energy-based clustering (s-MREC), 83–85, 88
Static power management (SPM), 120
Statistical en route filtering (SEF) mechanism, 204
Statistical method, 204
STEM protocol. See Sparse topology and energy management protocol
Stochastic scheme, 127–129
Storage unit, 119
StrongARM1100, 122
Strong statistic estimation model, 204
Student's t distribution test, 66
Supercapacitors, 143
Survivability, 182
SVM-based data aggregation tree, 172
Sweep coverage, 231
Sybil attack, 190
   data link layer attacks, 188
Symmetric key cryptography, 191, 192

Symmetric key management scheme
   entity-based schemes, 194–195
      exclusion basis system-based key predistribution
         schemes, 198
      matrix-based group key predistribution scheme,
         196–197
      pairwise key predistribution scheme, 195
      pure probabilistic key predistribution schemes, 196
      tree-based key predistribution schemes, 197–198
Synchronization through piggybacked reference
      timestamps (SPiRT), 227
Synchronous tree-based aggregation, tree-based
      approach, 171–172

T

TAG approach. *See* Tiny aggregation approach
Tampering attack, 186
Tamper-proofing packages, 190
Target coverage, 231
Task control block (TCB), 12
Task management (TM), 5
Task scheduling techniques, 130
TCB. *See* Task control block
TDMA. *See* Time-division multiple access
TDMA-based protocols, 29
TDMA MAC protocols, 35
TEG. *See* Thermoelectric generator
TEG-EH. *See* Thermo-electric generator energy
      harvesting
Telecoms cloud, 258
Terrestrial WSNs, 245–246
Thermal energy harvesting system, 147–149
Thermoelectric effect, 147, 155
Thermoelectric generator (TEG), 138, 148–149, 159
Thermo-electric generator energy harvesting (TEG-EH),
      148
Thermoelectric microgenerator, 155
Thevenin's equivalent circuit, 152
Thin-film solar cells, 142
3DES. *See* Triple DES
TICER. *See* Transmitter-initiated cycled receiver
Time-division multiple access (TDMA), 225
Time-out scheme, 126, 127
Time stamp, 226
Time synchronization protocols, 182, 225
      receiver-only-based protocols, 227
      receiver-receiver-based protocols, 226–227
Time sync protocol for sensor networks (TPSNs), 225,
      226
Tiny aggregation (TAG) approach, 170
TinyOS 2.0, 10–11
TinyOS codes, 14
"TinyOS" operating system, 7
TinyOS SIMulator (TOSSIM), 14
TinyOS 1.x, 10
TinySec, 201

TinySec key, 199
T-MAC, 35
Tmote Sky sensor node, 144
Topology management, 21
TOPSIS
      approach, 92–98
      ranking using fuzzy, 98–102
TOSSIM. *See* TinyOS SIMulator
Total residual energy, 29–30
TPSNs. *See* Time sync protocol for sensor networks
Traffic adaptive medium access (TRAMA) protocol, 32
Traffic-based scheduling MAC protocols, 32
TRAMA protocol. *See* Traffic adaptive medium access
      protocol
Transceiver, 216
Transceiver CC2420, 122
Transit-based attacks, information in, 185–186
Transmission media, 21
Transmit power, 223
Transmitter-initiated cycled receiver (TICER), 34
Transport layer
      attacks, 188–189, 191
      protocol stack for WSNs, 6
Tree-based aggregation, 171–172
Tree-based key predistribution schemes, 197–198
Triple DES (3DES), 192
TrueTime–PiccSIM extension, 16
Trusted third node-based scheme, 195
Turbo codes, 224
Two-dimensional grid, 233
Two-node data aggregation, 166
Two-storage device-based model, 145

U

UID. *See* Unique node identifier
Ultralow-power devices, 144
Unattended operation, security limitation, 183
Unauthorized admittance, 180
Unbalanced force, 150
Underground WSNs, 246
Underwater WSNs, 246–247
Unfairness, data link layer attacks, 192
UnificationEngine API, 259
Uniqueness of addresses, 228
Unique node identifier (UID), 227
Unique secret matrix, 197
Unix-like operating system, 11
Unreliable communication, security limitation, 183
Unscheduled MAC protocols, 33–35
Unweighted pair-group method using arithmetic
      averages (UPGMA), 72

V

Vehicular sensor network, 245
Verification, aggregation, 175

VIKOR, ranking using, 102–111
Viptos, 15
Viterbi algorithm, 224
Voltage multiplier, 156
Vortex shedding, 152
Vulnerable components
    BS security, 183
    sensor node security, 184–185

W

Wake-up mode, 242
W3C. *See* World Wide Web Consortium
Weaver API, 259
Web 1.0, 254
Web 3.0, 253–255
Web-centric software languages, 253
Web 2.0 era, 254
Web-hosted applications, 252
Web MIDI API, 259
Weighted pair-group method using arithmetic averages
    (WPGMA), 73
Wind energy harvesting system, 145–147
Wind generator, efficiency of, 146
Wind power, 146
Wind speed, 146
Wireless sensor networks (WSNs), 2, 3, 44, 138
    challenges for energy-constrained, 23
    and network, architecture of, 4

operating systems and networking environment,
    9–19
routing in. *See* routing
WiseMAC, 34
Wi-Se node for remote sensing, 147
Witness-based approach, 204
World Wide Web, 253
World Wide Web Consortium (W3C), 254
Wormhole attack, 188, 191
WPGMA. *See* Weighted pair-group method using
    arithmetic averages
WSNs. *See* Wireless sensor networks

X

XBee, 38
XBee DigiMesh, 38
Xbee sensor mote, 140, 144
XCTU, 38, 39
XLM. *See* Cross-layer module
XLP protocol, 235
xMatters API, 259

Z

ZigBee, 216, 229
    3.0, 37
    MAC module, 28, 37, 38
    protocol, 39
ZigBee-based home automation system, 244